About Island Press

Since 1984, the nonprofit organization Island Press has been stimulating, shaping, and communicating ideas that are essential for solving environmental problems worldwide. With more than 1,000 titles in print and some 30 new releases each year, we are the nation's leading publisher on environmental issues. We identify innovative thinkers and emerging trends in the environmental field. We work with world-renowned experts and authors to develop cross-disciplinary solutions to environmental challenges.

Island Press designs and executes educational campaigns, in conjunction with our authors, to communicate their critical messages in print, in person, and online using the latest technologies, innovative programs, and the media. Our goal is to reach targeted audiences—scientists, policy makers, environmental advocates, urban planners, the media, and concerned citizens—with information that can be used to create the framework for long-term ecological health and human well-being.

Island Press gratefully acknowledges major support from The Bobolink Foundation, Caldera Foundation, The Curtis and Edith Munson Foundation, The Forrest C. and Frances H. Lattner Foundation, The JPB Foundation, The Kresge Foundation, The Summit Charitable Foundation, Inc., and many other generous organizations and individuals.

APPLIED PANARCHY

APPLIED PANARCHY

Applications and Diffusion across Disciplines

Edited by
Lance H. Gunderson,
Craig R. Allen, and
Ahjond Garmestani

ISLANDPRESS | Washington | Covelo

Library of Congress Control Number: 2021942228

All Island Press books are printed on environmentally responsible materials.

Manufactured in the United States of America
10 9 8 7 6 5 4 3 2 1

Keywords: Island Press, adaptation, adaptive cycle, adaptive governance, adaptive management, biotechtonics, climate change, coerced regime, ecological organization, ecology, ecosystem, ecosystem management, ecosystem stewardship, environmental change, environmental governance, environmental law, environmental management, environmental policy, food security, food systems, multiscale, panarchy, pandemic, regime shift, resilience, scale, scale domain, social-ecological resilience, social-ecological systems, spatial regime, spatial resilience, system traps, transformation, transformative governance

For Buzz

CONTENTS

PREFACE

Applied Panarchy is the fourth book in a series of volumes addressing panarchy theory. Panarchy was proposed by systems ecologists who were interested in bringing together theory with the practice of managing natural resources. The theory was proposed to explain observations of the discontinuous structure and nonlinear dynamics of ecosystems. This framework was also developed to explain the surprising and unpredictable behavior in systems of people and their environment. We created the term *panarchy* by combining *Pan*, the mischievous Greek god of nature who scattered discord and chaos in mythology, with the Greek word *arkos*, "rules." However, since then we have discovered that in 1860, Paul Émile de Puydt coined the word *panarchy* to describe a form of governance that was universal and all inclusive. In this volume, *panarchy* does not mean "all rules," as described by de Puydt, but instead refers to "nature's rules" (Gunderson and Holling 2002).

The three previous volumes on panarchy developed and applied cross-scale theory to a wide range of resource systems from an interdisciplinary, integrative framework. The first volume, *Barriers and Bridges to the Renewal of Ecosystems and Institutions* (Gunderson, Holling, and Light 1995), introduced the word *panarchy*, imputed from case studies of complex managed systems. These cases all indicated patterns of increasing rigidity with development and the role of crises and instabilities in subsequent system trajectories. The second volume, *Panarchy: Understanding Transformations in Systems of Humans and Nature* (Gunderson and Holling 2002), resulted from a research project of the Beijer Institute for Ecological Economics and the University of Florida, and it demonstrated how economic theory, political theory, and social theory interacted with ecological theories of nonlinear change. That synthetic work led to a new generation of integrative scholarship on social-ecological system dynamics. The third volume on panarchy, *Practical Panarchy for Adaptive Water Governance* (Cosens and Gunderson 2018), examined how law, policy, and ecological dynamics interact to influence the governance of water resource systems.

To build on these previous volumes, this volume set out to explore how panarchy theory has diffused to and modified other fields of inquiry and scientific publications over the past two decades. We also describe how panarchy theory has influenced the practice, management, and governance of environmental stewardship. Three fundamental themes underlie our exploration of panarchy in this volume. One is the linkage between resilience scholarship and panarchy theory. The second theme of this work addresses the links between cross-scale discontinuous structures and processes and panarchy theory. Scholarship on cross-scale dynamics has received considerable attention (Allen and Holling 2008), and this aspect of panarchy theory has also given rise to numerous journal articles testing the theory and documenting the results of applying panarchy theory to resource management. The third theme of this volume is to document how concepts of panarchy theory have diffused into other scholarly areas and disciplinary discourses. These areas include governance and management of natural resources, law and policy, indigenous and traditional ecological knowledge, interdisciplinary approaches to agriculture and land management, and innovation, creativity, and novelty.

ACKNOWLEDGMENTS

This book is rooted in the original and creative theories of C.S. "Buzz" Holling. Buzz blended general systems theory with ecological theory to develop theories of change that remain relevant today. His 1973 publication on resilience theorized multiple stable states in ecosystems, which led ecologists and practitioners to move away from equilibrium centered assumptions about nature to adaptive management approaches. In one of the first volumes on sustainable development of the biosphere (Clark and Munn 1986), Buzz proposed the adaptive cycle to capture nonlinear cyclical ecosystem dynamics, which is one of the key ingredients of panarchy theory. In 1992, Buzz published a new paradigm describing the discontinuous nature and structure of ecosystems, further developing ideas that led to the development of panarchy with Lance Gunderson. Buzz passed away in 2019, during the early stages of this volume, and for that reason we dedicate this volume to him.

Financial support for this project was provided by Emory College of Arts and Sciences, Emory University, and the U.S. National Science Foundation. The material is based on work supported by the National Science Foundation under grant nos. DGE-1735362 and 1920938. Any opinions, findings, and conclusions or recommendations expressed in this material are those of the authors and do not necessarily reflect the views of the National Science Foundation.

The findings and conclusions in this book have not been formally disseminated by the U.S. Environmental Protection Agency and should not be construed to represent any agency determination or policy. Any use of trade names is for descriptive purposes only and does not imply endorsement by the U.S. government.

We thank Abbey Snyder, a graphic designer in the School of Natural Resources, University of Nebraska–Lincoln, who rendered all figures in this volume.

Literature Cited

Allen, C.R., and C.S. Holling (Eds.). 2008. *Discontinuities in Ecosystems and Other Complex Systems.* Columbia University Press, New York.

Clark, W.C., and R. E. Munn. 1986. *Sustainable Development of the Biosphere.* Cambridge University Press, Cambridge, UK.

Cosens, B., and L.H. Gunderson. 2018. *Practical Panarchy for Adaptive Water Governance: Linking Law to Social-Ecological Resilience*. Springer, New York.

Gunderson, L.H., and C.S. Holling. 2002. *Panarchy: Understanding Transformations in Human and Natural Systems.* Island Press, Washington, DC.

Gunderson, L.H., C.S. Holling, and S.S. Light. 1995. *Barriers and Bridges to the Renewal of Ecosystems and Institutions.* Columbia University Press, New York.

PART I

PANARCHY CONCEPTS

Chapter 1

Panarchy: Nature's Rules

Lance H. Gunderson, Ahjond Garmestani, and
Craig R. Allen

In late 2019, a previously unknown virus was detected in Hubei Province, China. Within a few months, the SARS-2 coronavirus had spread around world. The ensuing COVID-19 pandemic has led to concatenated environmental, social, and political crises (Walker et al. 2020). Irruptive phenomena, surprising in particulars but generally understood, generate drastic consequences for humanity but are of the same nature as many contagious ecological processes, such as forest fires or the spread of invasive species (Holling 1986, Garmestani et al. 2020). Contagious processes such as pandemics, fires, and political uprisings cover wide ranges of scales. In the case of the COVID-19 pandemic, the scale range spanned from the molecular (virus) to the planet. Studies of episodic crises (such as pandemics) and their relationship to cross-scale, nonlinear dynamics of ecological processes in natural resource systems led to the conceptualization and articulation of panarchy theory (Gunderson et al. 1995; Gunderson and Holling 2002; Cosens and Gunderson 2018).

The word *panarchy* is a scientific portmanteau, coined by combining the word *pan* with the root *archy* (Gunderson and Holling 2002). Pan was the Greek god of nature, part animal and part human, who scattered discord, chaos, and ensuing panic. *Archy* is derived from the Latin and Greek word for "rules" and is cognate with words such as *monarchy* ("one ruler") and *hierarchy* ("sacred rules"). Hence, *panarchy* describes the rules of nature or nature's rules. As Pan is a composite entity, part human and part animal, so are managed resource systems. That is, they are not just ecosystems, nor are they solely human systems, but rather they are coupled systems of people and nature (Gunderson and Holling 2002; Westley et al. 2002). Hence coupled, multiscale social-ecological systems (panarchies) reflect these characteristics of Pan (unpredictable nature) and the emergent rules by which these systems operate that generate abrupt, episodic, and nonlinear changes (Gunderson and Holling 2002).

Two decades ago, the book *Panarchy: Understanding Transformations in Human and Natural Systems* (Gunderson and Holling 2002) outlined an integrated theory of change in complex social-ecological systems (SESs). The ideas and theories were proposed to explain patterns of change in systems that were not ecological systems with human intervention, nor were they social systems attempting to dominate and control ecosystems (Westley et al. 2002). Such systems are now described as social-ecological systems (Anderies et al. 2004; Chapin et al. 2009; Garmestani and Allen 2014).

The motivations and organizing ideas for this new volume are twofold. In addition to improving our understanding of panarchy (Gunderson and Holling 2002) we focus on the application of panarchy concepts to assess and manage complex environmental issues across multiple scales. Because of the interdisciplinary nature of most environmental issues, our second goal was to document diffusion of panarchy concepts across fields of scholarship. Below, we review panarchy theory as described by Gunderson and Holling (2002), including nonlinear changes as depicted by adaptive cycles, cross-scale structure and functions, and cross-scale interactions. Then we introduce applications, expansions, and modifications of panarchy theory to provide a common set of ideas that are expanded in subsequent chapters. We conclude with a brief introduction to the diffusion of panarchy concepts into other scholarly disciplines, especially human geography, natural resource management, law and governance, economics, food systems, spatial management of land use, and ecosystem stewardship.

Panarchy Theory

A panarchy consists of dynamic, scale-specific processes and structures (Holling 1986; Holling et al. 2002; Allen et al. 2014). The distribution of process and structure varies across both scales and systems, creating breaks and discontinuities. Contagious processes, such as pandemics or forest fires, begin abruptly at small scales, then spread to broader scales until limited by spatial breaks (lack of connectivity) or by a shift in feedback mechanisms (Holling et al. 2002). Other cross-scale interactions occur as slow and broad processes and structures influence those that are faster and smaller, as typified in a hierarchy (Holling et al. 2002). These connections are often ephemeral, however, becoming dominant at certain times and dormant at others. Within a particular domain of scale, parallel but compartmentalized elements occur (such as different communities of similar size within an ecoregion or different habitat patches within an ecosystem). Complex adaptive systems, panarchies, are not static and equilibrium seeking; rather, they are dynamic and nonstationary.

Dynamics that occur within a particular scale have been described, within panarchy theory, as adaptive cycles.

Holling (1986) noted that many systems, ecological, economic, social, and political, are characterized by a number of paradoxes. One observed paradox was that systems undergoing changes do so at heterogeneous rates. Sometimes change is slow, such as the accumulation of plant biomass or diversity in a forest ecosystem, crops in an agricultural field, or wealth within a corporation. Sometimes changes are fast, such as a fire in a forest, a pest outbreak in agriculture, or collapse in a market. Another paradox identified by Holling (1986) is the emergence of periods of stability interspersed with periods of instability. Forest fires, pandemics, and political elections are all examples of periods of instability, or "creative destruction" (Schumpeter 1942, 83). A third key paradox proposed by Holling was the shift from viewing changes as linear and monotonic to one where patterns are viewed in phase space, in circular time, or as cycles. Yet another paradox for such systems is between conservation and novelty: what is conserved within systems and what are new and novel forms and structures in the systems that lead to system change. The adaptive cycle was developed to resolve these paradoxes (Holling 1986, 1992).

An adaptive cycle depicts a system that exhibits four distinct and usually sequential phases of change (figure 1.1): (1) an exploitation phase, (2) a conservation phase, (3) a release phase, and (4) a reorganization phase. An exploitation phase (labeled *r*) is a period of growth, fueled by inputs of energy and resources and results in the accumulation of structures, capital, and connections within the system. Over time, limits and constraints emerge to constrain growth, as the system matures. The conservation phase (K) is one in which materials are recycled and system energy flow increasingly goes to maintenance of structure accumulated during the growth phase. Moreover, the increased complexity, structure, and unachieved potential generally contributes to an increasing vulnerability to external variations or disturbances. The conservation phase is followed by an instability or release phase (Ω). Such instabilities may be triggered by fires, pest outbreaks, economic crisis, or political uprisings, for example, in which the slowly accumulated entities are quickly destroyed. Holling (1986) borrowed from Schumpeter (1942) and described the release phase as a period of creative destruction. The Ω phase is quickly followed by a reorganization (α) phase, where the system is renewed or a new system emerges, leading to a growth phase of a new cycle. The r phase may lead to a very similar system that redevelops around the same structures and functions,

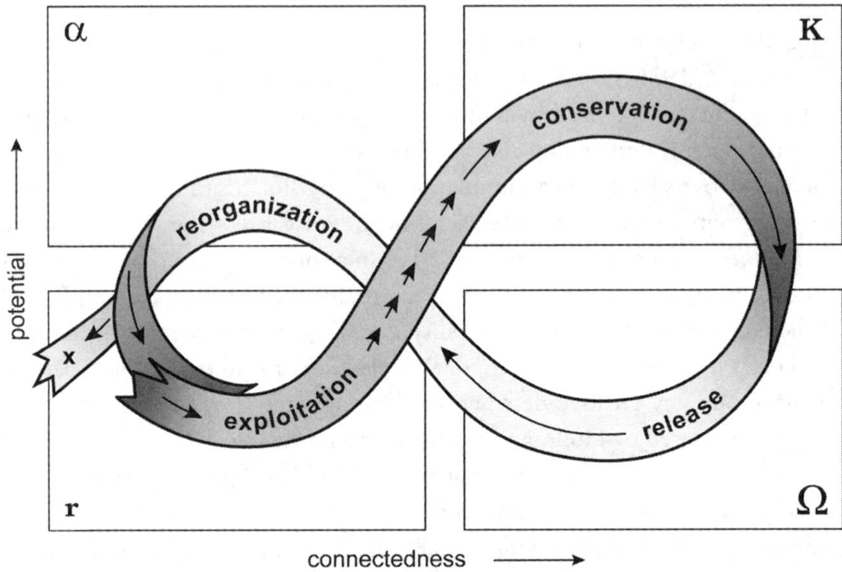

Figure 1.1. Observed phases in social-ecological systems. The four phases include exploitation (r phase), conservation (K phase), release (Ω phase), and reorganization (α phase). These phases are plotted along dimensions of connectedness (horizontal axis), or degree of connectivity between controlling system components, and potential (vertical axis) for accumulation of resources. Time flows unevenly through the cycle; systems slowly develop from r to K phases, then quickly through the release and reorganization phases. The exit pathway indicates a stage when the system is likely to transform into an alternative identity. Sequential movement of a system between the phases is called an adaptive cycle. (Adapted from Holling and Gunderson 2002.)

or it may be different, structured by new processes that establish during the r phase, giving rise to a new (or alternative) system state, driven by a new process regime. This recurring pattern of rapid, then slowing growth, swift destruction and reformation, has been observed in many systems (Gunderson et al. 1995; Gunderson and Holling 2002; Cosens and Gunderson 2018).

Scales and Scaling

"Think globally, act locally" has been a catchphrase of the environmental community for decades. The phrase captures two ideas that persist to date. One is the notion of Earth as a global system. The first photographs of Earth in 1972 helped humans conceptualize the planet as a blue marble, floating in space. That image showed Earth as one object bounded by the immensity of space. The image also reflected Earth as a system, or biosphere, made up of constituent systems such as the hydrosphere, atmosphere, and geosphere, which belied complexity at smaller scales. The second element of the phrase brings in the notion of human agency and action, at a scale appropriate to humans: the local. Yet as human populations have expanded, so has the scale of human activities, generating environmental changes at scales well beyond the local (Steffen et al. 2018). In turn, human agency has attempted to deal with environmental change at increasingly large scales, with mixed results and unforeseen consequences (Garmestani et al. 2020).

Key structures and processes that define a particular panarchy can be mapped by using grain and extent to define scale boundaries across dimensions of space and time. For example, vegetation components in forest systems consist of leaves, trees, stands, forests, landscapes, and biomes (Holling and Gunderson 2002; figure 1.2), each occurring at a discrete scale, each structured by a different process, and each possessing different structures. These features are dynamic. Leaves exhibit an annual cycle of new buds, leaf growth, senescence, and abscission. Patches of forest go through these phases of succession on cycles of decades, as indicated by periodicities of fire or pest outbreaks. The rich array of adaptive cycles within a system, interacting with other adaptive cycles at the same scale (e.g., patches of grass and forest of the same size) and interacting across scales (e.g., grass and forest patches interacting with larger-scale processes and structures), forms a panarchy. An atmospheric panarchy, for example, can consist of structures such as microbursts, thunderstorms, frontal waves, El Niño–Southern Oscillation, and global climate regimes (figure 1.2). Similar cross-scale structures are apparent in a wide variety of systems, including for governance institutions (Cash et al. 2006) and for cities (Eason and Garmestani 2012).

Holling and Gunderson (2002) made three observations about cross-scale structures and processes. The first is that different structures, objects, and processes appear and disappear with changes in spatial (and temporal) scale. For example, photographs of forest stands cannot capture the detail of smaller structures such as pine needles or flowers that are observable at smaller windows, nor can they capture larger patterns, such as landforms, watersheds, or plant communities. The second observation is that processes and structures cover different extents in space and time. Some processes such as forest fires range from scales of a square meter to thousands of square kilometers. Other processes such as changes in carbon dioxide concentration in the atmosphere are the result of aggregation of emission sources over twenty or so orders

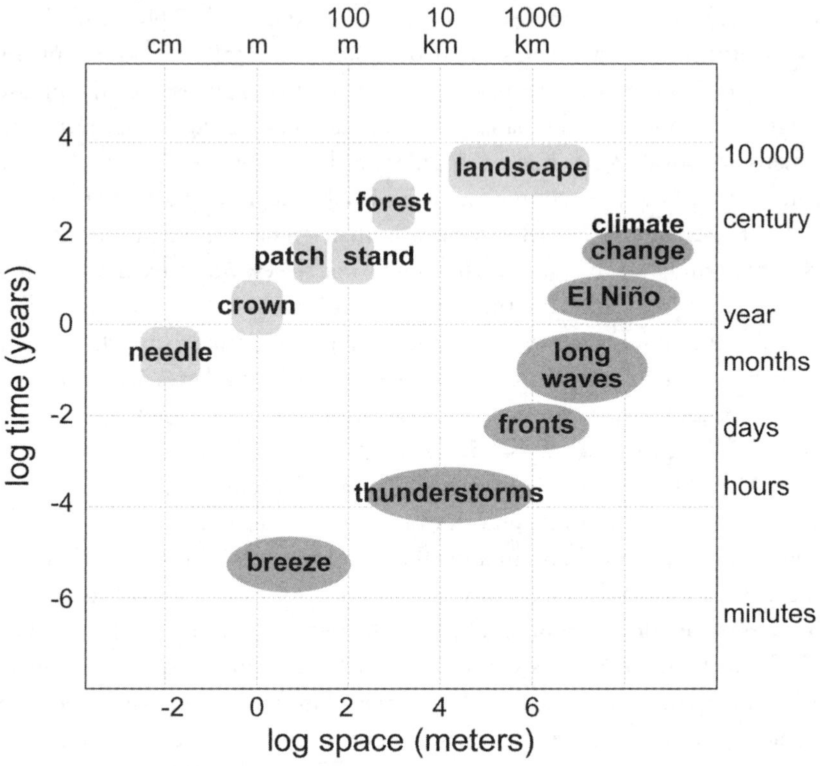

Figure 1.2. Cross-scale vegetation and atmospheric structures, plotted along dimensions of space (horizontal axis) and time (vertical axis). Each feature has characteristic size ranges and turnover times. The vegetation hierarchy consists of elements such as leaves, tree crowns, patches, forests, and landscapes, and the atmospheric hierarchy ranges from breezes to global climate change. (Modified from Holling et al. 2002.)

of magnitude. The third observation is that ecological systems consist of self-organized processes that are not scale invariant; that is, they are not self-similar across scales (as measured by a constant fractal dimension or fit of a common power law; Garmestani et al. 2005, 2009). Although many physical systems are self-similar or scale invariant, ecosystems are not because of the interaction between biotic and abiotic elements.

Cross-Scale Dynamics

The discontinuities and nonlinearities that characterize panarchies are caused by interactions of structures and processes across scales (Holling 1992;

Gunderson and Holling 2002; Allen and Holling 2008). Some of the organization within the system reflects scale-specific structure and hierarchical control. An example of hierarchical or top-down control is when slow, broad variables such as geology and soil types interact with faster, local climatic variables (temperature, photoperiod, rainfall) to determine the suite of plant and animal species that thrive within a panarchy. Many disturbance dynamics, such as forest fires or pandemics, are not the result of top-down control by slower variables but occur when faster, smaller variables control the system for periods of time (i.e., bottom-up change).

Panarchy also suggests that the slow and broad processes and structures influence those that are faster and smaller (i.e., the subsystems and downscale influences). But these connections are ephemeral, becoming dominant at certain times and dormant at others. There are potentially multiple connections between adaptive cycle phases at one level and phases at another level; four are discussed here. Two of these connections, revolt and remember, were introduced by Holling and Gunderson (2002). Revolts describe contagious disturbances, driven by positive feedback mechanisms, that propagate from small scales to larger (and longer) scales. Remember reflects larger to small-scale dynamics and top-down (large to small) interactions. Two other cross-scale connections, crisis and innovation, have been introduced into the panarchy framework (Allen and Holling 2010; Chaffin and Gunderson 2015; figure 1.3). These four cross-scale linkages influence two key phases of a focal system; they influence and drive periods of instability (Ω) and periods of reorganization (α).

Natural disasters occur when systems that originate from larger scales such as cyclones, tsunamis, or earthquakes influence local or smaller scale systems. Such instabilities can test the resilience of ecological and social systems at local scales. Other processes, described by the revolt arrow (figure 1.3), emerge from smaller scales and influence larger scales. However, when a level in the panarchy enters a phase of creative destruction and experiences a collapse, that collapse can cascade up to the next larger and slower level, particularly if that level is in the K phase, where resilience is low. Irruptive disease outbreaks and fires are two examples of such phenomena. The lighting of a match in a forest or a lightning strike can start a fire at a scale of a few square centimeters within seconds. Under some conditions that local fire is quickly extinguished, or a fire never begins. However, under certain conditions (such as droughts or low humidity), local ignitions can create a small ground fire that spreads to the crown of a tree, then to a patch in the forest and then to a stand of trees.

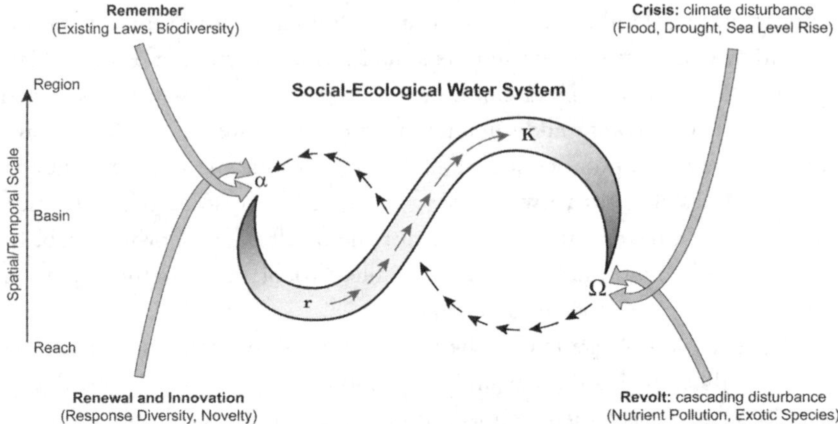

Figure 1.3. Cross-scale influences on a social-ecological hydrologic system. At basin scales, systems can change in patterns consistent with an adaptive cycle. Cross-scale influences that can result in a system instability or release (Ω phase) that originate from larger spatiotemporal scales are called crises; revolts originate at smaller scales and cascade to vulnerable systems at larger scales. Examples of crises include cyclones and tsunamis; examples of revolts include fires, disease, and pest outbreaks. Two other connections, remember and novelty, influence system reorganization (α phase). Remember describes how potential capital from broader scales can influence postreorganization pathways to drive renewal and recovery. Such larger-scale variables include evolved biota, constitutions in law, and world views and myths in social components of a system. New elements or combinations from smaller scales can also influence reorganization. (Modified from Chaffin and Gunderson 2015.)

Each step in that cascade moves the collapse to a larger spatiotemporal scale. Therefore, *revolt* is used to describe how fast and small events overwhelm slow and large ones. And that effect could cascade to still higher, slower levels if those levels have accumulated vulnerabilities and rigidities.

The downscale interactions in panarchy are captured by the word *remember*. This type of cross-scale interaction (figure 1.3) is important at times of change and renewal. Once a catastrophe is triggered at a given scale, the opportunities and constraints for the renewal of the cycle are strongly limited by the K phase of the next slower and larger scale. After a fire in an ecosystem, for example, processes and resources accumulated at a larger scale slow the leakage of nutrients that have been mobilized and released into the soil. And the options for renewal draw on the seed bank, physical structures and surviving species that form biotic legacies that have accumulated during the growth of the forest. It is as if this connection draws on the aspects of the capital accumulated during maturity, hence the choice of the word *remember*.

Below, we describe some of the ways in which panarchy theory has been applied to help elucidate and manage complex systems. We first describe the development of panarchy from the study of dramatic shifts in the form and function of SESs over time, through the lens of resilience (Holling 1986; Gunderson et al. 1995; Folke 2006, 2016). Then we depict how structure and processes vary across scales of space and time and the implications for those scale interactions on the dynamics of systems.

Resilience, Adaptation, and Transformation

Panarchy theory was inferred from a set of case studies of regional-scale resource management systems (Gunderson et al. 1995). The managed resource systems included a range of ecosystems such as the boreal forests of New Brunswick (Baskerville 1995), the Everglades of Florida (Light et al. 1995), Chesapeake Bay (Costanza and Greer 1995), the Columbia River (Lee 1995), the Great Lakes (Francis and Regier 1995), and the Baltic Sea (Jansson and Velner 1995). All of these SESs exhibited historical patterns consistent with an adaptive cycle. That is, periods of growth and development occurred as infrastructure, policies and institutions emerged to promote societal goals. For example, levees and canals were built in the Everglades in the 1950s to control unwanted flooding in urban and agricultural areas of the wetland (Light et al. 1995). Over time, policies and actions were successful at reducing urban flooding. However, other parts of the system were changing, as management agencies became focused on efficiency and cost control. The economic components in these systems became more dependent and vulnerable. Internal vulnerabilities intersected with external shocks, which in turn led to environmental crises. After these crises, the system reorganized and started new phases of growth and development. Other researchers have conducted reviews of natural resource management case studies (Walker et al. 2006; Walker and Salt 2006; Cosens and Gunderson 2018) and found similar patterns of policy development and implementation (r to K phases) and recurring environmental crises (Ω) that generate policy reformation in SESs.

In particular, two types of environmental crises, defined as unexpected or unforeseen system dynamics (Gunderson 2003), have led to systemic reformation in natural resource systems (Gunderson and Holling 2002; Chaffin and Gunderson 2015). One type of crisis occurs as a result of unforeseen variation in processes that operate at larger scales (figure 1.3). Such variation generates an external shock or disturbance to the system and can reveal systemic vulnerabilities (Adger et al. 2009). For example, extreme flooding due to excessive rainfall occurred in the Everglades in 1928 and 1948, leading to loss of life and property (Light et al. 1995). The excessive rainfall in these Everglades events was caused by the passage of tropical cyclones (depressions and hurricanes). Such weather events are part of the global climate system, driven in part by the differential heating of the scale of the planet. The other type of crisis or surprise studied in panarchical systems is related to ecological resilience.

C. S. Holling (1973) introduced the word *resilience* to describe systemic change in ecosystems. Ecosystems can shift from one qualitative state to another, changing both structure and function, and ecological resilience mediates such transitions or regime shifts (Holling 1996; Folke 2006). Since that time, hundreds of studies have demonstrated such state changes or regime shifts across a wide range of ecosystems (Gunderson and Pritchard 2002; Folke et al. 2004, 2010). These include marine ecosystems (Estes and Duggins 1995) and coral reef systems (Hughes et al. 2003), freshwater lakes (Scheffer et al. 2001), terrestrial grasslands (Walker and Salt 2006, 2012), drylands and deserts (Foley et al. 2003), temperate forests (Müller et al. 2016), and boreal forests (Chapin et al. 2004) subject to varying degrees of human intervention and management. Such regime shifts have been explored in a wide range of SESs as well (Holling and Gunderson 2002; Folke et al. 2010; Folke 2016).

To varying degrees, such regime shifts have been explained for these systems by a shift in system controls, through either changes in drivers (inputs) or shifts in feedbacks (internal influences) (Holling and Gunderson 2002). Such changes were also related to changes in controlling variables that operate at different spatial or temporal scales (Holling 1986). Thus, application of panarchy has become the focus of a growing body of research focused on improving operations (Wieland 2021), governance (Garmestani and Benson 2013; Chaffin et al. 2016), and management (Westley 2002) of complex SESs. Urban system research provides an example of applications of panarchy to a variety of complex adaptive systems.

Growth and Transformation in Cities

Urban systems have spatial heterogeneity in their social and ecological infrastructure (Grimm et al. 2000). The physical configuration of the environment (e.g., topography) can play a key role in city structure (e.g., San Francisco). Transportation networks of cities (e.g., rivers, rails, and roads) are important factors in the flow of people and commerce. In addition, education, power, property, status, and wealth manifest at different spatial and temporal scales and contribute to the manifestation of panarchy in cities (Pickett et al. 2001; Garmestani et al. 2009). This spatial heterogeneity can differ with scale and is influenced by factors originating at other scales. For example, zoning in cities is determined at the municipal scale by zoning regulations but is also affected by business interests and national economies (Garmestani et al. 2005). Furthermore, city growth rates are dependent on the size of the city, and cities (i.e., urban SESs) are influenced by hierarchical, historic, and stochastic factors that make their development differ from simple predictions that hold for physical systems.

Cities compete with one another for resources, and the differences between critical factors in cities can account for many of the differences in growth trajectories (Garcia et al. 2011). For example, the growth of cities in the southeastern United States is correlated with mean household income and the percentage of the population of a city with a college degree (Eason and Garmestani 2012). Cities are also known to exist in size-dependent aggregations, which have been interpreted as an objective manner to delineate scale in urban systems (Garmestani et al. 2009). Each of these scales is a level in a panarchy of urban systems, with U.S. cities largely remaining in their original size classes despite much growth during the past century. There are exceptions, however, as some cities have experienced tremendous growth (e.g., Phoenix), while others have experienced precipitous decline (e.g., Cleveland).

Panarchy also exists within an individual city, as variables at different scales exhibit different speeds, which in turn manifests as scale-dependent structure (Gunderson and Holling 2002). For urban systems, growth rate is a fast variable, governance is a medium variable, and infrastructure is a slow variable, with slow variables ultimately determining the resilience of a system (Allen and Holling 2008). For shrinking cities (e.g., Cleveland) that want to slow decline, catalyzing change in infrastructure (e.g., gray to green infrastructure) is a clear path to facilitate transformation (change to a more desirable state with human agency) in an urban system. An SES may be a candidate for transformation if it is in a degraded state (e.g., economically depressed city); transformation is the process whereby a system in an undesirable state has its resilience purposefully exceeded by human agency or experiences a collapse (e.g., hurricane impacts), and the reorganization phase is guided by humans to reorganize into a desirable state. Leadership, networks, learning, and trust are key aspects of the transformative capacity of an SES (Gunderson et al. 2006). However, cities are usually difficult to transform barring an unforeseen disaster, because of the massive investment in infrastructure that is challenging to overhaul (Brown et al. 2013).

For a shrinking city like Cleveland, the decline creates an opportunity to transform the city to a sustainable version of itself. In particular, a transition from gray to green infrastructure is possible because of an abundance of vacant parcels throughout the city (Green et al. 2016). Vacant parcels are the raw material for transformation in Cleveland and are being used for urban agriculture, wildlife corridors, recreational areas, and habitat for pollinators and to mimic the natural hydrologic cycle and reduce stormwater overflows (Green et al. 2016). Thus, transitioning from gray to green infrastructure is the manner by which Cleveland is transforming itself from a shrinking city to a sustainable city.

Maladaptive Trajectories

Many SESs exhibit rhythms of change, as exhibited in adaptive cycle dynamics and transformational changes (Gunderson et al. 1995; Walker and Salt 2006), but some do not. Systems can become trapped when they cannot or do not change, adapt to new conditions, or escape from a regime with an undesired trajectory. Holling and Gunderson (2002) referred to pathologic or maladaptive trajectories as system traps.

At least four different types of resource management traps can be identified (table 1.1). These different types are each defined by a combination of three properties: capital or potential, degree of connectivity, and level of resilience (Holling and Gunderson 2002). Although it is possible to have eight combinations of these three properties, four combinations have been identified (table 1.1). A system in a rigidity trap has high capital, high connectivity, and high resilience (Holling 2001). A system in a poverty trap has low levels or amounts of these three properties. A system caught in an eroding or lock-trap has low capital but high levels of connectivity and resilience. The fourth trap is the least well understood of these four and is called an isolation trap, as it has high capital or potential but is not highly coupled or resilient.

The descriptions and analyses of traps by various authors (Holling and Gunderson 2002; Allison and Hobbs 2004; Chapin et al. 2009; Cinner 2011; Carpenter and Brock 2008), characterize traps in terms of these three key variables, processes, and properties. They all note how slowly changing, structural variables influence the stability landscape of resilience. Rigidity traps are sustained by increasing control exerted by large-scale processes and minimizing

TABLE 1-1. *Levels of the three variables that characterize four system traps (Holling et al. 2002; Allison and Hobbs 2004; Angeler et al. 2020).*

	Capital	Connectivity	Resilience
Rigidity trap	High	High	High
Poverty trap	Low	Low	Low
Eroding trap	Low–eroding	High	High
Isolation trap	High	Low–none	Low

or eliminating small-scale processes. Rigidity traps are the result of sustained hierarchical controls (in the form of power, resources, and manipulation) that suppress innovation, diversity, and experimentation (Gunderson et al. 2018). In contrast, systems in poverty traps can be characterized by the lack of critical types of larger-scale inputs (memory, resources) and the inability to constrain or adapt to small-scale perturbations. When a system is in a poverty trap, small-scale disturbances lead to crises and reorganizations that sustain trajectories of continued poverty conditions.

The types of surprising and nonlinear changes exhibited by panarchical systems pose uncertainties for those managing or attempting to manage these systems (Gunderson and Holling 2002; Chapin et al. 2009). Laws and formal government organizations set up to pursue societal goals through ecosystem management are based primarily on linear, predictable system dynamics (Benson and Garmestani 2011; Garmestani and Allen 2014; Craig 2013, 2017). Adaptive and transformative approaches that use the panarchy framework are just beginning to emerge in resource management (Garmestani and Benson 2013; Auad et al. 2018; Garmestani et al. 2020) and governance (Green et al. 2016; Chaffin et al. 2016; Cosens and Gunderson 2018).

Applications and Diffusion of Panarchy Theory

In the preceding paragraphs, we have described key concepts and theories associated with panarchy (Gunderson and Holling 2002). The 2002 panarchy volume was a result of challenges to economic, political, and social theory interacting with ecological theories of nonlinear change. Work flowing from the first panarchy book has been published in dozens of books and tens of thousands of articles in scientific journals on resilience, adaptation, and transformation. The second thrust of scale related theoretical work has involved linking cross-scale discontinuous structures and processes with panarchy theory. Scholarship on cross-scale dynamics has achieved attention as well as numerous journal articles testing the theory and documenting the results of applying panarchy theory to resource management. Panarchy theory has informed a growing body of scholarship in which the concepts have diffused into other scholarly areas. These areas include governance and management of natural resources, law and policy, indigenous and traditional ecological knowledge, interdisciplinary approaches to agriculture and land management, innovation, creativity, and novelty (Allen and Holling 2010).

The remainder of this volume describes how panarchy theory has been modified and applied as a result of interdisciplinary scholarship and how panarchy theory has diffused into and modified other fields of inquiry and science, especially in applications of resource management and environmental stewardship. These two themes form the structure of the next two sections of this book. Part II, "Applications of Panarchy Theory," consists of brief chapters describing various applications, and Part III, "Diffusion of Panarchy Concepts," describes how panarchy concepts have diffused into scholarship in other disciplines.

Applications of panarchy theory

Panarchy was introduced as a theoretical construct two decades ago, and over this period its invocation in peer-reviewed literature has steadily increased, but its use remains primarily as a descriptive framework (Allen et al. 2014). As described in chapter 2 (Sundstrom et al., this volume), quantitative tests of hypotheses derived from panarchy theory can be addressed in terms of the discontinuous, cross-scale system attributes first described by Holling (1992). Holling (1992) and Allen and Holling (2008) developed the discontinuity hypothesis as a crucial link that connects adaptive cycles and panarchy through its depiction of how processes and patterns manifest in discrete system-scale domains and shape the ecological theater in which adaptive cycles play out. Sundstrom et al. (chapter 2) discuss how scale-specific structuring in complex systems manifests in discontinuous properties in a wide range of systems, including ecological, social, and social-ecological. The discontinuous nature of complex systems has provided insight on how resilience is generated and maintained by a nonrandom distribution of functional traits both within and across scales and how this characterization of complex systems is related to the response of complex systems to disturbance. System regime shifts and nonrandom phenomena are both associated with discontinuities, including variability at the individual, population, community, and system levels, sources of innovation and novelty, and how the cross-scale resilience model can help improve our management of complex systems of people and nature.

Sundstrom and Allen (chapter 3) describe the ubiquity of adaptive cycle dynamics in complex systems. They propose a set of ecological indicators that have been used to measure dynamics of change in ecosystems and are strong candidates for tracking system trajectory through the stages of an adaptive cycle. These indicators, when coupled with the explicit consideration of scales via the discontinuity hypothesis and panarchy, may minimize the need for multiple indicators to capture system dynamics and provide a richer picture of system trajectory than that offered by a single-scale analysis. Indicators are a generic concept broadly representing a system signal that can change over time (Nielsen and Jorgensen 2013). Sundstrom and Allen propose feasible ways in which researchers could systematically and quantitatively look for signatures of panarchical dynamics in ecosystems rather than relying on metaphorical and largely qualitative descriptions.

Management of systems of people and nature in times of accelerating social-ecological change has recently been translated into concepts that directly link management challenges, such as fostering and improving service provisioning for humans, to resilience theory. The concept of coerced resilience has been examined in a production ecosystem (agriculture, silviculture, fisheries) context and refers to the improvement of selected functions related to the production of food and fiber through substantial external physical and chemical subsidies (Angeler et al. 2019). A related concept, coerced regimes, refers to the creation and maintenance of artificial, non–self-sustaining system regimes (Angeler et al. 2020). Angeler and Allen (chapter 4) explore within-scale and cross-scale aspects of coercion, with special attention to how coercion may be implemented within and across the distinct phases of the adaptive cycles of a panarchy. They introduce the term *coerced panarchy* to show this control of hierarchical system dynamics through management. The Earth's climate is currently changing toward a catastrophic regime shift, which we call hothouse Earth. To avoid such a shift, the current climate regime must be maintained artificially through a series of mitigation interventions. Implementation of Earth stewardship measures within and across scales can potentially coerce the current climate regime.

Ongoing environmental changes at the scale of the planet can generate unpredictable changes at smaller geographic units. Under such global changes, extant spatial regimes in biomes may resist change, adapt via changes in processes or biota, move in space over time, or collapse and reorganize. The term *spatial regimes*, like the theories of resilience and panarchy from which it emerged, captures the multiscaled nature of complex systems but explicitly includes spatial and temporal components, which have not previously been linked. Such spatial social-ecological regimes respond to particular drivers in unpredictable ways; however, there is evidence that tracking spatial regime movement in space allows early warnings of regime change for any given location over time. Allen et al. (chapter 5) apply concepts of spatial regimes and present methods to identify spatial regimes that avoid the problem of arbitrary human-scale imposition. To account for nonstationary and multiscale processes in dynamic landscapes, the authors use the term *biotechtonics* to describe spatial regime changes. Consideration of the spatial dimensions of resilience and regime change has great potential for creating planning horizons that provide sufficient time for transformation or adaptation, as opposed to leading indicators of regime change that are based solely on temporal data, which often provide a warning only after it is too late to affect action.

Over the past few decades, resilience concepts have been increasingly applied by practitioners, engineers, and planners throughout North America (Cosens and Gunderson 2018). The concepts have been applied to the operation of extant systems and as a response to the increasing costs associated with natural disasters, especially in the United States. Some decision makers and policymakers acknowledge that managed systems cannot be universally robust or resilient. Disruptions will occur that will test system resilience and result in an SES either remaining in the same regime or experiencing a regime shift and emerging as a new state or norm. Pumo et al. (chapter 6) discuss different conceptualizations of resilience (i.e., what constitutes resilience for different systems and what configurations of system components contribute to or erode resilience and can lead to difficulties for system managers in context of panarchy theory). They discuss how reframing these concepts can better link planning and actions to resilience. The authors also discuss the implications of panarchy theory to a large U.S. federal agency, the U.S. Army Corps of Engineers, which is responsible for managing navigable waters and wetlands in the United States.

Twidwell et al. (chapter 7) bring together the principles of panarchy theory with the utility of screening and imaging for regime shifts across a range of spatial scales. Such visualizations, combined with the theory, can lead to more proactive management of undesired transitions by detecting and tracking regime shifts not just within but across multiple scales of organization. At the heart of panarchy is the adaptive cycle, and multiple adaptive cycles occur within a nested hierarchy of ecological organization. At each domain of scale, the forward loop in the adaptive cycle is consistent with ecological succession and emphasizes periods of growth and conservation. A critical transition in the adaptive cycle occurs as the system enters the backward loop, which describes a period of destruction and reorganization. In theory, this transition from the forward to the backward loop can manifest as a spatially explicit signal of a transition between alternative, neighboring spatial regimes. Imaging and screening for the leading edge of regime shifts captures this critical transition, highlighting signals of spatial irregularities (i.e., boundaries) where one regime ends, or collapses in geographic space, and another regime begins (Uden et al. 2019). This leading edge can then be tracked over time to provide insight into the directional movement of regime shifts, provide more advanced warning of ecological change, and identify heightened vulnerability to intact ecosystems long before regime shifts become problematic (Roberts et al. 2019).

Cross-disciplinary diffusion of panarchy

Olsson et al. (chapter 8) describe concepts of transformation in governance regimes in the context of the panarchy model. They propose a parallel panarchy model and describe how it can contribute to the understanding of complex interactions across scales and generate new insights on transformations toward safe and just futures. In the same way as for the panarchy model, the concepts of revolt and remembrance are central to the parallel panarchy model and are used to explore cross-scale interactions, through examples in governance transformations in Chile, Uzbekistan, and South Africa. They also apply the parallel panarchy to the emergence of neoliberalism and a market-based economy in order to understand and navigate large-scale transformations that are necessary for dealing with Anthropocene risks and challenges at the global level. An increase in the speed, magnitude, and direction of such transformations to sustainability are urgently needed to deal with Anthropocene challenges and risks.

Panarchy captures the reality that SESs continually adapt and transform in response to disturbances and cross-scale interactions (both temporal and spatial). The Anthropocene amplifies this behavior, as stressors such as climate change, pervasive pollution, and biodiversity loss provide constant disturbance and change. However, law has social stability as its ultimate goal. Achieving this goal over time requires a nuanced balance of legal predictability and legal flexibility that can and should vary by context: Law has to adapt to social-ecological changes, but it cannot be so continually and pervasively in flux that it becomes an impediment to social and economic investment and security (Craig et al., chapter 9). Thus, some aspects of law, such as the basics of criminal law, need to remain dependable; governments should deter, catch, and punish murderers regardless of social-ecological change. Other aspects of law, especially the facets of law that deal most directly with the ecological components of SESs (e.g., natural resources and environmental law), need to manage resilience and be more cognizant of cross-scale interactions in order to continue to be effective.

The original panarchy volume (Gunderson and Holling 2002) was developed in part through the Beijer International Institute of Ecological Economics in Sweden. The institute has a multidecade history of collaborative scholarship between ecologists and economists. Although the 2002 volume addressed the intersections between resilience theory and economic theory, panarchy theory was largely unaddressed. Farley and Egler (chapter 10) step up to fill this gap and review microeconomic, macroeconomic, and ecological economics through a panarchy lens. They demonstrate the need for economic panarchy that generates adaptive economic cycles at different scales (from the family level up to the global level) as a way of managing resilience in economic systems and generating antifragile strategies. They note that the impacts of economic growth can flip global SESs into alternative states with potentially deleterious impacts on human welfare (Daly 2014). They argue that both microeconomic and macroeconomic theory explain or address this potential. Thus there is a need for a new economic model based on current models of change in SESs. The authors argue that panarchy theory can play a critical role in the development of this economic model.

Hodbod and Wentworth (chapter 11) demonstrate that a focus on cross-scale drivers through a panarchy framing is not only feasible in food systems but is critical. Such an approach is well suited for food systems, given their complexities, but it requires mixed methods and data from multiple disciplines and benefits from a more engaged approach, such as integrating the community into research in a transdisciplinary manner. The result is a rich longitudinal analysis that links socioeconomic drivers to both food system activities and food security outcomes, and vice versa, outlining how poor food security outcomes affect social welfare and socioeconomic conditions. Integrating an equity focus into such an analysis illuminated the asymmetries within food systems and revealed how differential power dynamics can influence the trajectory of the food system's adaptive cycle, and in turn that of food insecure households. Although challenges arise when applying panarchy in urban food system contexts, given the diversity of social actors across multiple scales and their notions of desirability, Hodbod and Wentworth find panarchy to be a useful tool for elucidating the dynamics of change across these scales and recommend its use to food system scholars.

In coupled SESs, cross-scale interactions exert substantial influence on system trajectories and internal system dynamics (Chaffin, chapter 12). Impacts of panarchy include both disruption and entrenchment of critical system drivers at focal scales of investigation, as well as resulting changes in system processes and structures. These dynamics are visible across a range of nested levels within a broader SES, including at the individual (psychological), community or group (social), and governance (institutional and organizational) levels. As cogent actors within these systems, humans—at all levels—both wield and react to cross-scale interactions in an attempt to shape SES trajectories, mainly through adaptive and transformative responses to disturbances and regime shifts. The systematic analysis of these dynamics can reveal patterns of coupled human–environment relations that are essential to understand if sustainable SES trajectories are to be achieved. Chaffin surveys a breadth of examples of human responses to cross-scale dynamics in SESs and identifies key capacities for adaptation and transformation at the individual, group, and societal levels. Such capacities address one of the most significant challenges to operationalizing concepts of panarchy toward the achievement of sustainable SES trajectories: bridging the cognitive and normative barriers of society toward the personal, social, technological, and eventually social-ecological adaptation and transformation necessary to displace entrenched forms of environmental governance and provide space for innovation in SESs.

Stewardship traditionally focused on local efforts to sustain the health of discrete ecosystems or landscapes to foster the benefits (ecosystem services) they provide to society (Chapin et al. 2009). Historically, such stewardship was implemented in top-down fashion to meet the goals of landowners and managers. Increases in the mobility of information (the internet), goods (international trade), pollutants (climate change), and people (migration) have rendered this stewardship model insufficient to meet either the local or global needs of nature and society or to address the inequitable distribution of vulnerabilities, privileges, and needs. Chapin et al. (chapter 13) discuss how stewardship must be reframed to meet the panarchy of these accelerating cross-scale interdependencies. Social-ecological stewardship (i.e., ecological solidarity) develops most readily at a single (usually local) scale and relies on trust that emerges when there is respectful engagement, dialogue, and power sharing. Cross-scale stewardship is less common and often requires new configurations of social networks and institutions and knowledge coproduction processes that are linked to innovations in policy and practice. This cross-scale integration is necessary to cultivate the trust that fosters buy-in and collaboration between groups that have not traditionally interacted. Examples of (sometimes) successful cross-scale stewardship include international development that fosters grassroots community empowerment and endogenous development, transnational businesses that invest in sustainable and equitable supply chains, international nongovernment organizations that engage local communities in designing and implementing conservation programs on working landscapes, and global change scientific programs that respectfully engage local and traditional knowledge. Cross-scale, social-ecological stewardship would benefit from greater emphasis on understanding how power mediates cross-scale integration, interdependencies, and decision-making processes across cultures, sectors, and disciplines, which is essential to address complex real-world problems that are clouded by inevitable uncertainty.

Literature Cited

Adger, W.N., H. Eakin, and A. Winkels. 2009. Nested and teleconnected vulnerabilities to environmental change. *Frontiers in Ecology and the Environment* 7: 150–157.

Allen, C.R., D.G. Angeler, A.L. Garmestani, L.H. Gunderson, and C.S. Holling. 2014. Panarchy: Theory and application. *Ecosystems* 17: 578–589.

Allen, C.R., and C.S. Holling. 2008. *Discontinuities in ecosystems and other complex systems.* Columbia University Press, New York.

Allen, C.R., and C.S. Holling. 2010. Novelty, adaptive capacity, and resilience. *Ecology and Society* 15(3): 24.

Allison, H.E., and R.J. Hobbs. 2004. Resilience, adaptive capacity, and the "lock-in trap" of the Western Australian agricultural region. *Ecology and Society* 9(1): 3.

Anderies, J. M., M. A. Janssen, and E. Ostrom. 2004. A framework to analyze the robustness of social-ecological systems from an institutional perspective. *Ecology and Society* 9(1): 18.

Angeler, D.G., B.C. Chaffin, S.M. Sundstrom, A. Garmestani, K.L. Pope, D. Twidwell, D. Uden, and C.R. Allen. 2020. Coerced regimes: Managing artificial feedbacks to navigate the Anthropocene. *Ecology and Society* 25(1): 4.

Angeler, D.G., H. Peterson, C.R. Allen, A. Garmestani, D. Twidwell, W. Chuang, V.M. Donovan, T. Eason, C.P. Roberts, S.M. Sundstrom, and C.L. Wonkka. 2019. Adaptive capacity in ecosystems. *Advances in Ecological Research* 60: 1–24.

Auad, G., J. Blythea, K. Coffman, and B.D. Fath. 2018. A dynamic management framework for socio-ecological system stewardship: A case study for the United States

Bureau of Ocean Energy Management. *Journal of Environmental Management* 225: 32–45.

Baskerville, G.L. 1995. The forestry problem: Adaptive lurches of renewal. In *Barriers and Bridges to the Renewal of Ecosystems and Institutions*, 37–102. ed. L.H. Gunderson, C.S. Holling, and S.S. Light. Columbia University Press, New York.

Benson, M.H., and A.S. Garmestani. 2011. Embracing panarchy, building resilience and integrating adaptive management through a rebirth of the National Environmental Policy Act. *Journal of Environmental Management* 92: 1420–1427.

Brown, R.R., M.A. Farrelly, and D.A. Loorbach. 2013. Actors working the institutions in sustainability transitions: The case of Melbourne's stormwater management. *Global Environmental Change* 23: 701–718.

Carpenter, S. R., and W. A. Brock. 2008. Adaptive capacity and traps. *Ecology and Society* 13(2): 40.

Cash, D.W., W. Adger, F. Berkes, P. Garden, L. Lebel, P. Olsson, and O. Young. 2006. Scale and cross-scale dynamics: Governance and information in a multilevel world. *Ecology and Society* 11(2): 8.

Chaffin, B., A. Garmestani, L. Gunderson, M. Benson, D. Angeler, C. Arnold, B. Cosens, R. Craig, J.B. Ruhl, and C.R. Allen. 2016. Transformative environmental governance. *Annual Review Environment and Resources* 41: 399–423.

Chaffin, B., and L. Gunderson. 2015. Emergence, institutionalization and renewal: Rhythms of adaptive governance in complex social-ecological systems. *Journal of Environmental Management* 165: 81–87.

Chapin, F.S., G. Kofinas, and C. Folke, Eds. 2009. *Principles of Ecosystem Stewardship*. Springer, New York.

Chapin, F.S., G. Peterson, F. Berkes, T.V. Callaghan, P. Angelstam, M. Apps, C. Beier, A.-S.C.Y. Bergeron, K. Danell, T. Elmqvist, C. Folke, B. Forbes, N. Fresco, G. Juday, J. Niemelä, A. Shvidenko, and G. Whiteman. 2004. Resilience and vulnerability of northern regions to social and environmental change. *Ambio* 336: 344–349.

Cinner, J.E. 2011. Social-ecological traps in reef fisheries. *Global Environmental Change* 21: 835–839.

Cosens, B., and L.H. Gunderson, Eds. 2018. *Practical Panarchy for Adaptive Water Governance: Linking Law to Social-Ecological Resilience*. Springer, New York.

Costanza, R., and J. Greer. 1995. The Chesapeake Bay and its watershed: A model for sustainable ecosystem management. In *Barriers and Bridges to the Renewal of Ecosystems and Institutions*, ed. L.H. Gunderson, C.S. Holling, and S.S. Light, 169–213. Columbia University Press, New York.

Craig, R.K. 2013. Learning to think about complex environmental systems in environmental and natural resource law and legal scholarship: A twenty-year retrospective. *Fordham Environmental Law Review* 24: 87–102.

Craig, R.K. 2017. Putting resilience theory into practice: The example of fisheries management. *Natural Resources and the Environment* 31: 3–7.

Daly, H.E. 2014. *From Uneconomic Growth to a Steady-State Economy*. Edward Elgar, New York.

Eason, T., and A.S. Garmestani. 2012. Cross-scale dynamics of a regional urban system through time. *Region et Developpement* 36: 55–77.

Estes, J.A., and D.O. Duggins. 1995. Sea otters and kelp forests in Alaska: Generality and variation in a community ecological paradigm. *Ecological Monographs* 651: 75–100.

Foley, J. A., M. T. Coe, M. Scheffer, and G. L. Wang. 2003. Regime shifts in the Sahara and Sahel: Interactions between ecological and climatic systems in northern Africa. *Ecosystems* 66: 524–539.

Folke, C. 2006. Resilience: The emergence of a perspective for social–ecological systems analyses. *Global Environmental Change* 16: 253–267.

Folke, C. 2016. Resilience. In *Subject: Framing Concepts in Environmental Science. Oxford Research Encyclopedias, Environmental Science*, ed. H. Shugart. Oxford University Press, New York.

Folke, C., S. Carpenter, B. Walker, M. Scheffer, T. Chapin, and J. Rockstrom. 2010. Resilience thinking: Integrating resilience, adaptability and transformability. *Ecology and Society* 15(4): 20.

Folke, C., S. Carpenter, B. Walker, M. Scheffer, T. Elmqvist, L. Gunderson, and C.S. Holling. 2004. Regime shifts, resilience, and biodiversity in ecosystem management. *Annual Review of Ecology Evolution and Systematics* 35: 557–581.

Francis, G.R., and H. Regier. 1995. Barriers and bridges to the restoration of the Great Lakes Basin ecosystem. In *Barriers and Bridges to the Renewal of Ecosystems and Institutions*, ed. L.H. Gunderson, C.S. Holling, and S.S. Light, 239–291. Columbia University Press, New York.

Garcia, J.H., A.S. Garmestani, and A.T. Karunanithi. 2011. Threshold transitions in a regional urban system. *Journal of Economic Behavior & Organization* 78: 152–159.

Garmestani, A., and C.R. Allen. 2014. *Social-Ecological Resilience and Law*. Columbia University Press, New York.

Garmestani, A.S., C.R. Allen, and K.M. Bessey. 2005. Time-series analysis of clusters in city size distributions. *Urban Studies* 42: 1507–1515.

Garmestani, A.S., C.R. Allen, and L. Gunderson. 2009. Panarchy: discontinuities reveal similarities in the dynamic system structure of ecological and social systems. *Ecology and Society* 14(1): 15.

Garmestani, A.S., and M.H. Benson. 2013. A framework for resilience-based governance of social-ecological systems. *Ecology and Society* 18(1): 9.

Garmestani, A., D. Twidwell, D.G. Angeler, S.M. Sundstrom, C. Barichievy, B.C. Chaffin, T. Eason, N.A.J. Graham, D. Granholm, L. Gunderson, M. Knutson, K.L. Nash, R. Nelson, M. Nystrom, T.L. Spanbauer, C.A. Stow, and C.R. Allen. 2020. Panarchy: Opportunities and challenges for ecosystem management. *Frontiers in Ecology and the Environment* 18: 576–583.

Green, O.O., A.S. Garmestani, S. Albro, N. Ban, A. Berland, C. Burkman, M.M. Gardiner, L.H. Gunderson, M.E. Hopton, M. Schoon, and W.D. Shuster. 2016. Adaptive governance to promote ecosystem services in urban green spaces. *Urban Ecosystems* 19: 77–93.

Grimm, N.B., J.M. Grove, S.T.A. Pickett, and C.L. Redman. 2000. Integrated approaches to long-term studies of urban ecological systems. *BioScience* 50: 571–593.

Gunderson, L. H. 2003. Adaptive dancing: Interactions between social resilience and ecological crises. In *Navigating Social-Ecological Systems: Building Resilience for Complexity and Change*, ed. F. Berkes, J. Colding, and C. Folke, 33–52. Cambridge University Press, Cambridge, UK.

Gunderson, L.H., S.R. Carpenter, C. Folke, P. Olsson, and G.D. Peterson. 2006. Water RATs (resilience, adaptability, and transformability) in lake and wetland social-ecological systems. *Ecology and Society* 11(1): 16.

Gunderson, L. A. Garmestani, K. Rizzardi, J.B. Ruhl, and A. Light. 2018. Social, legal and ecological capacity for adaptation and transformation in the Everglades. In *Practical Panarchy, Linking Law, Resilience and Adaptive Water Governance of Regional Scale Social-Ecological Systems*, ed. B. Cosens and L. Gunderson, 65–82. Springer, Switzerland.

Gunderson, L.H., and C.S. Holling. 2002. *Panarchy: Understanding Transformations in Human and Natural Systems*. Island Press, Washington, DC.

Gunderson, L.H., C. S., Holling, and S.S. Light. 1995. *Barriers and Bridges to the Renewal of Ecosystems and Institutions*. Columbia University Press, New York.

Gunderson, L.H., and L. Pritchard. 2002. *Resilience and the Behavior of Large-Scale Systems*. Island Press, Washington, DC.

Holling, C.S. 1973. Resilience and stability of ecological systems. *Annual Review of Ecology and Systematics* 4: 1–23.

Holling, C.S. 1986. The resilience of terrestrial ecosystems: Local surprise and global change. In *Sustainable Development of the Biosphere*, ed. W. C. Clark and R. E. Munn, 292–317. Cambridge University Press, Cambridge, UK.

Holling, C.S. 1992. Cross-scale morphology, geometry, and dynamics of ecosystems. *Ecological Monographs* 62: 447–502.

Holling, C. S. 1996. Engineering resilience versus ecological resilience. In *Engineering within Ecological Constraints*, ed. P. Schulze, 31–43. National Academies Press, Washington, DC.

Holling, C.S. 2001. Understanding the complexity of economic, ecological, and social systems. *Ecosystems* 4: 390–405.

Holling, C.S., and L.H. Gunderson. 2002. Resilience and adaptive cycles. In *Panarchy: Understanding Transformations in Human and Natural Systems*, ed. L.H. Gunderson and C.S. Holling, 25–62. Island Press, Washington, DC.

Holling, C.S., L.H. Gunderson, and G.D. Peterson. 2002. Sustainability and panarchies. In *Panarchy: Understanding Transformations in Systems of Humans and Nature*, ed. L.H. Gunderson and C.S. Holling, 63–101. Island Press, Washington, DC.

Hughes, T.P., A.H. Baird, D.R. Bellwood, M. Card, S.R. Connolly, C. Folke, R. Grosberg, O. Hoegh-Guldberg, J.B.C. Jackson, J. Kleypas, J.M. Lough, P. Marshall, M. Nystrom, S.R. Palumbi, J.M. Pandolfi, B. Rosen, and J. Roughgarden. 2003. Climate change, human impacts, and the resilience of coral reefs. *Science* 301: 929–933.

Jansson, B.O., and H. Velner. 1995. The Baltic: The sea of surprises. In *Barriers and Bridges to the Renewal of Ecosystems and Institutions*, ed. L.H. Gunderson, C.S. Holling, and S.S. Light, 292–374. Columbia University Press, New York.

Lee, K.N. 1995. Deliberately seeking sustainability in the Columbia River Basin. In *Barriers and Bridges to the Renewal of Ecosystems and Institutions*, ed. L.H. Gunderson, C.S. Holling, and S.S. Light, 214–238. Columbia University Press, New York.

Light, S.S., L.H. Gunderson, and C.S. Holling. 1995. The Everglades: Evolution of management in a turbulent ecosystem. In *Barriers and Bridges to the Renewal of Ecosystems and Institutions*, ed. L.H. Gunderson, C.S. Holling, and S.S. Light, 103–168. Columbia University Press, New York.

Müller, F., M. Bergmann, R. Dannowski, J.W. Dippner, A. Gnauck, P. Haase,

M.C. Jochimsen, P. Kasprzak, I. Kröncke, R. Kümmerlin, M. Küster, G. Lischeid, H. Meesenburg, C. Merz, G. Millat, J. Müller, J. Padisák, C.G. Schimming, H. Schubert, M. Schult, G. Selmeczy, T. Shatwell, S. Stoll, M. Schwabe, T. Soltwedel, D. Straile, and M. Theuerkauf. 2016. Assessing resilience in long-term ecological data sets. *Ecological Indicators* 65: 10-43.

Nielsen, S.N., and S.E. Jorgensen. 2013. Goal functions, orientors and indicators (GoFOrIt's) in ecology. Application and functional aspects—Strengths and weaknesses. *Ecological Indicators* 28: 31–47.

Pickett, S.T.A., M.L. Cadenasso, J.M. Grove, C.H. Nilon, R.V. Pouyat, W.C. Zipperer, and R. Costanza. 2001. Urban ecological systems: linking terrestrial ecological, physical, and socioeconomic components of metropolitan areas. *Annual Review of Ecology and Systematics* 32: 127–157.

Roberts, C.P., C.R. Allen, D.G. Angeler, and D. Twidwell. 2019. Shifting avian spatial regimes in a changing climate. *Nature Climate Change* 9: 562–568.

Scheffer, M., S. Carpenter, J.A. Foley, C. Folke, and B. Walker. 2001. Catastrophic shifts in ecosystems. *Nature* 413: 591–596.

Schumpeter, J. A. 1942. *Capitalism, Socialism and Democracy*. Harper and Brothers, New York.

Steffen, W., J. Rockstrom, K. Richardson, T.M. Lenton, C. Folke, D. Liverman, and H.J. Schellnhuber. 2018. Trajectories of the Earth system in the Anthropocene. *Proceedings of the National Academy of Sciences* 115: 8252–8259.

Uden, D.R., D. Twidwell, C.R. Allen, M.O. Jones, D.E. Naugle, J.D. Maestas, and B.W. Allred. 2019. Spatial imaging and screening for regime shifts. *Frontiers in Ecology and Evolution* 7: 407.

Walker, B., S.R. Carpenter, C. Folke, L. Gunderson, G.D. Peterson, M. Scheffer, M. Schoon, and F.R. Westley. 2020. Navigating the chaos of an unfolding global cycle. *Ecology and Society* 25(4): 23.

Walker, B., L. Gunderson, A. Kinzig, C. Folke, S. Carpenter, and L. Schultz. 2006. A handful of heuristics and some propositions for understanding resilience in social-ecological systems. *Ecology and Society* 11(1): 13.

Walker, B., and D. Salt. 2006. *Resilience Thinking: Sustaining Ecosystems and People in a Changing World.* Island Press, Washington, DC.

Walker, B., and D. Salt. 2012. *Resilience Practice: Building Capacity to Absorb Disturbance and Maintain Function*. Island Press, Washington, DC.

Westley, F. 2002. The devil in the dynamics: Adaptive management on the front lines. In *Panarchy: Understanding Transformations in Human and Natural Systems*, ed. L.H. Gunderson and C.S. Holling, 333–360. Island Press, Washington, DC.

Westley, F., S.R. Carpenter, W.A. Brock, C.S. Holling, and L. Gunderson. 2002. Why systems of people and nature are not just social and ecological systems. In *Panarchy: Understanding Transformations in in Human and Natural Systems*, ed. L.H. Gunderson and C.S. Holling, 103–120. Island Press, Washington, DC.

Wieland, A. 2021. Dancing the supply chain: Towards transformative supply chain management. *Journal of Supply Chain Management* 57: 1–16.

PART II

APPLICATIONS OF PANARCHY THEORY

Chapter 2

Panarchy, Cross-Scale Resilience, and Discontinuous Structures and Processes

Shana M. Sundstrom, Craig R. Allen, and David G. Angeler

When C. S. Holling published his seminal article on ecological resilience in 1973 (Holling 1973), it is unlikely he anticipated that by 2020, more than 80,000 articles would be published on resilience (Web of Science topic search). Resilience science has transformed ecology and other fields by expanding our understanding of system dynamics and behavior away from a reductionist view of ecosystems as operating on a single, fixed attractor with equilibrium dynamics and linear behavior to a complex system view that embraces nonlinear change and multiple regimes. Resilience science has been successful in part because Holling and colleagues developed a rich and compelling narrative of system behavior that quite simply describes reality better than previous scientific worldviews of linked systems of human and nature (Gunderson and Holling 2002). Holling's ability to observe the same phenomena as other scientists, such as avian predation on budworm infestations of spruce trees, but use those observations to transform how we view the dynamics and behavior of ecosystems was unique (Holling 1988). In fact, individual facets of resilience science are often their own fields of study; we refer to concepts such as regime shifts and threshold behaviors, adaptive management, resilience and law, transformative governance, adaptive cycles, and the mechanisms by which resilience emerges (e.g., cross-scale resilience, adaptive capacity, and ecological memory) (Holling 1978; Scheffer 2009; Garmestani and Allen 2014; Chaffin et al. 2016b; Angeler et al. 2019b; Sundstrom and Allen 2019). Although resilience has had an extraordinary amount of exposure in the scientific community, there remains an untold story that rests at the core of resilience science. Or, perhaps more accurately, individual pieces of this narrative have been explored, but their relationship to each other remains underdeveloped (Allen et al. 2014; Nash et al. 2014a; Sundstrom and Allen 2019). This chapter links those narratives, around concepts of scale, cycles, and the structure of complex systems. It focuses on the resilience concepts of discontinuities, adaptive cycles and panarchy, their relationship to each other, and the implications for more rigorous assessments of resilience in complex systems.

Complex systems have in common a set of attributes that differentiate them from systems that are merely complicated. For example, a diesel combustion engine is complicated rather than complex because its behavior is equal to the sum of its parts, abides by a well-understood set of deterministic rules, and is thus predictable. In contrast, complex systems are inherently unpredictable and abide by a broad set of rules defining dynamics and behavior that emerge from basic laws of thermodynamics and can include chaotic behavior and emergent phenomena. Attributes shared across all types of complex systems include operating out of equilibrium, having thresholds and nonlinearities, and containing multiple scales of structure. Hierarchical structure emerges in order to degrade energy, and the more complex a system is, the more hierarchical levels are presumably present (Schneider and Kay 1994). *Hierarchical structure* in common parlance refers to a nested hierarchy, envisioned as a food pyramid, or a company's organizational chart, but this usage conflates organizational level with hierarchical structure. For example, ecosystems are commonly understood to contain the following levels of organization: species < population < community < ecosystem < landscape < biome. The assumption too often made is that these convenient levels of organization are also the scales at which pattern and process manifest. To a certain extent it is true, in that biological and other processes have generated structure, in the form of a species, and species represent a level of structure in a hierarchy that degrades energy. However, this typology of organization does not capture the totality of the relevant scales of pattern and process. In resilience science, *hierarchical structure* refers to the spatial and temporal scales at which pattern and process manifest. A very small species, such as a mouse, does not convert as much organic energy as a moose. Both the spatial and temporal scales at which a mouse's life history plays out are profoundly different from those of a moose. So although the "species" level of organization as represented by a mouse and a moose is clearly a useful approach for some research, it is also incomplete. A major contribution of Holling (1986, 1992) to our understanding of pattern, process, and scale in ecosystems and other types of complex systems was to introduce three concepts—the discontinuity hypothesis, adaptive cycles, and panarchy—which provide a more objective and empirical approach to understanding scaling, structure, and the emergence of resilience in complex systems.

The term *scale* refers specifically to the spatiotemporal domains at which pattern and process manifest. Scale domains are the range of spatial and temporal scales that compartmentalize a particular set of ecological patterns and processes (Wiens 1989). Furthermore, scale domains are separated by nonlinear and abrupt transitions from one set of processes that generate pattern and structure to another set; in other words, scale domains are separated by discontinuities (Wiens 1989; Holling 1992). The discontinuity hypothesis describes the mechanism by which scale domains are generated and provides the theoretical underpinning for objectively and quantitatively identifying them. A panarchy describes the nested nature of multiple scale domains that collectively make up an ecosystem or other complex system and that can be structurally deduced from discontinuities (Gunderson and Holling 2002; Nash et al. 2014a).

Identifying scales in ecosystems has been a prominent issue for decades (Levin 1992). Historically, spatial scale has been defined by grain and extent, where extent is the size of the geographic area under consideration, and grain is the minimum size of individual units of observation (Turner et al. 1989). Although these definitions are useful, the choice of grain and extent is up to each researcher and often is constrained by nonbiological or nonecological factors, such as funding issues, geopolitical boundaries, or time constraints. In ecology, the conversation about scaling phenomena fundamentally changed with a seminal article by Wiens (1989), who pointed out that the importance of scale is not confined to sampling methods in field work (e.g., using sampling scales appropriate for the geographic distribution of a species or the scales at which processes play out) but is "fundamental to all ecological investigations." Furthermore, it is a key aspect of managing complex systems, because management that is not matched to the scale of the problem is limited in its ability to meet management objectives over the long term (Garmestani and Benson 2013; Green et al. 2014, 2015). Even under best-case scenarios, where policies and management are carefully matched to the spatial and temporal scales of the system, species, or process of concern, rapid and unexpected changes can occur when changes at either smaller or larger spatial and temporal scales cascade up (or down), driving unwanted changes at the scale of concern. Despite the critical importance of scale for linked systems of humans and nature, scale is still often dealt with in a cursory fashion because of the difficulties in objectively identifying the critical scales in a system. Too often, this means defaulting to identifying a focal scale of concern and then taking into consideration an arbitrary (though not scientifically defined) scale above and scale below. Moving beyond the arbitrary delineation of scale and toward assessments that incorporate both spatial and temporal axes of scale, is critical for ecology and for understanding complex social-ecological systems.

Scale: The Key to Understanding Panarchy

A panarchy can be simplistically defined as a set of nested adaptive cycles (see chapters 1 and 7 for discussion of adaptive cycles). Implicit in the idea of panarchy as a model of complex system organization is that there is scale-specific structure occurring at specific scale domains and that most strong interactions and feedbacks are largely compartmentalized within a scale domain. Furthermore, those within-scale dynamics can be described as adaptive cycles, or cycles of growth, conservation, collapse, and renewal. Adaptive cycles are separated from one another by thresholds or discontinuities between scale domains (Holling and Gunderson 2002). Processes occurring at larger and slower scale domains constrain smaller ones, whereas smaller and faster adaptive cycles have the flexibility to test, invent, experiment, and occasionally trigger change at larger scales (Holling et al. 2002). That is, in a panarchy information can flow both from the bottom up and the top down across adaptive cycles, thereby steering the dynamics of the overall system.

This separation of distinct scale domains has several important implications and effects. First, it means that key variables within systems are distributed discontinuously. Second, it indicates that self-organizing interactions and processes, such as community-level interactions for organisms (including interactions such as competition), are compartmentalized by scale (Allen et al. 2015). Therefore, similarly sized entities, whether they be animals, human communities, or firms, are more likely to strongly interact with each other than entities of grossly different sizes, although exceptions occur (e.g., with predation, where bigger things tend to eat smaller things). The compartmentalization of systems along an axis of scale provides rich opportunities for experimentation within scale domains without putting the system itself at risk because risk is largely confined to that scale domain. For example, disturbance in ecosystems is scale specific, manifested in a much higher vulnerability to hail showers at the scale of seedling patches compared with the scale of the entire forest (Angeler et al. 2018). This compartmentalization by scale also leads to the development of high levels of diversity within systems (O'Neill et al. 1986), because ecological niches occur along temporal as well as spatial axes, allowing more species to coexist despite using similar resources. Discontinuities between scale domains can be objectively identified and used as a starting basis for quantitative assessments of system scales and nested adaptive cycle dynamics and cross-scale feedbacks (Angeler et al. 2015).

Discontinuities in Structures and Processes

Panarchy has been widely used as a conceptual model by which to understand and envision the organization of complex systems (Apeldoorn et al. 2011; Porto 2012; Ruhl 2012; Randle et al. 2014; Chaffin and Gunderson 2016; Garmestani et al. 2020). It can also be used as a model of system dynamics, which gives rise to concrete and testable hypotheses about the functioning of complex systems (e.g., ecosystems, urban systems, social systems, and linked social-ecological systems) (Allen et al. 2014). When combined with the discontinuity hypothesis, it can move from the conceptual to quantitative realm (Allen and Holling 2008; Angeler et al. 2016).

Components in ecosystems and other complex systems interact to create structures in time and space. Interactions that are reinforced over time persist, whereas those that are not fade away, as is manifested in the turnover of species in ecological assemblages. This structure provides the core "memory" of a system in that its structure is resilient to change (Allen and Holling 2002). This is important for humans because ecosystems often appear stable, such that we can expect reasonably predictable dynamics and the relatively constant provision of ecosystem goods and services. This system-level conservativeness is maintained by processes of self-organization due in part to the interaction of biotic and abiotic elements, where processes create ecological patterns in ecological structure, but species interact with that structure in a way that reinforces patterns beneficial to them. When ecological processes consistently and persistently occur at distinct spatial and temporal scales, they in turn create scaled structure. For example, the spatial and temporal scales of forest structure are shaped by photosynthesis (small and fast scales), patterns in insectivory and herbivory (moderate spatial and temporal scales), fire (larger scales), and geoclimatic processes of soil, temperature, and precipitation (largest spatial and temporal scales). The scaled nature of forest structure therefore provides resource opportunities to species at particular ranges of scale, and species with a body mass that can take advantage of the scales of available resources will be more successful in the long term (Holling 1992). Many animal life history traits are strongly allometric with size, so animal body mass represents the scales at which species perceive and interact with their environment (Peters 1983; Haskell et al. 2002). Thus, just as ecological structure is scaled into distinct scale domains at which pattern in the form of structure manifests, so too should animal body masses. This was the original discontinuity hypothesis, and it has since been strongly validated with empirical evidence (Forys and Allen 2002; Allen and Holling 2008; Wardwell et al. 2008; Sundstrom et al. 2012). However, animals are not just passively interacting with their environment. In exploiting their environments, animals, including humans, often change ecological (and nonecological) structures in ways that are favorable for their own well-being and success. For example, large herbivores can alter the dynamics of succession (and competition between grasses, bushes, and trees) such that the habitat is, in some sense of the word, optimal for them. Self-organization involves other biotic system elements as well. For example, many grasses worldwide are pyrophilic and highly flammable. In the absence of fire, succession would often eliminate these grasslands from the ecosystems they occupy. However, the presence of these grasses encourages fire, which favors their spread and excludes competitors. The end result of these interactions is nested scale domains separated by discontinuities, in both the structure of the environment and the entities interacting with that environment. This physically manifests as highly conservative size classes (e.g., Garmestani et al. 2009) in both structure and the entities interacting with that structure. Species' size classes represent the scales at which they interact with their environment and thus the availability of resources at those spatial and temporal scales. These are also the scale domains at which adaptive cycles manifest as a result of the scaled interactions between species, ecological processes, and environmental structure. This should hold true in any complex system, not just ecosystems (Sundstrom et al. 2014).

Holling first proposed the discontinuity hypothesis in 1992 (Holling 1992). A decade later, the full relationship between discontinuities, adaptive cycles, and panarchy was described. Holling et al. (2002) presented five proximate mechanisms that could explain the discontinuous distributions of body masses he had identified in multiple taxa across multiple ecosystems. There were three slow processes—the discontinuity hypothesis, phylogeny, and the founder effect—and two fast processes, competition and trophic relationships. A decade had passed since the original proposition, and evidence had accumulated not only in support of the discontinuity hypothesis but also disproving some of the other possible mechanisms (Sendzimir 1998; Allen et al. 1999; Forys and Allen 1999, 2002; Sendzimir et al. 2002). In Allen et al. (1999), phylogeny was disproven as a structuring mechanism, because it couldn't explain the size classes found in south Florida herpetofauna, bird, or mammal body mass distributions. More recently, competitive processes such as emergent neutrality were tested as a structuring mechanism and authors found that, when applied to simulated and plankton communities, emergent neutrality could generate multimodal size classes separated by discontinuities (Scheffer and van Nes 2006; Vergnon et al. 2012). However, the locations of the discontinuities were not stable, contrary to what has been found in real ecosystems where the location of size classes has been highly robust and conservative over time (Lambert and Holling 1998; Raffaelli et al. 2000; Garmestani et al. 2005; but see Angeler et al. 2019b). Allen et al. (2006) discuss the different spatial and temporal scales over which the discontinuity hypothesis remains explanatory relative to other possible mechanisms and conclude that it is the dominant mechanism producing discontinuous body distributions within an ecosystem boundary but is less explanatory when data from multiple ecosystems are aggregated. Perhaps more importantly, the highly conservative nature of the location of scale domains and the discontinuities separating them has continued to be demonstrated in multiple types of complex systems and over long time scales, from ecosystems, to city sizes, and the size of economies (Garmestani et al. 2005; Nash et al. 2013, 2014a; Spanbauer et al. 2016; Sundstrom et al. 2020). It has also been empirically demonstrated that ecological structure (e.g., coral reef) is strongly scaled and correlated to scale domains in the reef fish community, something that had been difficult to assess because of the challenge of measuring "size" in ecological structure (Nash et al. 2013).

In the 30 years since Holling (1992) first proposed the discontinuity hypothesis, overwhelming evidence has accumulated that complex systems are not only hierarchical, they contain discrete, nested scale domains that can be objectively identified. Pattern and process within each scale domain both shape and are shaped by the entities that interact with each other and with their environment at those scales. Not only that, the scale domains are separated by discontinuities, or nonlinearities in processes, patterns, and entities in that system. There are several implications of this depiction of the hierarchical, scaled nature of complex systems: It provides a more objective approach for understanding and using scale in complex systems; it has a strong bearing on system resilience; it provides a framework for understanding how disturbance, change, and cross-scale feedbacks can potentially cascade up or down scale domains; it highlights the importance of management actions being explicitly embedded in a clear understanding of the scales at which pattern and process manifest and interact; it suggests that scale-specific structure is a key defining element of any complex system; and it provokes interesting questions about what kind of entities and behaviors have evolved to operate close to a discontinuity, where pattern and process are less stable, as opposed to entities and behaviors that take advantage of more stable patterns and processes far from the discontinuities.

Quantifying Panarchies

Objectively identifying the scales present in a panarchy has occurred primarily via a suite of methods developed to conduct discontinuity analyses, including cluster analysis, a kernel density estimator called the Gap Rarity Index (Restrepo et al. 1997), and Bayesian classification and regression trees (Chipman et al. 1998; Stow et al. 2007). More recently, the Gap Rarity Index has been modified to be fully transparent and free of subjective constants, and the revised method is called the Discontinuity Detector (Barichievy et al. 2018). These methods can be adjusted to account for differences in the life history traits of organisms (e.g., species with determinate vs. indeterminate growth (Nash et al. 2014b). All early tests used body mass distributions of a census of all species present in an ecosystem, sorted by taxa (Holling 1992). Later studies expanded to other types of complex systems, using the size distributions of key system elements such as size of a city, firms, or an economy (Garmestani et al. 2005, 2007, 2008; Garcia et al. 2011; Sundstrom et al. 2014, 2020).

Size is not the only attribute that can be used to identify discontinuities. Consider that the discrete and persistent spatial and temporal scales of processes are what shape ecological structure; therefore, the frequency or temporal scale of an entity's interaction with its environment should be effective and manifested in other variables (e.g., species densities and abundances) as well (Angeler et al. 2014a). For example, analysis of long-term data has revealed discrete groups of lake invertebrate species that exhibited distinct temporal frequencies. Some invertebrate species responded to slow environmental variables, and others responded to fast variables (Angeler et al. 2013).

Resilience and Panarchies

Arguably the most useful application of discontinuity analysis and panarchy theory thus far has been the cross-scale resilience model, a quantitative method for assessing system resilience. This model posits that system resilience is predicated, in part, on the nonrandom distribution of functions within and across the scale domains of a panarchy (Peterson et al. 1998). The model provides a framework for the analysis of functions within and across scales, and discontinuity analysis provides a method for objectively identifying scales present in a system and assessing resilience (Allen et al. 2005).

In ecosystems, species have functional traits that maintain system processes and feedbacks, such as insectivores, which control insect predation, or detritivores, which contribute to decomposition and nutrient cycling. A resilient system will have diverse functions present within a scale domain and a redundancy of function distributed across scale domains, which provides buffering against scale-specific disturbances (Peterson et al. 1998). Within-scale diversity should be imbricated, with overlap in functional and response diversity, allowing a rich response to disturbance or perturbation. Redundancy of function is also critical, because so-called redundant functions occur at different spatial and temporal scales, thus filling different ecological niches because of their separation by scale. This represents an important aspect of resilience: the ability to limit the potential cascade of collapse given perturbations that could otherwise scale up, such as insect outbreaks, and the ability to provide compensatory function for reductions in species that provide similar functionality at other scales (Holling 1988; Winfree and Kremen 2009; Mouillot et al. 2013). Research has shown that species' functions are indeed distributed in a nonrandom manner that matches the predictions of the cross-scale resilience model (Forys and Allen 2002; Wardwell et al. 2008). This distribution has been shown to contribute to recovery from disturbance and to be a valuable indicator to managers of system resilience and ability to recover from environmental disturbances (Nash et al. 2015). Furthermore, in a nondegraded system (defined here as containing the full complement of native species with few losses via extirpation), the nonrandom distribution of functions appears to be strongly resilient to the loss of species in the system, although there are clear thresholds of species loss beyond which resilience is rapidly eroded (Sundstrom et al. 2012; Mouillot et al. 2014). The model remains an underused but valuable tool for objectively assessing a key measure of resilience (distribution of function across nonarbitrary system-scale domains) and could be valuable for tracking how resilience changes in response to both an immediate perturbation (and how it changes over longer time frames) and more persistent disturbances such as climate change. The cross-scale resilience model could help guide management; for example, management actions could target the reinforcement of critical ecological functions at scales that are relatively free from stress, in order to buffer against the potential loss of functions at affected scales, thereby fostering ecosystem resilience (Angeler et al. 2013).

Cross-scale resilience reflects systemic resilience across all scales present within a system, and resilience emerges from the distribution of functions within and across system-scale domains identified by a discontinuity analysis. Dynamics and behavior within a scale domain are described by the adaptive cycle: cycles of growth, conservation, collapse, and renewal that play out at each scale domain within a system. The cross-scale resilience model does not directly address within-scale resilience, which is largely a function of the strength of feedbacks within a scale. This is important because it explains, in part, how adaptive cycles at the largest scales (an extreme example is the global climate system) don't collapse despite the lack of redundancy or diversity at those largest scales. In fact, managing for resilience at a particular scale domain within a system can be at odds with system-level resilience. As an example, consider a grassland that has a regime defined by a dominance of grasses, as opposed to another ecosystem type (e.g., woodland) that could occupy the same geographic space. At the system level we may want it to remain a grassland rather than convert to a desert or woody shrubland. Resilience emerges from pattern and process playing out at multiple spatial and temporal scales within the system. If we were to consider the resilience of small patches of grass within the system, only to see them become bare ground after a fire occurrence, the collapse of the grass patches from fire would be deemed a regime shift at that scale domain. However, a heterogeneous mosaic of burn patches throughout a grassland is critical for system-level resilience and persistence. Similarly, although the maintenance of functions as provided by species is critical for system-level resilience, both the presence or absence and abundances of species at all scales can be highly variable, and that variability provides critical adaptive capacity and flexibility to the system as it responds to disturbances. A focus on the resilience of one scale could be either positive or negative with regard to system resilience: positive if it means maintaining functionality that simply isn't present at other scales or is especially critical, as in a keystone species, or negative if it eliminates the variability necessary for system-level resilience by attempting to keep that scale domain within a particular regime. Thus it is more customary to discuss cross-scale resilience at the system level as a function of factors occurring at smaller-scale domains, as opposed to a narrower focus on within-scale resilience. A primary goal of understanding system-level resilience is to understand the vulnerability of a system to a regime shift, where the system reorganizes around a new set of defining structures, processes, and functions, often to the detriment of both human and other species.

Applications of Identifying Discontinuities in Panarchies

Identifying discontinuities in a panarchy is a powerful entry point for a deeper understanding of central resilience concepts. Regime shift research has focused primarily on the system level, for example, without consideration of the full panarchy. Spatial regime research also benefits from a discontinuity and panarchy approach, because they can be used to detect the movement of regimes in space. Finally, the emergence of novelty, both beneficial and otherwise, appears to be linked to discontinuities in a panarchy.

Regime shifts

Regime shifts occur when a system shifts from one characteristic set of controlling processes, structures, and functions to another. These shifts can also be described as critical transitions or movement to an alternative basin of attraction and are discussed almost exclusively at the ecosystem level or the focal scale of interest of the researcher. Thus far, humans have understandably defined both systems of interest and the scales of interest in ways that are tractable and most relevant to human concerns. This means our focus has been on managing and understanding systems at meso scales, with geographic extents of tens of meters to thousands of meters, and temporal frequencies of weeks to decades. As a result, our best understanding of regime shifts and resilience comes from studies of freshwater lakes, which are systems with defined physical boundaries, and have processes and cycles that occur from weeks to a few decades (Scheffer 2009). We have substantial knowledge gaps about regime shifts that occur at other scales, and this lack of knowledge becomes fraught with risk as the human global footprint continues to expand. Furthermore, because of our singular focus on subjective scales of human interest, we know very little about how change at smaller spatial and temporal scales can cascade up to drive collapse at larger scales (Hughes et al. 2013), or what collapse at global levels means for smaller-scale domains on Earth (Rockström et al. 2009a; Steffen et al. 2018). Discussing regime shifts in the context of panarchy theory expands the conversation in critical ways and takes into account the full social and ecological complexity of systems. It should enrich our ability to mechanistically understand and anticipate critical transitions in complex systems. At least four areas of interest need further work: the consequences of collapse at the global scale, the conditions under which smaller-scale regime shifts can cascade up, the implications of moving between the phases of the adaptive cycle at a single scale domain, and the implications of moving from one scale domain to another.

A system-level regime shift occurs when the level of interest within a panarchy collapses and reorganizes into a new state. Because panarchies are hierarchical systems, the collapse of a given level also collapses lower levels. Although we often focus on ecosystems or smaller ecological units as the level of our concern, climate change is a global disturbance, and the risk of crossing other, non–carbon-based planetary thresholds (Rockström et al. 2009b) means that the planet, as the highest level in the panarchy, ought to be a major focus of human concern given that collapse at global scales poses such risk for all the scale domains below it. Even less is known about how collapse at smaller scales can cascade up. Collapse and renewal at smaller spatial and temporal scales plays an important role in system-level resilience, as it provides the flexibility and adaptability that buffer disturbances and prevents them from cascading up. However, under certain conditions it is possible for smaller-scale collapse to cascade up and drive larger scale shifts, as is currently happening in the central plains as invasive eastern red cedar (*Juniperus virginiana*) has transitioned from a localized disturbance to the cause of the large-scale transformation of regional grasslands into woody shrublands (Twidwell et al. 2013; chapter 7). A further example is the Arab Spring, in which authoritarian regimes in North African countries were toppled as a result of local revolts.

Regime shifts that percolate up or cascade down scales may be less well understood than system-level shifts, but two other aspects of panarchy theory and discontinuities are rarely, if ever, discussed. The first is that shifts between each phase in an adaptive cycle are shifts in process and behavior, which can help us understand the regimes themselves. The phases of exploitation, conservation, collapse, and reorganization can be defined by different goal functions of the system (Fath et al. 2004). Goal functions are used by theoretical ecologists to measure changes in thermodynamic expression as a system develops or responds to a disturbance (Schneider and Kay 1994; Nielsen and Jorgensen 2013). For example, exergy storage, which is the amount of work a biological system can perform, is expected to rise throughout system development, whereas entropy production (maximum exergy dissipation) is expected to peak partway through the exploitation phase (Sundstrom and Allen 2019).

Understanding the fundamental differences in system behavior between the different stages of the adaptive cycle and at multiple scales would allow researchers to track when system behavior is deviating from normal ranges of behavior that may not be noticeable from other types of monitoring data. However, it is also important to note that anytime an adaptive cycle at any scale domain goes through the reorganization phase, a hysteretic regime shift is possible, even if the scales involved are small. This can occur because the conditions that supported the previous regime may have changed, and the system can no longer reorganize around the same set of processes and structures (Aguiar et al. 1996). For example, when cloud forest is lost, it is difficult for forest patches to regenerate because the precipitation patterns necessary for renewal were a function of the forest itself; with the forest structure removed, precipitation patterns shift, and forest species cannot reestablish themselves (Hildebrandt and Eltahir 2006, 2008). Thus, movement from one phase of the adaptive cycle to another is normal but is also a point at which irreversible change can occur.

The second point is that movement from one scale domain or adaptive cycle to another also constitutes a regime shift as there is a rapid change in characteristic processes and structure between scale domains (Allen et al. 2014). Adaptive cycles play out at each of the nested scale domains within a panarchy and are separated by the spatial and temporal scale differences between scale domains. Consider the annual growth and senescence of deciduous leaves compared with the millennial or longer processes structuring where forest biomes occur. Short of larger-scale processes changing (such as climate change) or system resilience being eroded, collapse and renewal processes within scale domains tend to result in a return to a similar configuration (we recognize these are two substantial caveats, so substantial, in fact, that most systems are probably experiencing both conditions). Understanding that changing scales also changes the regime in question is important, because it means that the structures, processes, and species affected are different and require different management. This approach and understanding is in stark contrast to fitting power laws to complex systems; power laws are general fits that ignore the rich dynamics captured by patterns of deviations from those power laws (i.e., discontinuities). For all these reasons and more, locating within-system thresholds in processes and patterns could be critically useful in understanding how or in what circumstances regime shifts at one scale domain may affect other scales in the systems. In fact, although identifying discontinuities in body mass distributions has been the primary method for identifying scale domains, thresholds such as those discussed here provide an alternative method for identifying discontinuities between scale domains.

Research on discontinuities in a panarchy has also demonstrated that variability in species abundances is higher for species whose body masses place them near a discontinuity (Wardwell and Allen 2009). This fits with theory that argues that the discontinuities between scale domains, which are nonlinearities in pattern and process, are also areas with heightened variance (Wiens 1989; Holling 1992). Because increasing variance is also an indicator of impending ecological transitions (Carpenter and Brock 2006), it seems likely that it is possible to identify the variables that are most likely to exhibit increased variability, and monitor them at much smaller scales, before a system experiences a regime shift. This would greatly improve our ability to detect whether there is an increased probability of a system-level regime shift.

Regime movement

Finally, in a novel application of discontinuities and regime shift identification, discontinuities in forty-five years of Breeding Bird Survey community data were used to demonstrate that both northern and southern boundaries of bird community spatial regimes have shifted northward, with the northern boundary shifting farther and at a faster rate (Roberts et al. 2019). The application of regime shift detection approaches to spatial data has revealed interesting dynamics and has the potential to improve management. For example, the analysis by Roberts et al. (2019), showed the southern spatial regime to be moving at a slower rate than the northern regime, consistent with arctic amplification. Time series data such as those presented in Roberts et al. (2019) also allows the determination of rates of spatial change, which in turn has the potential to provide decades of early warning; an early warning indicator that signals before eminent collapse has been the holy grail for identifying and using leading indicators of regime shifts. Advance notice of regime shifts provides managers time in which to proactively act rather than merely react.

Novelty

Theory and recent empirical analysis suggest that the thresholds between scale domains are the location of novelty and innovation as a result of variable dynamics at these transition zones. That means these regions are critical sources of adaptive capacity as well as sources of vulnerability. Heightened variability at the species, population, and community levels has been observed at these transitions between scales (Allen et al. 1999; Allen and Saunders 2002, 2006; Gunderson et al. 2007; Skillen and Maurer 2008; Wardwell and Allen 2009; Angeler et al. 2014b), as indicated by phenomena such as invasion, extinction, population variability and fitness, nomadism, and migration. Resources are thought to be more variable at the edge of a scale domain because structuring processes are more variable. Studies on migratory and nomadic avian species that take advantage of variability in resources shows that they have body masses that place them closer to a discontinuity than expected by chance (Allen and Saunders 2002; Alai 2010). The heightened variability in resource availability provides both risk and opportunity. For example, invasive species are more successful when they have body masses that place them near a discontinuity, and species close to a discontinuity are also more likely to be declining or extinct (Allen et al. 1999; Skillen and Maurer 2008; Sundstrom and Allen 2014).

We suspect the heightened variability at transitions between scale domains also plays a role in the production of novelty. Without novelty, systems become stagnant and rigid and lack the capacity to respond to new disturbances. Novelty and innovation are a source of diversity, a key component of adaptive capacity. Adaptive capacity is a property of ecosystems that allows them to respond to disturbances in a manner that allows them to stay in the same regime; it is core to system resilience (Angeler et al. 2019b). Diversity at all levels of organization provides flexibility and alternative pathways for a system so that functions are maintained under a broad range of circumstances. Although invasive species are often viewed negatively for their impact on native species, they are also a source of renewal and may replace lost function or help systems adapt to changing biophysical conditions (Chaffin et al. 2016a).

Conclusions

The successful management of any complex system, including ecosystems, requires a thorough understanding of how pattern and process manifest within and across system scales. Panarchy theory, along with the closely related concepts of discontinuities and adaptive cycles, provides a robust framework for understanding dynamics in multiscaled, hierarchical systems. This framework includes ways to objectively identify system scales (discontinuity analysis) (Nash et al. 2014a), ways to model within-scale and across-scale dynamics and feedbacks (adaptive cycle dynamics within and across scales) (Angeler et al. 2015), a richer understanding of the risk of cascading regime shifts, tools to quantify resilience (cross-scale resilience model and expansion thereof) (Peterson et al. 1998; Sundstrom et al. 2018), and sources of both positive and harmful innovation and novelty and their relationship to system resilience (Allen and Holling 2010). Although each of these concepts has received individual attention in the literature, the integration of discontinuities, adaptive cycles, and panarchy, as Holling originally intended (Holling et al. 2002), remains a fruitful and underdeveloped pathway for understanding complex systems. It is a powerful story that provides researchers a rigorous way to identify critical scales and to model how pattern and dynamics at multiple scale domains interact and shape system-level resilience. Our history of ecosystem management is littered with examples of how managing at a single scale can lead to a loss of resilience and unpleasant surprises, such as the collapse of ecosystems (Crépin et al. 2012). Panarchy provides a way to understand the dynamics of complex systems at multiple scales, and discontinuities provide an approach to objectively identify those scales.

Acknowledgments

This material is based on work supported by the National Science Foundation under grant nos. DGE-1735362 and 1920938. Any opinions, findings, and conclusions or recommendations expressed in this material are those of the authors and do not necessarily reflect the views of the National Science Foundation.

Literature Cited

Aguiar, M.R., J.M. Paruelo, O.E. Sala, and W.K. Lauenroth. 1996. Ecosystem responses to changes in plant functional type composition: An example from the Patagonian steppe. *Journal of Vegetation Science* 7: 381–390.

Alai, A.L. 2010. The *Textural Discontinuity Hypothesis and Its Relation to Nomadism, Migration, Decline, and Competition*. MS thesis, University of Nebraska–Lincoln.

Allen, C.R., D.G. Angeler, A.S. Garmestani, L.H. Gunderson, and C.S. Holling. 2014. Panarchy: Theory and application. *Ecosystems* 17: 578–589.

Allen, C.R., D.G. Angeler, M.P. Moulton, and C.S. Holling. 2015. The importance of scaling for detecting community patterns: Success and failure in assemblages of introduced species. *Diversity* 7: 229–241.

Allen, C.R., E.A. Forys, and C.S. Holling. 1999. Body mass patterns predict invasions and extinctions in transforming landscapes. *Ecosystems* 2: 114–121.

Allen, C.R., A.S. Garmestani, T.D. Havlicek, P.A. Marquet, G.D. Peterson, C. Restrepo, C.A. Stow, and B.E. Weeks. 2006. Patterns in body mass distributions: Sifting among alternative hypotheses. *Ecology Letters* 9: 630–643.

Allen, C.R., L. Gunderson, and A.R. Johnson. 2005. The use of discontinuities and functional groups to assess relative resilience in complex systems. *Ecosystems* 8: 958–966.

Allen, C.R., and C.S. Holling. 2002. Cross-scale structure and scale breaks in ecosystems and other complex systems. *Ecosystems* 5: 315–318.

Allen, C.R., and C.S. Holling (Eds.). 2008. *Discontinuities in Ecosystems and Other Complex Systems*. Island Press, New York.

Allen, C.R., and C.S. Holling. 2010. Novelty, adaptive capacity, and resilience. *Ecology and Society* 15: 24.

Allen, C.R., and D.A. Saunders. 2002. Variability between scales: Predictors of nomadism in birds of an Australian Mediterranean-climate ecosystem. *Ecosystems* 5: 348–359.

Allen, C.R., and D.A. Saunders. 2006. Multimodel inference and the understanding of complexity, discontinuity, and nomadism. *Ecosystems* 9: 694–699.

Angeler, D.G., C.R. Allen, H.E. Birgé, S. Drakare, B.G. McKie, and R.K. Johnson. 2014a. Assessing and managing freshwater ecosystems vulnerable to environmental change. *Ambio* 43: 113–125.

Angeler, D.G., C.R. Allen, A.S. Garmestani, L.H. Gunderson, O. Hjerne, and M. Winder. 2015. Quantifying the adaptive cycle. *PLoS One* 10: 1–17.

Angeler, D.G., C.R. Allen, A.S. Garmestani, L.H. Gunderson, and I. Linkov. 2016. Panarchy use in environmental science for risk and resilience planning. *Environment Systems and Decisions* 36: 225–228.

Angeler, D.G., C.R. Allen, A.S. Garmestani, K.L. Pope, D. Twidell, and M. Bundschuh. 2018. Resilience in environmental risk and impact assessment: Concepts and measurement. *Bulletin of Environmental Contamination and Toxicology* 101: 543–548.

Angeler, D.G., C.R. Allen, and R.K. Johnson. 2013. Measuring the relative resilience of subarctic lakes to global change: Redundancies of functions within and across temporal scales. *Journal of Applied Ecology* 50: 572–584.

Angeler, D.G., C.R. Allen, D. Twidwell, and M. Winder. 2019a. Discontinuity analysis reveals alternative community regimes during phytoplankton succession. *Frontiers in Ecology and Evolution* 7: 139.

Angeler, D.G., C.R. Allen, A. Vila-Gispert, and D. Almeida. 2014b. Fitness in animals correlates with proximity to discontinuities in body size distributions. *Ecological Complexity* 20: 213–218.

Angeler, D.A., H.B. Fried-Petersen, C.R. Allen, A.S. Garmestani, and D. Twidwell. 2019b. Adaptive capacity in ecosystems. *Resilience in Complex Socioecological Systems* 60: 1–24.

Apeldoorn, D.F., K. Kok, M.P. Sonneveld, and T.A. Veldkamp. 2011. Panarchy rules? Rethinking resilience of agroecosystems, evidence from Dutch dairy-farming. *Ecology and Society* 16: 39.

Barichievy, C., D.G. Angeler, T. Eason, A.S. Garmestani, K.L. Nash, C.A. Stow, S. Sundstrom, and C.R. Allen. 2018. A method to detect discontinuities in census data. *Ecology and Evolution* 8: 9614–9623.

Carpenter, S.R., and W.A. Brock. 2006. Rising variance: A leading indicator of ecological transition. *Ecology Letters* 9: 311–318.

Chaffin, B., A.S. Garmestani, D.G. Angeler, D.L. Herrmann, C.A. Stow, M. Nyström,

J. Sendzimir, M.E. Hopton, J. Kolasa, and C.R. Allen. 2016a. Biological invasions, ecological resilience and adaptive governance. *Journal of Environmental Management* 183: 399–407.

Chaffin, B.C., A.S. Garmestani, L. Gunderson, M. Harm Benson, D.G. Angeler, C.A. Arnold, B. Cosens, R.K. Craig, J.B. Ruhl, and C.R. Allen. 2016b. Transformative environmental governance. *Annual Review of Resources and Environment* 41: 399–423.

Chaffin, B.C., and L.H. Gunderson. 2016. Emergence, institutionalization and renewal: Rhythms of adaptive governance in complex social-ecological systems. *Journal of Environmental Management* 165: 81–87.

Chipman, H.A., E. George, and R.E. Mcculloch. 1998. Bayesian CART model search. *Journal of the American Statistical Association* 93: 935–948.

Crépin, A.-S., R. Biggs, S. Polasky, M. Troell, and A. de Zeeuw. 2012. Regime shifts and management. *Ecological Economics* 84: 15–22.

Fath, B.D., S.E. Jørgensen, B.C. Patten, and M. Straškraba. 2004. Ecosystem growth and development. *BioSystems* 77: 213–228.

Forys, E.A., and C.R. Allen. 1999. Biological invasions and deletions: Community change in south Florida. *Biological Conservation* 87: 341–347.

Forys, E.A., and C.R. Allen. 2002. Functional group change within and across scales following invasions and extinctions in the Everglades ecosystem. *Ecosystems* 5: 339–347.

Garcia, J.H., A.S. Garmestani, and A.T. Karunanithi. 2011. Threshold transitions in a regional urban system. *Journal of Economic Behavior and Organization* 78: 152–159.

Garmestani, A.S., and C.R. Allen. 2014. *Social-Ecological Resilience and Law.* Columbia University Press, New York.

Garmestani, A.S., C.R. Allen, and K.M. Bessey. 2005. Time-series analysis of clusters in city size distributions. *Urban Studies* 42: 1507–1515.

Garmestani, A.S., C.R. Allen, and C.M. Gallagher. 2008. Power laws, discontinuities and regional city size distributions. *Journal of Economic Behavior and Organization* 68: 209–216.

Garmestani, A.S., C.R. Allen, C.M. Gallagher, and J.D. Mittelstaedt. 2007. Departures from Gibrat's law, discontinuities and city size distributions. *Urban Studies* 44: 1997–2007.

Garmestani, A.S., C.R. Allen, and L. Gunderson. 2009. Panarchy: Discontinuities reveal similarities in the dynamic system structure of ecological and social systems. *Ecology and Society* 14: 15.

Garmestani, A.S., and M.H. Benson. 2013. A framework for resilience-based governance of social-ecological resilience. *Ecology and Society* 18: 9.

Garmestani, A.S., D. Twidwell, D.G. Angeler, S. Sundstrom, C. Barichievy, B.C. Chaffin, T. Eason, N. Graham, D. Granholm, L. Gunderson, M. Knutson, K.L. Nash, R.J. Nelson, M. Nystrom, T.L. Spanbauer, C.A. Stow, and C.R. Allen. 2020. Panarchy: Opportunities and challenges for ecosystem management. *Frontiers in Ecology and the Environment* 18: 576–583.

Green, O.O., A.S. Garmestani, L.H. Gunderson, C.R. Allen, J.B. Ruhl, C.A. Arnold, N.A.J. Graham, B. Cosens, D.G. Angeler, B. Chaffin, and C.S. Holling. 2015. Barriers and bridges to the integration of social-ecological resilience and law. *Frontiers in Ecology and the Environment* 13: 332–337.

Green, O.O., A.S. Garmestani, M.E. Hopton, and M.T. Heberling. 2014. A multi-scalar examination of law for sustainable ecosystems. *Sustainability* 6: 3534–3551.

Gunderson, L., C.R. Allen, and D. Wardwell. 2007. Temporal scaling in complex systems: Resonant frequencies and biotic variability. In *Temporal Dimensions in Landscape Ecology: Wildlife Responses to Variable Resources*, ed. J.A. Bissonette and I. Storch, 78–89. Springer, New York.

Gunderson, L.H., and C.S. Holling (Eds.). 2002. *Panarchy: Understanding Transformations in Human and Natural Systems*. Island Press, Washington, DC.

Haskell, J.P., M.E. Ritchie, and H. Olff. 2002. Fractal geometry predicts varying body size scaling relationships for mammal and bird home ranges. *Nature* 418: 527–530.

Hildebrandt, A., and E.A.B. Eltahir. 2006. Forest on the edge: Seasonal cloud forest in Oman creates its own ecological niche. *Geophysical Research Letters* 33: 2–5.

Hildebrandt, A., and E.A.B. Eltahir. 2008. Using a horizontal precipitation model to investigate the role of turbulent cloud deposition in survival of a seasonal cloud forest in Dhofar. *Journal of Geophysical Research* 113: 1–11.

Holling, C.S. 1973. Resilience and stability of ecological systems. *Annual Review of Ecological Systems* 4: 1–23.

Holling, C.S. (Ed.). 1978. *Adaptive Environmental Assessment and Management.* John Wiley and Sons, Chichester, UK.

Holling, C.S. 1986. The resilience of terrestrial ecosystems: Local surprise and global change. In *Sustainable Development of the Biosphere*, ed. W.C. Clark and R.E. Munn, 292–317. Cambridge University Press, Cambridge, UK.

Holling, C.S. 1988. Temperate forest insect outbreaks, tropical deforestation and migratory birds. *Memoirs of the Entomological Society of Canada* 120: 21–32.

Holling, C.S. 1992. Cross-scale morphology, geometry, and dynamics of ecosystems. *Ecological Monographs* 62: 447–502.

Holling, C.S., and L.H. Gunderson. 2002. Resilience and adaptive cycles. In *Panarchy: Understanding Transformations in Human and Natural Systems*, ed. L.H. Gunderson and C. S. Holling, 25–62. Island Press, Washington, DC.

Holling, C.S., L.H. Gunderson, and G.D. Peterson. 2002. Sustainability and panarchies. In *Panarchy: Understanding Transformations in Human and Natural Systems*, ed. L.H. Gunderson and C.S. Holling, 63–102. Island Press, Washington, DC.

Hughes, T.P., S. Carpenter, J. Rockström, M. Scheffer, and B. Walker. 2013. Multiscale regime shifts and planetary boundaries. *Trends in Ecology and Evolution* 28: 389–395.

Lambert, W.D., and C.S. Holling. 1998. Causes of ecosystem transformation at the end of the Pleistocene: Evidence from mammal body-mass distributions. *Ecosystems* 1: 157–175.

Levin, S.A. 1992. The problem of pattern and scale in ecology. *Ecology* 73: 1943–1967.

Mouillot, D., D.R. Bellwood, C. Baraloto, J. Chave, R. Galzin, M. Harmelin-Vivien, M. Kulbicki, S. Lavergne, S. Lavorel, N. Mouquet, C.E.T. Paine, J. Renaud, and W. Thuille. 2013. Rare species support vulnerable functions in high-diversity ecosystems. *PLoS Biology* 11: e1001569.

Mouillot, D., D.R. Bellwood, C. Baraloto, J. Chave, R. Galzin, M. Harmelin-Vivien, M. Kulbicki, S. Lavergne, S. Lavorel, N. Mouquet, C.E.T. Paine, J. Renaud, and W. Thuiller. 2014. Functional over-redundancy and high functional vulnerability in global fish faunas on tropical reefs. *Proceedings of the National Academy of Sciences* 111: 13757–13762.

Nash, K.L., C.R. Allen, D. Angeler, C. Barichievy, A.S. Garmestani, N. A. J. Graham, D. Granholm, M. Knutson, J. Nelson, M. Nyström, S. Riley, C.A. Stow, and S.M. Sundstrom. 2014a. Discontinuities, cross-scale patterns and the organization of ecosystems. *Ecology* 95: 654–667.

Nash, K.L., C.R. Allen, C. Barichievy, M. Nystrom, S. Sundstrom, and N.A.J. Graham. 2014b. Habitat structure and body size distributions: Cross-ecosystem comparison for taxa with determinate and indeterminate growth. *Oikos* 123: 971–983.

Nash, K.L., N.A.J. Graham, S. Jennings, S.K. Wilson, and D.R. Bellwood. 2015. Herbivore cross-scale redundancy supports response diversity and promotes coral reef resilience. *Journal of Applied Ecology* 53: 646–655.

Nash, K.L., N. Graham, S. Wilson, and D.R. Bellwood. 2013. Cross-scale habitat structure drives fish body size distributions on coral reefs. *Ecosystems* 16: 478–490.

Nielsen, S.N., and S.E. Jorgensen. 2013. Goal functions, orientors and indicators (GoFOrIt's) in ecology. Application and functional aspects: Strengths and weaknesses. *Ecological Indicators* 28: 31–47.

O'Neill, R.V., D.L. DeAngelis, J.B. Waide, and T.F.H. Allen. 1986. *A Hierarchical Concept of Ecosystems*. Princeton University Press, Princeton, NJ.

Peters, R. 1983. *The Ecological Implications of Body Size*. Cambridge University Press, Cambridge, UK.

Peterson, G.D., C.R. Allen, and C.S. Holling. 1998. Ecological resilience, biodiversity, and scale. *Ecosystems* 1: 6–18.

Porto, L.E. 2012. The spiraling propensities of mind: Towards an ecological theory of human meaning systems within a panarchy of adaptive cycles of human consciousness. *Proceedings of the 56th Annual Meeting of the ISSS*. International Society for the Systems Sciences, San Jose, CA.

Raffaelli, D., S. Hall, C. Emes, and B. Manly. 2000. Constraints on body size distributions: An experimental approach using a small-scale system. *Oecologia* 122: 389–398.

Randle, J.M., M.L. Stroink, and C.H. Nelson. 2014. Addiction and the adaptive cycle: A new focus. *Addiction Research and Theory* 6359: 1–8.

Restrepo, C., L. Renjifo, and P. Marples. 1997. Frugivorous birds in fragmented neotropical montane forests: Landscape pattern and body mass distribution. In *Tropical Forest Remnants: Ecology, Management and Conservation of Fragmented Communities*, ed. W.F. Laurance and R. Bierregaard, 171–189. University of Chicago Press, Chicago.

Roberts, C.P., C.R. Allen, D.G. Angeler, and D. Twidwell. 2019. Shifting avian spatial regimes in a changing climate. *Nature Climate Change* 9: 562–568.

Rockström J., W. Steffen, K. Noone, A. Persson, and F.S. Chapin. 2009a. A safe operating space for humanity. *Nature* 461: 472–475.

Rockström, J., W. Steffen, K. Noone, A. Persson, F.S. Chapin, E. Lambin, and T.M. Lenton. 2009b. Planetary boundaries: Exploring the safe operating space for humanity. *Ecology and Society* 14: 32.

Ruhl, J.B. 2012. Panarchy and the law. *Ecology and Society* 17: 31.

Scheffer, M. 2009. *Critical Transitions in Nature and Society*. Princeton University Press, Princeton, NJ.

Scheffer, M., and E.H. van Nes. 2006. Self-organized similarity, the evolutionary emergence of groups of similar species. *Proceedings of the National Academy of Sciences* 103: 6230–6235.

Schneider, E., and J.K. Kay. 1994. Life as a manifestation of the second law of thermodynamics. *Mathematical and Computer Modelling* 19: 25–48.

Sendzimir, J.P. 1998. *Patterns of Animal Size and Landscape Complexity: Correspondence Within and across Scales*. Dissertation, University of Florida, Gainesville.

Sendzimir, J.P., C.R. Allen, L.H. Gunderson, and C. Stow. 2002. Implications of body mass patterns. In *Landscape Ecology and Resource Management: Linking Theory with Practice*, ed. J. Bissonette and I. Storch, 125–152. Island Press, Washington, DC.

Skillen, J.J., and B.A. Maurer. 2008. The ecological significance of discontinuities in body-mass distributions. In *Discontinuities in Ecosystems and Other Complex Systems*, ed. C.R. Allen and C.S. Holling, 193–218. Columbia University Press, New York.

Spanbauer, T.L., C.R. Allen, D.G. Angeler, T. Eason, S.C. Fritz, A.S. Garmestani, K.L. Nash, J.R. Stone, C.A. Stow, and S.M. Sundstrom. 2016. Body size distributions signal a regime shift in a lake ecosystem. *Proceedings of the Royal Society B* 283: 20160249.

Steffen, W., J. Rockström, K. Richardson, T.M. Lenton, C. Folke, D. Liverman, C.P. Summerhayes, A.D. Barnosky, S.E. Cornell, M. Crucifix, J.F. Donges, I. Fetzer, S.J. Lade, M. Scheffer, R. Winkelmann, and H.J. Schellnhuber . 2018. Trajectories of the Earth system in the Anthropocene. *Proceedings of the National Academy of Sciences* 115: 8252–8259.

Stow, C., C.R. Allen, and A.S. Garmestani. 2007. Evaluating discontinuities in complex systems: Toward quantitative measures of resilience. *Ecology and Society* 12: 26.

Sundstrom, S.M., and C.R. Allen. 2014. Complexity versus certainty in understanding species' declines. *Diversity and Distributions* 20: 344–355.

Sundstrom, S.M., and C.R. Allen. 2019. The adaptive cycle: More than a metaphor. *Ecological Complexity* 39: 100767.

Sundstrom, S.M., C.R. Allen, and D.G. Angeler. 2020. Scaling and discontinuities in the global economy. *Journal of Evolutionary Economics* 30: 319–345.

Sundstrom, S.M., C.R. Allen, and C. Barichievy. 2012. Species, functional groups, and thresholds in ecological resilience. *Conservation Biology* 26: 305–314.

Sundstrom, S.M., D.G. Angeler, C. Barichievy, T. Eason, A.S. Garmestani, L.H. Gunderson, M. Knutson, K.L. Nash, T. Spanbauer, C. Stow, and C.R. Allen. 2018. The distribution and role of functional abundance in cross-scale resilience. *Ecology* 99: 2421–2432.

Sundstrom, S., D.G. Angeler, A.S. Garmestani, J. García, and C.R. Allen. 2014. Transdisciplinary application of cross-scale resilience. *Sustainability* 6: 6925–6948.

Turner, M.G., R.V. O'Neill, R.H. Gardner, and B.T. Milne. 1989. Effects of changing spatial scale on the analysis of landscape pattern. *Landscape Ecology* 3: 153–162.

Twidwell, D., W.E. Rogers, S.D. Fuhlendorf, C.L. Wonkka, D.M. Engle, J.R. Weir, U.P. Kreuter, and C.A. Taylor. 2013. The rising Great Plains fire campaign: Citizens' response to woody plant encroachment. *Frontiers in Ecology and the Environment* 11: e64–e71.

Vergnon, R., E.H. van Nes, and M. Scheffer. 2012. Emergent neutrality leads to multimodal species abundance distributions. *Nature Communications* 3: 1–6.

Wardwell, D.A., and C.R. Allen. 2009. Variability in population abundance is associated with thresholds between scaling regimes. *Ecology and Society* 14: 42.

Wardwell, D.A., C.R. Allen, G.D. Peterson, and A.J. Tyre. 2008. A test of the cross-scale resilience model: Functional richness in Mediterranean-climate ecosystems. *Ecological Complexity* 5: 165–182.

Wiens, J.A. 1989. Spatial scaling in ecology. *Functional Ecology* 3: 385–397.

Winfree, R., and C. Kremen. 2009. Are ecosystem services stabilized by differences among species? A test using crop pollination. *Proceedings of the Royal Society B* 276: 229–237.

Chapter 3

The Adaptive Cycle: More than a Metaphor

Shana M. Sundstrom and Craig R. Allen

The adaptive cycle and its contribution to panarchy were proposed as a metaphor and conceptual tool for understanding long-term dynamics of change in complex adaptive systems (CASs) such as ecosystems and social-ecological systems (Holling 1986; Gunderson and Holling 2002). As such, the concept has had uptake by researchers from a variety of fields (Odum et al. 1995; Bunce et al. 2009; Burkhard et al. 2011; Randle et al. 2014; Fath et al. 2015; Kharrazi et al. 2016; Thapa et al. 2016) despite the limited empirical evidence demonstrating adaptive cycles in real data (though see Carpenter et al. 1999; Angeler et al. 2015). Other CAS work, such as self-organized criticality (SOC), edge of chaos, regime shifts, sustainability, resilience, punctuated equilibrium, game theory, and thermodynamics (Langton 1990; Lindgren and Nordahl 1994; Kauffman 1995; Bak and Boettcher 1997; Ulanowicz 1997; Aronson and Plotnick 2001; Jorgensen et al. 2007; Lockwood and Lockwood 2008; Scheffer 2009), suggests that the adaptive cycle emerges from endogenous processes of self-organization, evolution, feedbacks, and exogenous inputs of energy and materials.

We propose that adaptive cycle dynamics are ubiquitous in CASs and discuss a set of indicators developed from literature on ecological and general systems that are strong candidates for tracking system trajectory through phases of an adaptive cycle. These indicators, when coupled with the explicit consideration of scales via the discontinuity hypothesis and panarchy, may minimize the need for multiple indicators to capture system dynamics and provide a richer picture of a system's trajectory than that offered by a single-scale analysis. We use the word *indicator* as a generic concept broadly representing a system signal that can change over time (Nielsen and Jorgensen 2013). Our emphasis is on feasible ways in which researchers could systematically and quantitatively look for signatures of panarchical dynamics in ecosystems rather than relying on metaphor and largely qualitative descriptions. In short, we propose that adaptive cycles are real dynamics of real systems and not just handy conceptual metaphors, and we identify the simplest indicators likely to signal phases in an adaptive cycle dynamic.

Adaptive Cycles and the Discontinuity Hypothesis

An adaptive cycle describes system movement through a three-dimensional state space defined by system potential, connectedness, and resilience (figure 3.1) (Holling 1986; Gunderson and Holling 2002). A key insight from Holling (1986) was that system progression through phases of an adaptive cycle is driven by endogenous variables interacting with variables operating at different scale domains. Although many cross-scale linkages occur throughout the phases of an adaptive cycle, two key cross-scale linkages, revolt and remember, define a panarchy (figure 3.2). A core hypothesis of panarchy is that the key processes that structure ecosystems occur at different ranges of spatial and temporal scales, often separated by orders of magnitude (chapter 2). Thus the adaptive cycle of a pine tree in a boreal forest occurs at smaller spatial and shorter temporal scales than the broader and longer processes that drive boreal forest dynamics at the scale of physiographic regions.

The discontinuity hypothesis (Holling 1992) was proposed to test whether systems had identifiable scales at which adaptive cycles unfold (Holling, pers. comm.). This hypothesis demonstrates that ecosystems contain distinct spatiotemporal domains over which key processes, ecological structures, and resources either do not change or change monotonically (Wiens 1989; Holling 1992; Nash et al. 2014). Scale domains are separated by discontinuation of structuring processes, or scale breaks, and the transition to a new set of structuring processes. For example, biochemical processes such as photosynthesis and respiration structure pine needle dynamics, whereas herbivory and fire drive forest patch dynamics. Thus, adaptive cycles occur at each scale domain within the system, resulting in complex systems with multiple and nested domains of scales at different phases of the adaptive cycle. At smaller and faster scale domains within a larger ecosystem, disturbances and processes of self-organization can drive cycling dynamics that are confined to those scales, whereas processes of renewal and regeneration depend on system memory at larger scales. Occasionally, disturbances can cascade up to larger and slower spatiotemporal scales, especially if those larger scale domains are in the conservation phase of their own adaptive cycle and smaller scale cycles approach

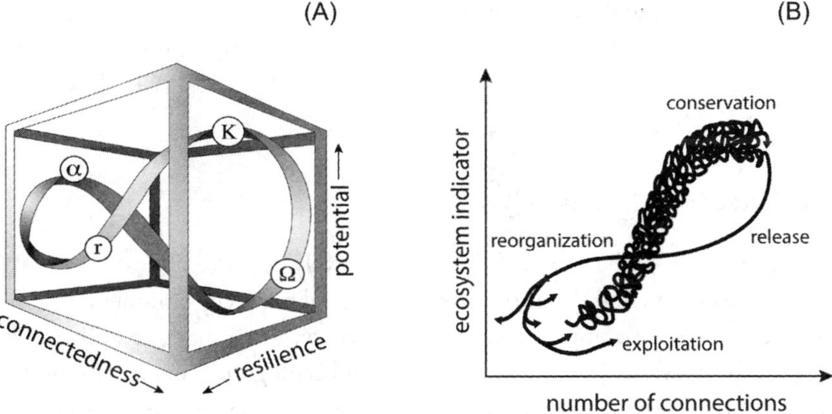

Figure 3.1. (A) Depiction of an adaptive cycle along different dimensions, indicating system trajectory through phases of exploitation (r), conservation (K), release (Ω), and reorganization (α), along dimensions of connectedness, potential, and resilience. (Adapted from Gunderson and Holling 2002.) (B) Two-dimensional system trajectory, rotated 45° so that there is no increase in resilience at the end of the release phase, and to more realistically capture non-monotonic growth during the exploitation and conservation phases (Adapted from Burkhard et al. 2011)

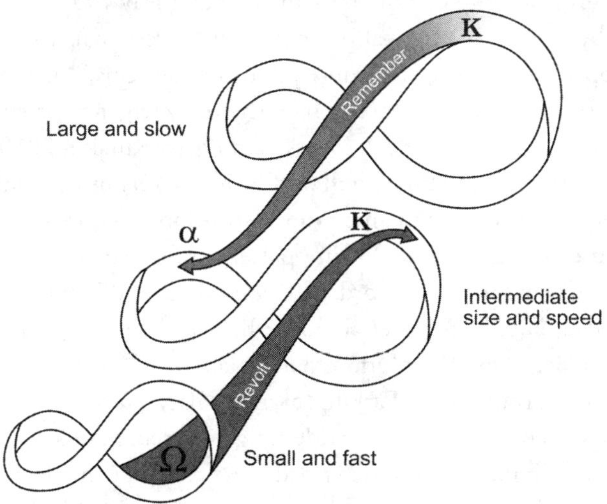

Figure 3.2. Panarchy showing three nested adaptive cycles with bidirectional, cross-scale influences. The three cycles cross a wide range of spatial and temporal scales of system processes and dynamics, from small and fast to large and slow. (Reproduced from Gunderson and Holling 2002.)

synchrony, because the accumulation of system potential in the form of standing biomass and bound nutrients shaped by high connection between system elements renders it more vulnerable to cascading effects (Gunderson and Holling 2002).

Evidence for Adaptive Cycles

Descriptions of cyclic system behavior intuitively apply to almost any living, adaptable system, and the authors of the original panarchy volume both push back against the compulsion to see adaptive cycles everywhere and use examples from a variety of economic, social, and ecological systems (Gunderson and Holling 2002). Although work on the adaptive cycle and panarchy has occurred largely within resilience science, research in nonrelated fields suggests that the fundamental dynamics of system development, maturation, collapse, and reorganization captured by the adaptive cycle are endogenous dynamics of CASs that reflect internal processes of self-organization and evolution over time (Burkhard et al. 2011).

CASs arise when systems are nonisolated (Jørgensen et al. 1999) or driven by energy inputs and exporting entropy. System structure and complexity emerge from processes that arise to dissipate energy, generating the essential characteristics of a complex system (Schneider and Kay 1994). When systems are driven by the same "physical principles and forces that drive self-organization in open, inorganic, far-from-equilibrium systems," such as nonequilibrium thermodynamics, then the patterns of emergence of structure and process should be generalizable across system types (Kurakin 2010, 5).

Predator–prey interactions, bull and bear markets in economic systems, collapse and renewal of companies and institutions, and evolution in social systems and ecosystems show shifting patterns of dynamics that parallel the adaptive cycle (Lindgren and Nordahl 1994; Foster and Wild 1999; Jain and Krishna 2002a, 2002b; Becks et al. 2005; Beinhocker 2006). In archaeology, a dynamic model of social evolution that explains the "cycles of consolidation, expansion, and dissolution" (Parkinson and Galaty 2007, 117) of geopolitical states has been shown to fit a wide range of archaeological communities (Marcus 1998). Tainter (1988) described what can happen when the collapse phase occurs at the level of the panarchy, rather than at smaller-scale adaptive cycles (i.e., collapse of civilizations). Adaptive cycles do indeed appear to be ubiquitous, but that does not explain the underlying principles that drive such system dynamics.

Early pioneers in complex system science argued that there are a limited number of system states, such as limit points, limit cycles, chaos, strange attractors, and long transients (Wolfram 1984). These states strongly parallel the four phases of the adaptive cycle and demonstrated that given too much or too little order, systems can be trapped in states that lead to collapse because they are either too static or too chaotic to support the processes necessary to sustain self-organized, persistent structural hierarchies that can adapt and evolve under changing conditions (Wolfram 1984; Langton 1986; Kauffman 1993; Ulanowicz 1997). Systems are thought to evolve to the edge of chaos, where they avoid the trap of frozen order and stay poised at the cusp of chaos where they can maximize information, entropy rate, and adaptation (Langton 1986; Latora et al. 2000). Ulanowicz (1997) called the edge of chaos a window of vitality and defined this window by a range or region of parameter space (also see Zorach and Ulanowicz 2003; Nakajima and Haruna 2011; Benincà et al. 2015). A similar system state is described by theories of SOC.

SOC argues that systems can self-organize to a region of state space that is at the cusp of order and chaos. When operating in this region, systems exhibit spatial patterns that exhibit constant fractal dimensions or temporal frequencies that follow power law distributions (Kauffman 1995; Bak and Boettcher 1997; Jørgensen et al. 1998; Pascual and Guichard 2005; Jørgensen et al. 2007). Schneider and Kay (1994) proposed that life itself is a far-from-equilibrium dissipative structure, poised at the cusp between low and high entropy (order and disorder); this tension between opposing forces is mirrored in ecology, where the tension and trade-off between diversity and redundancy (chaos and order) play out in evolution, community assembly, and ecological resilience (Page 2010). Studies from a broad range of fields have found edge of chaos dynamics and self-organized criticality in complex systems (Bonabeau 1997; Jørgensen et al. 1998; Latora et al. 2000; Li 2000; Turchin and Ellner 2000; Lansing 2003; Kurakin et al. 2007; Kitzbichler et al. 2009; Kong et al. 2009; Upadhyay 2009; Nakajima and Haruna 2011; Salem 2011; Chua et al. 2012; Benincà et al. 2015), but interestingly, not all systems stay there, as some systems show such dynamics for only a range of parameter space, or for a limited duration of time, or only larger system scales stay poised at criticality (Li 2000; Upadhyay 2009; Medvinsky et al. 2015; Lansing et al. 2017) whereas smaller scales experience collapse and renewal.

If systems are tuned to evolve to criticality where even small events can trigger a collapse or phase transition (Bak and Paczuski 1985; de Oliveira 2001; Pascual and Guichard 2005), then it is not obvious that all spatial and temporal scales of a complex adaptive system could be at a criticality point or even within a narrow range of criticality concurrently. This would generate severe instability because even small disturbances would constantly cascade up and down system scales. Yet many large-scale ecosystems exhibit stable or quasi-equilibrium behavior, which can persist long enough that it led earlier ecologists to assume that deterministic successional behavior and equilibrium dynamics were the norm (Clements 1936). In fact, there is robust evidence for stability in larger-scale patterns such as biomass even while community composition and abundance can be highly variable and even chaotic (Ernest and Brown 2001; Scheffer et al. 2003; Hatton et al. 2015; Vallina et al. 2017; Sundstrom et al. 2018). Brunk (2002) argued that systems need time to rebuild the structure that allows the transmission of disturbance. In forests, for example, it takes time to regrow the biomass that becomes the fuel load that can spread fire throughout the forest, making it likely that the region of parameter space encompassing SOC for a mature system is relatively broad and *is a result of* the higher frequency of regular collapses at smaller spatial and temporal scales (recall the power law behavior of disturbance events that defines self-organized criticality) that prevent disturbances from cascading up to the largest scales of the system. We hypothesize that the conservation (K) phase of an adaptive cycle may well operate at SOC or the edge of chaos, but only if cycles of collapse and renewal occur with sufficient frequency at smaller spatial and temporal scales (Brunk 2002; Gunderson and Holling 2002). Therefore, the timescales for a power law distribution of disturbance size and frequency that include the collapse of an entire mature system will necessarily be very long, unless a system is gradually pushed out of the parameter space in which a system can maintain SOC. For example, a slow-changing variable such as climate shrinks the region of criticality, making it easier for disturbances to trigger a system-wide collapse. If SOC is a feature of the evolution of complex systems over time (Bolliger et al. 2003), then this suggests that younger systems, systems that experience disturbance frequency rates that prevent the generation of order and complex features, and adaptive cycles at all scale domains of the system will show a range of evidence for the hallmarks of SOC—namely, differing degrees of power law behavior in the size distribution of disturbances and the spatial pattern of clusters of vegetation (Kéfi et al. 2014). In short, we suggest that within an adaptive cycle, power law structure and edge of chaos dynamics will increase from none during a collapse phase, to weak in the reorganization phase, and then should steadily increase to strong as the system moves through the exploitation and conservation phases (figure 3.3)

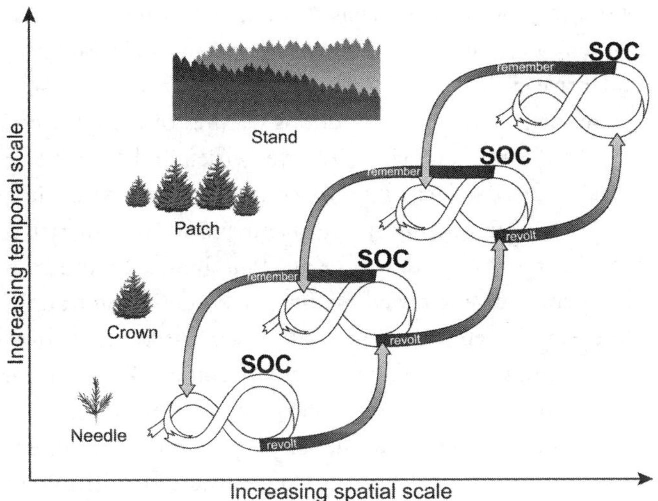

Figure 3.3. A conceptual diagram of nested adaptive cycles for a pine-dominated ecosystem. Self-organized criticality (SOC) should peak in the conservation phase of the adaptive cycle. Because the spatial and temporal scales increase at each level of the hierarchy, the forest system at the highest level in the panarchy should spend the most time at self-organized criticality. (Adapted from Allen et al. 2014.)

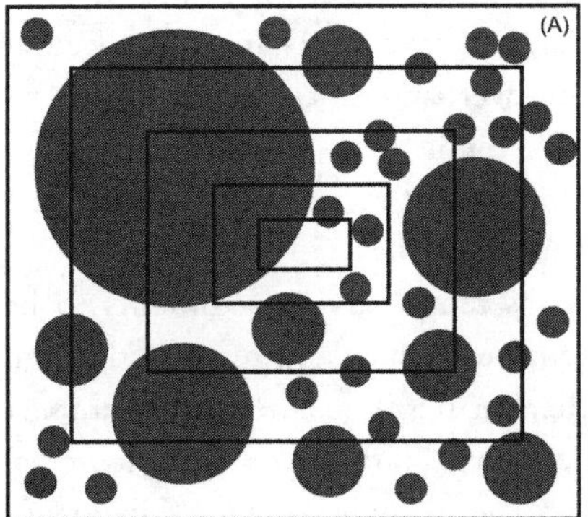

Figure 3.4. Stylized representation of vegetation patches (circles), with size of the circle corresponding to the size of a patch and window of observation (area) represented by rectangles (A). The distribution of patch sizes fits a power law across a range of observation windows, as is characteristic of self-organized criticality. This is not mutually exclusive with (B), where patches operating at different spatial and temporal scales (represented by shades of gray) may be in different phases of the adaptive cycle or different stages of self-organized criticality.

(Brunk 2002). We note that power law behavior in system features such as spatial organization or temporal frequencies of events is not mutually exclusive with asynchrony in adaptive cycle dynamics within a panarchy (figure 3.4).

Adaptive cycles may be generic and ubiquitous features of complex systems (Schneider and Kay 1994; Ulanowicz 1997; Jørgensen and Fath 2004; Beinhocker 2006; Kurakin 2011). Such systems accrue complexity over time as processes of self-organization generate more hierarchical layers of structure that dissipate more energy (Schneider and Kay 1994; Jørgensen and Fath 2004). System development is characterized by increased order, organization, and storage of usable energy in sequential phases that see first biomass, then networks, and finally information (in terms of genetic complexity) increase (Fath et al. 2004; Jørgensen et al. 2016). Complexity and order evolve from relative simplicity and disorder under the influence of "periodic but transient setbacks in the form of organization relaxations and restructuring" (Kurakin 2011, 1). These authors describe detailed dynamics of change in CASs in terms of exergy, infrared thermal measurements, and electron and proton transport in autocatalytic processes (Schneider and Kay 1994; Jørgensen and Fath 2004; Kurakin 2011) that fully align with adaptive cycles and panarchy but places less emphasis on the collapse and renewal phases, and scale breaks. They argued that setbacks to this trajectory of increasing complexity have occurred at all spatiotemporal scales (e.g., from a small forest fire to mass extinctions to the fall of prior civilizations), but have not changed the fundamental trajectory of increasing complexity over time; the players may come and go, but the organization of relationships tends to be preserved and evolve (Kurakin 2011). Setbacks in this trajectory toward increased complexity are therefore temporary and of little importance.

We argue that externalizing setbacks as temporary impediments to be overcome rather than critical for long-term persistence and renewal through innovation and adaptation is problematic. Furthermore, the relative impact of setbacks depends on the timescale under consideration. For example, social and economic systems at the global scale are increasingly complex (Ulanowicz 1997), but many ecological systems are at risk of simplification as anthropogenic degradations accumulate, rapid species extinctions reduce the diversity necessary for systems to retain and build complexity, and climate change drives an increased risk of system-level regime shifts. Alternative regimes can be simpler, more homogenous systems as a result of missing crucial elements that allow them to reorganize into a similar state after a disturbance. When viewed at geological timescales, it seems likely that processes of thermodynamics, self-organization, and evolution will resume the inexorable march to increasing complexity, but this may provide small comfort for humans in the twenty-first century. Regardless of alternative views of archetypic system dynamics, a set of signals or indicators is needed to track such nonlinear changes demonstrated by the adaptive cycle.

Indicators for Tracking Systems through Phases of Adaptive Cycle Dynamics

A range of thermodynamic indicators have been proposed to track system change over time (Ulanowicz 1997; Fath et al. 2001, 2004; Müller 2005; Jørgensen et al. 2007; Burkhard et al. 2011). However, such indicators have been developed to reflect dynamics of systems at a particular range of spatio-temporal scales (Fath et al. 2001; Müller 2005) and are best used in combination with each other to sufficiently capture the complexity of system changes over broader domains. We describe the potential of a new approach: applying the indicators *at scale domains within the system*, which reflect an adaptive cycle. This approach may reduce the need to measure multiple indicators because partitioning dynamics by scale domains within the system will account for some of the issues that necessitate the use of multiple indicators. It will also provide a sensible way to address the importance of the collapse and reorganization phases at smaller scales in maintaining system dynamics in the conservation stage at larger spatial and temporal scales. To be fair, it is possible the literature on these indicators has not dealt substantively with the collapse and renewal phases because the emphasis has been on system-level dynamics, where collapse is fairly rare (Ulanowicz 1997; Fath et al. 2004; Coscieme et al. 2013). Finally, the indicators are not easily used by general ecologists or land managers. They are intimidating in their language, their complexity, and their daunting data requirements (Müller 2005). We discuss a subset of indicators that have varying degrees of feasibility; we think it is important to discuss ascendency despite its difficulty of use because it is one of the more comprehensive and well-developed thermodynamic indicators. However, our ultimate objective is to focus on indicators that simplify tracking dynamics of change within an ecosystem by explicitly accounting for panarchy dynamics and that are feasible for ecologists to use because they lower the bar for the volume and detail of data needed.

We identified a set of system indicators from literature by leading scholars in the field of ecological and general systems. We discuss indicators of emergy

(Odum 1996), eco-exergy (Fath and Cabezas 2004), entropy production and infrared (Schneider and Kay 1994), ascendency (Ulanowicz 1997), Fisher information (Fath and Cabezas 2004; Mayer et al. 2006; Karunanithi et al. 2008), connectance (Woodward et al. 2005), and biomass (Odum 1969). The indicators are ultimately all thermodynamic in origin in that they track flows and stores of energy in a system. Many of these indicators have been derived from the study of ecosystems, sometimes by comparing successional stages within or across the seasons (Bastianoni and Marchettini 1997; Lu et al. 2015) or comparing systems with different degrees of degradation by processes such as lake eutrophication (Patrício et al. 2004; Ludovisi 2006), but few indicators have been evaluated by researchers for their performance during the release and reorganization phases, over long temporal spans, or at multiple scales. The following section develops a rationale for each indicator, beginning with emergy.

Emergy

Emergy and the maximum empower principle were developed by Odum (1996) as a way to quantify energy inputs and storages and flows across energy hierarchies, such as trophic levels of food webs. Emergy represents the amount of useful work that is stored at different hierarchal levels (such as autotrophs, primary or secondary consumers) in relation to the energy input (solar radiation) that supports that level in a hierarchy (Cai et al. 2004; Nielsen and Jorgensen 2013), hence the term *embodied energy*, or *emergy*. For example, the emergy inputs into a managed forest may include the sun, evapotranspiration, runoff, human labor, and litter (Lu et al. 2006). These energy inputs are transformed via photosynthesis into tree biomass, which in turn is transformed by primary consumers, and transformed again by secondary consumers. Because the energy pathway to the secondary consumers is longer than the path to the tree (in terms of more energy transformations), the secondary consumers have higher emergy than tree biomass (Bastianoni and Marchettini 1997). Emergy increases with ecosystem development because the energy pathway to produce a more complex system is necessarily longer (Burkhard et al. 2011). Emergy metrics out of context are uninformative with regard to how a system uses its emergy inputs or feedbacks (Burkhard et al. 2011; Coscieme et al. 2013). They also need a formidable amount of information and can be difficult to calculate (Brown and Ulgiati 2010). As a result, practitioners have developed emergy/exergy ratios to better understand the state of a system or how emergy is distributed across a system's structure, further increasing the data demands

(Bastianoni and Marchettini 1997; Bastianoni et al. 2006; Lu et al. 2015), but a similar pattern of increase during development is supported. Calculating emergy at scale domains within an ecosystem might alleviate the need for more data-intensive emergy/exergy metrics because of the relationship between organism size and complexity (Ludovisi 2006; Campbell and Garmestani 2012).

Eco-exergy

Eco-exergy measures the amount of work a biological system can perform, or the difference in entropy between a system at equilibrium and its actual state (Fath et al. 2004). There are two primary hypotheses regarding the change in exergy over the course of ecosystem development. The first focuses on the maximization of total amount of exergy dissipated, and it has been argued to be the functional equivalent of maximizing entropy production (Lin et al. 2018). The second hypothesis focuses on the maximization of exergy storage (Fath et al. 2004). Because ecosystems operate far from equilibrium, ecosystem exergy storage reflects the total amount of available energy stored in organic structures (Ludovisi 2009). Eco-exergy equations take into account information, structure, and concentration, and they indirectly account for the manner in which biological matter is distributed between ecosystem compartments, typically by using carbon as the energy currency and genetic complexity as the information (Ulanowicz 1997; Scharler 2012; Lu et al. 2015). More biomass of more complex organisms will cause eco-exergy to rise (Fath and Cabezas 2004). Both theory and empirical data support the hypothesis that exergy storage increases throughout ecosystem development, first from increases in biomass, then increases in network connections and cycling of materials, and finally from an increase in genetic complexity (Marques and Jorgensen 2002; Fath et al. 2004; Jørgensen and Fath 2004; Jørgensen et al. 2007). This does not incorporate the back-loop stages of the adaptive cycle, where release and renewal should decrease exergy storage.

The detailed data necessary to calculate eco-exergy for an equilibrium system (used for comparison and necessary for the calculation) as well as the system of interest is daunting. Eco-exergy by itself is disconnected from a broader understanding of resilience, as the extinction of lower-order species with less genetic complexity and an increase in abundance of higher-order organisms with more complexity can drive an increase in exergy (Fath and Cabezas 2004) but will also reduce functional redundancy and future adaptive capacity. From the perspective of the adaptive cycle, an increase in eco-exergy in this scenario would equate to reduced resilience and reduced potential; thus, when eco-exergy is calculated at the system level, it does not sufficiently capture critical system features. Various combinations of eco-exergy and emergy exist to address the limitations of each individual indicator (Bastianoni et al. 2006; Coscieme et al. 2013; Lu et al. 2015), but this also increases the information or data demands of the analysis. However, were eco-exergy to be calculated for each scale domain of the complete nested set of adaptive cycles, it would partition eco-exergy by the complexity of organisms because organisms such as mice, found at smaller and faster scales, are necessarily simpler than organisms that operate at longer and larger scales, such as wolves. Understanding changes in eco-exergy at scales may provide a sufficiently nuanced view of change that it reduces the need for multiple indicators to adequately capture system change and development through time.

Entropy production

Maximum entropy production proposes that nonequilibrium systems will evolve to steady states at which entropy production is maximized (Schneider and Kay 1994; Fath et al. 2004; Kleidon 2010; Skene 2017). It is considered the equivalent of maximum exergy dissipation for a given fixed temperature (Kleidon 2010; Lin et al. 2018; though see Ludovisi et al. 2012) and has been explored through a variety of formulations, making it difficult to understand how broadly applicable the principle is across ecosystem types and stages of development (Meysman and Bruers 2010). Theory suggests it is valid as a local principle of behavior but not necessarily for the nonlocal case or over longer timescales (Martyushev 2013; Skene 2017). Both modeling and empirical data indicate that the local asymptote of maximum entropy production is constrained by other levels of organization and a given environmental context (Quijano and Lin 2015; Skene 2017). It has also been demonstrated that maximum entropy production occurs as a function of the stage of system development and peaks during the early to middle stages of successional development (Fath et al. 2004; Ludovisi et al. 2005, 2012; Maes et al. 2011), which translates in the adaptive cycle to somewhere between the exploitation phase and conservation phase.

Few scientists directly consider entropy production at scales or levels of organization within the system (though see Dewar 2010; Skene 2017), although many do so indirectly by incorporating information on the structural complexity of biota such as considerations of body size, *r* versus *K* species, or complexity of a food web (Ludovisi et al. 2005; Ludovisi 2006, 2009; Meysman and Bruers 2010). Some of these indicators address the efficiency of the system in terms of maximizing storage of exergy per unit entropy produced and find that it increases both over the course of successional development and across lakes in different development states in a consistent manner (Ludovisi et al. 2005; Ludovisi 2006). In other words, as a system matures, more exergy is stored per entropy produced because of the increased complexity of species along the successional gradient. However, Meysman and Bruers (2010) found that food webs as a whole behave differently from individual trophic compartments, suggesting that the scale at which entropy production is evaluated matters. Much of the research on entropy production has used shallow lake systems and calculated biological entropy production as a function of phytoplankton photosynthesis, because phytoplankton directly convert incoming solar radiation via photosynthesis (Ludovisi 2006). Data on species presence, size, and abundance is therefore readily available from aquatic sampling. Such data are far less available for terrestrial systems, which as a result have been analyzed via exergy dissipation in the form of radiation (Lin et al. 2011). Calculating entropy across objectively identified scale domains might facilitate a more nuanced understanding of the pattern of maximal entropy production over time and across scales and provide a clearer expectation for the relationship between entropy production, or an entropy production/exergy storage ratio, and the stages of exploitation, conservation, release, and renewal in ecosystem dynamics over time.

Infrared

Infrared is a thermodynamic metric proposed by Schneider and Kay (1994) as a test of their proposition that complex systems such as ecosystems should increase their total dissipation over time and become more complex, whereas simple or degraded systems should dissipate less energy. They argued that "more mature systems should degrade incoming solar radiation into lower quality exergy" (Schneider and Kay 1994, 2), resulting in lower reradiated temperatures (lower airborne infrared thermal measurements). However, as Fath et al. (2004, 215) pointed out, "ecosystems are complex adaptive systems, and as such one would expect the thermodynamic properties of the ecosystem to change during development." It is now understood that although more mature systems dissipate more exergy than less mature or degraded systems, exergy dissipation eventually plateaus while system maturation continues and is reflected in other thermodynamic metrics (Aoki 1995; Fath et al. 2004; Ludovisi 2014; Stoy et al. 2014).

Measuring infrared has been explored primarily in terrestrial systems as a viable way to measure exergy dissipation and the degree of self-organization across different stages of ecological development. Recent refinements suggest that canopy surface temperature is highly correlated with exergy dissipation and entropy production and is easier to measure (Lin et al. 2011, 2016, 2018; Lin 2015). As with the other indicators, understanding infrared canopy surface temperature in terms of nested adaptive cycles may address some of the shortcomings; Schneider and Kay's (1994) hypothesis that maturing ecosystems will continually increase total dissipation as expressed by infrared may be correct only for the renewal and exploitation phases of the adaptive cycle. Tracking changes in infrared within scale domains could be an effective signal of ecosystem change over time because it would account for the differing spatial and temporal scales at which processes of exergy dissipation play out. Furthermore, tracking changes in infrared canopy surface temperature at scales directly incorporates the differences in structural information between short-lived, high-turnover vegetative species and longer-lived species.

Ascendency

Ascendency theory (Ulanowicz 1997) quantifies change in system dynamics by using information theory to measure growth and structure in food webs, where growth is an increase in system activity or total system throughput, and structure is the mutual information contained in the trophic flow. In the absence of major perturbations, ecosystems are expected to increase in ascendency over time (Ulanowicz 1986; Ulanowicz et al. 2006). Ascendency, when coupled with overhead, which captures system redundancies and the material for adaptive capacity, is a process of growth and maturation in ecosystems that fully parallels panarchy. In fact, Ulanowicz (1997) used ascendency theory to test the adaptive cycle and concluded that they are fundamentally telling the same story. However, panarchy explicitly addresses the notion of cycling dynamics occurring at multiple domains of scale and via the discontinuity hypothesis provides a method for detecting scale domains (Holling 1992; Nash et al. 2014), whereas ascendency theory only touches on feedbacks across levels in the hierarchy and does not explicitly model or account for them. Furthermore, it fails to substantively treat collapse and renewal as integral and necessary processes that are both unavoidable and critical for system resilience and persistence;

the stages of maturation only briefly acknowledge that there are "temporary setbacks" and downplay the possibility of collapse occurring at large scales when a system has reached senescence (Holling's K phase). Failing to treat collapse and renewal as integral, endogenously driven processes that are critical for system resilience at larger spatiotemporal scales constrains its ability to explain and predict future behavior. Relegating collapse and renewal to minor roles can also influence research choices that in turn can lead to misleading results; for example, Matutinovic et al. (2016) trimmed their data to exclude major disturbances.

Although Ulanowicz's (1997) rigorous and quantitative ascendency theory captures system development, the data demands of fully realized food webs are intense and provide only a snapshot of ecosystems at one point in time. Comparisons between aquatic systems with differing degrees of degradation based on a suite of ascendency-related measures in a network analysis are common (Patrício et al. 2004; Heymans et al. 2014; Meddeb et al. 2018), and temporal analyses are less so (Scharler 2012); some, but not all, support the contention that ascendency is higher in less disturbed or stressed systems. Modeling and simulated networks are also used in the search for patterns of behavior for ascendency and its related measures (Grami et al. 2011; Saint-Béat et al. 2013; Brinck and Jensen 2017; Ludovisi and Scharler 2017), but robust patterns have not emerged. For example, over evolutionary timescales, ascendency and average mutual information increase over time and drop during periods of "hectic transitions" or high-disturbance periods (Brinck and Jensen 2017), whereas Ludovisi et al. (2017) found that average mutual information and diversity of interactions between ecosystem components systematically increased along food web succession, whereas ascendency did not have a consistent trend. Although tracking system development or dynamics of change via ascendency, overhead, and related indicators at scales within a system to elucidate expected patterns of change in the context of the adaptive cycle could provide rich insight, the data demands remain highly challenging for nonaquatic systems, and the necessary temporal data are missing for highly resolved aquatic food webs.

Fisher information

Fisher information is a measure of the amount of disorder contained in any given parameter or system characteristic, and it is based on the probability of observing a particular system state (Fath and Cabezas 2004; Mayer et al. 2006). It has been used to detect spatial and temporal regime shifts in ecosystems (Karunanithi et al. 2008; Eason et al. 2014; Spanbauer et al. 2014; Sundstrom et al. 2017). This measure has the potential to track system change as ecosystems move between the phases of the adaptive cycle because the degree of order can be reflected in patterns of species richness, abundance, functional richness, growth rate, connectivity, and complexity, all of which are anticipated to change in systematic ways among the phases of the adaptive cycle. Whereas a drop in Fisher information indicates a loss of order or pattern in the data from unstable dynamics and a loss of resilience, as we would expect during the collapse phase of the adaptive cycle, a rise in order indicates less change and possible movement to more consistent patterns, as we would expect in the growth phases, and a stable value for Fisher information ought to occur during the conservation phase where the system spends the most time in a stable regime (Fath and Cabezas 2004). Because Fisher information can handle any kind and amount of multivariate data, there is the opportunity to exploit a variety of data types that characterize system dynamics in order to explore changes in Fisher information over time and within scale domains. For example, Müller (2005) presented a list of simplified thermodynamic variables (number of species, index of abiotic heterogeneity, primary production, respiration per biomass, transpiration per evapotranspiration, and loss of nutrients) that could be adapted for use in a Fisher information analysis to generate a holistic view of ecosystem change over time and at scales.

Connectance

Network theory can be used to depict ecological networks, where each species is typically a node in the network, and the relationship between nodes is captured via topographical features such as connectance (the number of other species to which a species is connected) or flows of energy or matter (Woodward et al. 2005; Ludovisi and Scharler 2017), which has significant data requirements (Fath et al. 2007). Scale is often only an indirect feature of network analyses, either when species are classified by trophic levels, which can be a crude classification of scale (see O'Gorman et al. 2012), or when organism body size is embedded in the network (Woodward et al. 2005). Although Holling and Gunderson (2002) did not reference network studies in his explication of adaptive cycles and panarchy, connectedness is an axis in the graphical depiction of an adaptive cycle. Their depiction of connectedness is more akin to topological studies than flow network studies such as ascendency theory (Ulanowicz 1997), because it focuses on connectivity between system elements, and its relationship to the degree to which system elements are influenced by external variables. Low connectivity between elements means their behavior is controlled primarily by external variability in processes, whereas high connectivity between system elements can strongly mediate and buffer external variability. However, high connectivity between system elements also renders the system more vulnerable to collapse (Ulanowicz 1997), because the degree to which elements such as nutrients are bound up in existing pathways and relationships between elements reduce the system's ability to "sample alternative and potentially better configurations" (Kurakin 2007, 17). It is expected that connectance will increase during the renewal and exploitation phase, peak during the conservation phase, and decrease during the release phase. Unlike some of the thermodynamic indicators that are expected to be maximized, a less-than-maximum degree of connectance appears to be optimal for system stability and resilience (Wagensberg et al. 1990; Zorach and Ulanowicz 2003; Fath et al. 2004; Burkhard et al. 2011; Ulanowicz et al. 2014).

More recent studies suggest that highly complex food webs (networks) can be simplified to just a handful of functional groups that describe the types of direct and indirect interactions species have and that these functional groups are well predicted by body mass (Kéfi et al. 2015, 2016). Because the size of a species is directly related to the spatial and temporal scales at which it interacts with its environment, this is a promising outcome. Therefore, it may be possible to understand changes in simple network metrics such as connectance or the number of trophic and nontrophic interaction types within scale domains in order to track system change over time. Because many interactions naturally cross scale domains (in general, predators are larger than prey), it is not immediately clear how to calculate these metrics when partitioned by scale domains. Furthermore, as with calculating flows in metrics such as ascendency, the data needs are fairly prohibitive because of the extent of monitoring and expert knowledge needed to populate these food webs (for example, see Kéfi et al. 2015, 2016). Such efforts are likely to have a high reward; however, how to integrate network theory, which is focused on relationships between network elements and therefore accounts for scale only indirectly, and resilience theory via the discontinuity hypothesis, which is focused on the scales at which species and processes operate, is an open topic for research. Jørgensen et al. (2016) argued for the need to integrate vertical and horizontal topology in network studies by bringing together hierarchy theory with thermodynamic theory via networks, but his conceptualization of hierarchy focused on levels of organization rather than the more objectively defined scale domains driven by pattern and process that underpin discontinuity theory (Holling 1992). Tracking connectedness at each scale domain and across time may be a robust signal of changing dynamics, but it may also fail to sufficiently capture critical dynamics without also incorporating thermodynamic relationships in network flows *sensu* Ludovisi et al. (2017).

Biomass

Perhaps the simplest possible signal of dynamics of change is biomass. Biomass accumulates rapidly during the reorganization and exploitation phases and approaches a limit during the conservation phase, because energy flows are being diverted to system maintenance rather than growth or increase in biomass (Odum 1969; Fath et al. 2004; Holdaway et al. 2010; Lu et al. 2015). In a resilient system, system-level biomass should remain stable while collapse and reorganization phases play out at smaller spatial and temporal scales, resulting in increased variability in biomass at the particular spatial and temporal scales of the disturbance as compensation processes occur; empirical data support this contention (Ernest and Brown 2001; Hatton et al. 2015). Changes in biomass of both flora and fauna ought to reflect movement between the phases of the adaptive cycle within scale domains, and biomass could be converted to a measure of carbon similar to exergy analyses in order to have common currency for the modeling of stocks of carbon in vegetation and animal species between scale domains and across time (Scharler 2012). Understanding how biomass changes across scale domains, such as rate of increase in biomass, or a change in the distribution of biomass across functional groups captures the basic thermodynamic changes that drive system growth, development, collapse, and renewal (Kurakin 2010), and can be an indicator of a regime shift at the system level (White et al. 2004; Sundstrom et al. 2018). Other features of resilience, such as functional diversity and functional redundancy, which mirror Ulanowicz's "overhead" and provide the buffering capacity that prevents system-level regime shifts, can be easily incorporated into models of changing biomass at scale domains, merely by partitioning biomass within scale domains into functional groups (Peterson et al. 1998; Forys and Allen 2002).

Conclusions

We have discussed a variety of indicators, some of which may be unfeasible because of data limitations and some of which may operate in the sweet spot of complexity science: simple enough to be feasible but complex enough to capture system patterns. We also make the case that evaluating these indicators at scale domains within the system may increase the complexity of system dynamics that are captured while reducing the need for multiple indicators. We propose that the adaptive cycle reflects the inevitable dynamics of CASs as a result of the internal processes of self-organization and evolution over time. The qualitatively similar system dynamics described in a variety of systems through diverse approaches are the result of system development in a thermodynamically nonisolated system and as such ought to manifest in signals of development and change that can be tracked across the spatial and temporal dimensions of a system in accordance with panarchy theory. None of the indicators proposed here are based on a fixed species identity or a particular community structure beyond how species identity is related to rates of energy consumption, functional role, or type of interactions with other species. As Kurakin (2009, 10) explained, "What is preserved are the spatiotemporal relationships between individual components, i.e., a certain organizational structure—a form—but not individual components. Members come and go, but the organization persists." Our interest is in system-level properties that remain stable because of, not in spite of, dynamics of change at smaller and faster spatial and temporal scales. Although ecosystems do transform over geological timescales in response to global change, it is reasonable to expect many ecosystems to remain in an exploitation and conservation phase for human timescales of decades to centuries, if not longer. Maintaining resilience of systems that are currently in desirable states requires signals that track dynamics of change at explicit and objective scales and use data that are realistic to acquire.

Acknowledgments

S.M.S. was supported by the University of Nebraska–Lincoln School of Natural Resources, the Natural Sciences and Engineering Research Council of Canada, and Emory University.

Literature Cited

Allen, C R., D.G. Angeler, A.L Garmestani, L.H. Gunderson, and C.S. Holling. 2014. Panarchy: Theory and application. *Ecosystems* 17: 578–589.

Angeler, D.G., C.R. Allen, A.S. Garmestani, L.H. Gunderson, O. Hjerne, and M. Winder. 2015. Quantifying the adaptive cycle. *PLoS One* 10: 1–17.

Aoki, I. 1995. Entropy production in living systems: From organisms to ecosystems. *Thermochimica Acta* 250: 359–370.

Aronson, R.B., and R.E. Plotnick. 2001. Scale-independent interpretations of macroevolutionary dynamics. In *Biodiversity Dynamics: Turnover of Populations, Taxa, and Communities*, ed. M.L. McKinney and J.A. Drake, 430–450. Columbia University Press, New York.

Bak, P., and S. Boettcher. 1997. Self-organized criticality and punctuated equilibria. *Physica D* 107: 143–150.

Bak, P., and M. Paczuski. 1985. Complexity, contingency, and criticality. *Proceedings of the National Academy of Sciences* 92: 6689–6696.

Bastianoni, S., and N. Marchettini. 1997. Emergy/exergy ratio as a measure of the level of organization of systems. *Ecological Modelling* 99: 33–40.

Bastianoni, S., F.M. Pulselli, and M. Rustici. 2006. Exergy versus emergy flow in ecosystems: Is there an order in maximizations? *Ecological Indicators* 6: 58–62.

Becks, L., F.M. Hilker, H. Malchow, K. Jürgens, and H. Arndt. 2005. Experimental demonstration of chaos in a microbial food web. *Nature* 435: 1226–1229.

Beinhocker, E.D. 2006. *The Origin of Wealth: Evolution, Complexity, and the Radical Remaking of Economics*. Harvard Business School Press, Boston, MA.

Benincà, E., B. Ballantine, S.P. Ellner, and J. Huisman. 2015. Species fluctuations sustained by a cyclic succession at the edge of chaos. *Proceedings of the National Academy of Sciences* 112: 6389–6394.

Bolliger, J., J.C. Sprott, and D.J. Mladenoff. 2003. Self-organization and complexity in historical landscape patterns. *Oikos* 100: 541–553.

Bonabeau, E. 1997. Flexibility at the edge of chaos: A clear example from foraging in ants. *Acta Biotheoretica* 45: 29–50.

Brinck, K., and H.J. Jensen. 2017. The evolution of ecosystem ascendency in a complex systems based model. *Journal of Theoretical Biology* 428: 18–25.

Brown, M.T., and S. Ulgiati. 2010. Emergy indices of biodiversity and ecosystem dynamics. In *Handbook of Ecological Indicators for Assessment of Ecosystem Health*, ed. S.E. Jorgensen, F.-L. Xu, and R. Costanza, 89–112. Taylor and Francis, New York.

Brunk, G.G. 2002. Why do societies collapse? *Journal of Theoretical Politics* 14: 195–230.

Bunce, M., L. Mee, L.D. Rodwell, and R. Gibb. 2009. Collapse and recovery in a remote small island-A tale of adaptive cycles or downward spirals? *Global Environmental Change* 19: 213–226.

Burkhard, B., B.D. Fath, and F. Müller. 2011. Adapting the adaptive cycle: Hypotheses on the development of ecosystem properties and services. *Ecological Modelling* 222: 2878–2890.

Cai, T.T., T.W. Olsen, and D.E. Campbell. 2004. Maximum (em)power: A foundational principle linking man and nature. *Ecological Modelling* 178: 115–119.

Campbell, D.E., and A.S. Garmestani. 2012. An energy systems view of sustainability: Emergy analysis of the San Luis Basin, Colorado. *Journal of Environmental Management* 95: 72-97.

Carpenter, S.R., W.A. Brock, and P. Hanson. 1999. Ecological and social dynamics in simple models of ecosystem management. *Conservation Ecology* 3: 4.

Chua, L., V. Sbitnev, and H. Kim. 2012. Neurons are poised near the edge of chaos. *International Journal of Bifurcation and Chaos* 22: 1250098.

Clements, F.E. 1936. Nature and structure of the climax. *Journal of Ecology* 24: 252–284.

Coscieme, L., F.M. Pulselli, S.E. Jørgensen, S. Bastianoni, and N. Marchettini. 2013. Thermodynamics-based categorization of ecosystems in a socio-ecological context. *Ecological Modelling* 258: 1–8.

de Oliveira, P.M.C. 2001. Why do evolutionary systems stick to the edge of chaos. *Theory in Biosciences* 120: 1–19.

Dewar, R. C. 2010. Maximum entropy production and plant optimization theories. *Philosophical Transactions of the Royal Society B* 365: 1429–1435.

Eason, T., A.S. Garmestani, and H. Cabezas. 2014. Managing for resilience: Early detection of regime shifts in complex systems. *Clean Technologies and Environmental Policy* 16: 773–783.

Ernest, S.M., and J.H. Brown. 2001. Homeostasis and compensation: The role of species and resources in ecosystem stability. *Ecology* 82: 2118–2132.

Fath, B.D., and H. Cabezas. 2004. Exergy and Fisher Information as ecological indices. *Ecological Modelling* 174: 25–35.

Fath, B.D., C.A. Dean, and H. Katzmair. 2015. Navigating the adaptive cycle: An approach to managing the resilience of social systems. *Ecology and Society* 20: 24.

Fath, B.D., S.E. Jørgensen, B C. Patten, and M. Straškraba. 2004. Ecosystem growth and development. *BioSystems* 77: 213–228.

Fath, B.D., B.C. Patten, and J.S. Choi. 2001. Complementarity of ecological goal functions. *Journal of Theoretical Biology* 208: 493–506.

Fath, B.D., U.M. Scharler, R E. Ulanowicz, and B. Hannon. 2007. Ecological network analysis: Network construction. *Ecological Modelling* 208: 49–55.

Forys, E.A., and C R. Allen. 2002. Functional group change within and across scales following invasions and extinctions in the Everglades ecosystem. *Ecosystems* 5: 339–347.

Foster, J., and P. Wild. 1999. Detecting self-organisational change in economic processes exhibiting logistic growth. *Journal of Evolutionary Economics* 9: 109–133.

Grami, B., S. Rasconi, N. Niquil, M. Jobard, B. Saint-Béat, and T. Sime-Ngando. 2011. Functional effects of parasites on food web properties during the spring diatom bloom in Lake Pavin: A linear inverse modeling analysis. *PLoS One* 6: e23273.

Gunderson, L.H., and C S. Holling (Eds.). 2002. *Panarchy: Understanding Transformations in Human and Natural Systems*. Island Press, Washington, DC.

Hatton, I.A., K.S. McCann, J.M. Fryxell, T.J. Davies, M. Smerlak, A.R.E. Sinclair, and M. Loreau. 2015. The predator–prey power law: Biomass scaling across terrestrial and aquatic biomes. *Science* 349: aac6284.

Heymans, J.J., M. Coll, S. Libralato, L. Morissette, and V. Christensen. 2014. Global patterns in ecological indicators of marine food webs: A modelling approach. *PLoS One* 9: e95845.

Holdaway, R.J., A.D. Sparrow, and D.A. Coomes. 2010. Trends in entropy production during ecosystem development in the Amazon Basin. *Philosophical Transactions of the Royal Society B* 365: 1437–1447.

Holling, C.S. 1986. The resilience of terrestrial ecosystems: Local surprise and global change. In *Sustainable Development of the Biosphere*, ed. W.C. Clark and R.E. Munn, 292–317. Cambridge University Press, Cambridge, UK.

Holling, C.S. 1992. Cross-scale morphology, geometry, and dynamics of ecosystems. *Ecological Monographs* 62: 447–502.

Holling, C.S., and L.H. Gunderson. 2002. Resilience and adaptive cycles. In *Panarchy: Understanding Transformations in Human and Natural Systems*, ed. L.H. Gunderson and C.S. Holling, 25–62. Island Press, Washington, DC.

Jain, S., and S. Krishna. 2002a. Crashes, recoveries, and "core shifts" in a model of evolving networks. *Physical Review E* 65: 026103.

Jain, S., and S. Krishna. 2002b. Large extinctions in an evolutionary model: The role of innovation and keystone species. *Proceedings of the National Academy of Sciences* 99: 2055–2060.

Jørgensen, S.E., and B.D. Fath. 2004. Application of thermodynamic principles in ecology. *Ecological Complexity* 1: 267–280.

Jørgensen, S.E., B.D. Fath, S. Bastianoni, J.C. Marques, F. Muller, S.N. Nielsen, B.C. Patten, E. Tiezzi, and R.E. Ulanowicz. 2007. *A New Ecology: Systems Perspective*. Elsevier, Amsterdam, The Netherlands.

Jørgensen, S.E., H. Mejer, and S.N. Nielsen. 1998. Ecosystem as self-organising critical systems. *Ecological Modelling* 111: 261–268.

Jørgensen, S.E., S.N. Nielsen, and B.D. Fath. 2016. Recent progress in systems ecology. *Ecological Modelling* 319: 112–118.

Jørgensen, S.E., B.C. Patten, and M. Stras. 1999. Ecosystems emerging: 3. Openness. *Ecological Modelling* 117: 41–64.

Karunanithi, A.T., H. Cabezas, B.R. Frieden, and C.W. Pawlowski. 2008. Detection and assessment of ecosystem regime shifts from Fisher information. *Ecology and Society* 13: 22.

Kauffman, S.A. 1993. *The Origins of Order: Self-Organization and Selection in Evolution*. Oxford University Press, Oxford.

Kauffman, S. 1995. *At Home in the Universe: The Search for Laws of Complexity*. Oxford University Press, New York.

Kéfi, S., E.L. Berlow, E.A. Wieters, L.N. Joppa, S.A. Wood, U. Brose, and S.A. Navarrete. 2015. Network structure beyond food webs: Mapping non-trophic and trophic interactions on Chilean rocky shores. *Ecology* 96: 291–303.

Kéfi, S., V. Guttal, W.A. Brock, S.R. Carpenter, A.M. Ellison, V.N. Livina, D.A. Seekell, M. Scheffer, E.H. Van Nes, and V. Dakos. 2014. Early warning signals of ecological transitions: Methods for spatial patterns. *PLoS One* 9: 10–13.

Kéfi, S., V. Miele, E. A. Wieters, S.A. Navarrete and E.L. Berlow. 2016. How structured is the entangled bank? The surprisingly simple organization of multiplex ecological networks leads to increased persistence and resilience. *PLoS Biology* 14: e1002527.

Kharrazi, A., B. Fath, and H. Katzmair. 2016. Advancing empirical approaches to the concept of resilience: A critical examination of panarchy, ecological information, and statistical evidence. *Sustainability* 8: 935.

Kitzbichler, M.G., M.L. Smith, S.R. Christensen, and E. Bullmore. 2009. Broadband criticality of human brain network synchronization. *PLoS Computational Biology* 5: 1–13.

Kleidon, A. 2010. A basic introduction to the thermodynamics of the Earth system far from equilibrium and maximum entropy production. *Philosophical Transactions of the Royal Society B* 365: 1303–1315.

Kong, S.-G., W.-L. Fan, H.-D. Chen, J. Wigger, A. Torda, and H. Lee. 2009. Quantitative measure of randomness and order for complete genomes. *Physical Review E* 79: 1–11.

Kurakin, A. 2007. The universal principles of self-organization and the unity of Nature and knowledge. *The Soft* 1–34. http://alexeikurakin.org/main/soft.html

Kurakin, A. 2009. Scale-free flow of life: On the biology, economics, and physics of the cell. *Theoretical Biology and Medical Modelling* 6: 6. doi:10.1186/1742-4682-6-6

Kurakin, A. 2010. Order without design. *Theoretical Biology and Medical Modelling* 7: 12. doi:10.1186/1742-4682-7-12

Kurakin, A. 2011. The self-organizing fractal theory as a universal discovery method: The phenomenon of life. *Theoretical Biology and Medical Modelling* 8: 4. doi:10.1186/1742-4682-8-4

Kurakin, A., A. Swistowski, S.C. Wu, and D.E. Bredesen. 2007. The PDZ domain as a complex adaptive system. *PLoS One* 2: e953.

Langton, C.G. 1986. Studying artificial life with cellular automata. *Physica D* 22: 120–149.

Langton, C.G. 1990. Computations at the edge of chaos: Phase transitions and emergent computation. *Physica D* 42: 12–37.

Lansing, J.S. 2003. Complex adaptive systems. *Annual Review of Anthropology* 32: 183–204.

Lansing, J.S., S. Thurner, N.N. Chung, A. Coudurier-Curveur, A. Kurt, and L.Y. Chew. 2017. Adaptive self-organization of Bali's ancient rice terraces. *Proceedings of the National Academy of Sciences* 114: 6504–6509.

Latora, V., M. Baranger, A. Rapisarda, and C. Tsallis. 2000. The rate of entropy increase at the edge of chaos. *Physics Letters A* 273: 97–103.

Li, B.-L. 2000. Fractal geometry applications in description and analysis of patch patterns and patch dynamics. *Ecological Modelling* 132: 33–50.

Lin, H. 2015. Thermodynamic entropy fluxes reflect ecosystem characteristics and succession. *Ecological Modelling* 298: 75–86.

Lin, H., M. Cao, and Y. Zhang. 2011. Self-organization of tropical seasonal rain forest in southwest China. *Ecological Modelling* 222: 2812–2816.

Lin, H., Z. Fan, L. Shi, A. Arain, H. McCaughey, D. Billesbach, M. Siqueira, R. Bracho, and W. Oechel. 2016. The cooling trend of canopy temperature during the maturation, succession, and recovery of ecosystems. *Ecosystems* 20: 406–415.

Lin, H., H. Zhang, and Q. Song. 2018. Transition from abstract thermodynamic concepts to perceivable ecological indicators. *Ecological Indicators* 88: 37–42.

Lindgren, K., and M. G. Nordahl. 1994. Evolutionary dynamics of spatial games. *Physica D* 75: 292–309.

Lockwood, D.R., and J.A. Lockwood. 2008. Grasshopper population ecology: Catastrophe, criticality, and critique. *Ecology and Society* 13: 34.

Lu, H.F., D.E. Campbell, Z.A. Li, and H. Ren. 2006. Emergy synthesis of an agro-forest restoration system in lower subtropical China. *Ecological Engineering* 27: 175–192.

Lu, H., F. Fu, H. Li, D.E. Campbell, and H. Ren. 2015. Eco-exergy and emergy based self-organization of three forest plantations in lower subtropical China. *Scientific Reports* 5: 15047.

Ludovisi, A. 2006. Use of thermodynamic indices as ecological indicators of the development state of lake ecosystems: Specific dissipation. *Ecological Indicators* 6: 30–42.

Ludovisi, A. 2009. Exergy vs information in ecological successions: Interpreting community changes by a classical thermodynamic approach. *Ecological Modelling* 220: 1566–1577.

Ludovisi, A. 2014. Effectiveness of entropy-based functions in the analysis of ecosystem state and development. *Ecological Indicators* 36: 617–623.

Ludovisi, A., P. Pandolfi, and M.I. Taticchi. 2005. The strategy of ecosystem development:

Specific dissipation as an indicator of ecosystem maturity. *Journal of Theoretical Biology* 235: 33–43.

Ludovisi, A., L. Roselli, and A. Basset. 2012. Testing the effectiveness of exergy-based tools on a seasonal succession in a coastal lagoon by using a size distribution approach. *Ecological Modelling* 245: 125–135.

Ludovisi, A., and U.M. Scharler. 2017. Towards a sounder interpretation of entropy-based indicators in ecological network analysis. *Ecological Indicators* 72: 726–737.

Maes, W.H., T. Pashuysen, A. Trabucco, F. Veroustraete, and B. Muys. 2011. Does energy dissipation increase with ecosystem succession? Testing the ecosystem exergy theory combining theoretical simulations and thermal remote sensing observations. *Ecological Modelling* 222: 3917–3941.

Marcus, J. 1998. The peaks and valleys of ancient states: An extension of the Dynamic Model. In *Archaic States*, ed. G.M. Feinman and J. Marcus, 59–94. School of American Research Press, Sante Fe, NM.

Marques, J.C., and S.E. Jorgensen. 2002. Three selected ecological observations interpreted in terms of a thermodynamic hypothesis. Contribution to a general theoretical framework. *Ecological Modelling* 158: 213–221.

Martyushev, L. 2013. Entropy and entropy production: Old misconceptions and new breakthroughs. *Entropy* 15: 1152–1170.

Matutinovic, I., S.N. Salth, and R.E. Ulanowicz. 2016. The mature stage of capitalist development: Models, signs and policy implications. *Structural Change and Economic Dynamics* 39: 17–30.

Mayer, A.L., C.W. Pawlowski, and H. Cabezas. 2006. Fisher Information and dynamic regime changes in ecological systems. *Ecological Modelling* 195: 72–82.

Meddeb, M., B. Grami, A. Chaalali, M. Haraldsson, N. Niquil, O. Pringault, and A. S. Hlaili. 2018. Plankton food-web functioning in anthropogenically impacted coastal waters (SW Mediterranean Sea): An ecological network analysis. *Progress in Oceanography* 162: 66–82.

Medvinsky, A.B., B.V. Adamovich, A. Chakraborty, E.V. Lukyanova, T.M. Mikheyeva, N.I. Nurieva, N.P. Radchikova, A.V. Rusakov, and T.V. Zhukova. 2015. Chaos far away from the edge of chaos: A recurrence quantification analysis of plankton time series. *Ecological Complexity* 23: 61–67.

Meysman, F.J.R., and S. Bruers. 2010. Ecosystem functioning and maximum entropy production: A quantitative test of hypotheses. *Philosophical Transactions of the Royal Society B* 365: 1405–1416.

Müller, F. 2005. Indicating ecosystem and landscape organisation. *Ecological Indicators* 5: 280–294.

Nakajima, K., and T. Haruna. 2011. Self-organized perturbations enhance class IV behavior and 1/f power spectrum in elementary cellular automata. *BioSystems* 105: 216–224.

Nash, K.L., C.R. Allen, D. Angeler, C. Barichievy, A.S. Garmestani, N. A. J. Graham, D. Granholm, M. Knutson, J. Nelson, M. Nyström, S. Riley, C.A. Stow, and S.M. Sundstrom. 2014. Discontinuities, cross-scale patterns and the organization of ecosystems. *Ecology* 95: 654–667.

Nielsen, S.N., and S.E. Jorgensen. 2013. Goal functions, orientors and indicators (GoFOrIt's) in ecology. Application and functional aspects—Strengths and weaknesses. *Ecological Indicators* 28: 31–47.

Odum, E.P. 1969. The strategy of ecosystem development. *Science* 164: 262–270.
Odum, H.T. 1996. *Environmental Accounting: Emergy and Decision Making*. John Wiley and Sons, New York.
Odum, W.E., E.P. Odum, and H.T. Odum. 1995. Nature's pulsing paradigm. *Estuaries* 18: 547–555.
O'Gorman, E.J., D.E. Pichler, G. Adams, J.P. Benstead, H. Cohen, N. Craig, W.F. Cross, B.O.L. Demars, N. Friberg, G.M. Gíslason, R. Gudmundsdóttir, A. Hawczak, J.M. Hood, L.N. Hudson, L. Johansson, M.P. Johansson, J.R. Junker, A. Laurila, J.R. Manson, E. Mavromati, D. Nelson, J.S. Ólafsson, D.M. Perkins, O.L. Petchey, M. Plebani, D.C. Reuman, B.C. Rall, R. Stewart, M.S.A. Thompson, and G. Woodward. 2012. Impacts of warming on the structure and functioning of aquatic communities: Individual- to ecosystem-level responses. *Advances in Ecological Research* 47: 81–176.
Page, S.E. 2010. *Diversity and Complexity* (Primers in Complex Systems). Princeton University Press, Princeton, NJ.
Parkinson, W.A., and M.L. Galaty. 2007. Secondary states in perspective: An integrated approach to state formation in the prehistoric Aegean. *American Anthropologist* 109: 113–129.
Pascual, M., and F. Guichard. 2005. Criticality and disturbance in spatial ecological systems. *Trends in Ecology and Evolution* 20: 88–95.
Patrício, J., R. Ulanowicz, M.A. Pardal, and J.C. Marques. 2004. Ascendency as an ecological indicator: A case study of estuarine pulse eutrophication. *Estuarine, Coastal and Shelf Science* 60: 23–35.
Peterson, G.D., C.R. Allen, and C.S. Holling. 1998. Ecological resilience, biodiversity, and scale. *Ecosystems* 1: 6–18.
Quijano, J.C., and H. Lin. 2015. Is spatially integrated entropy production useful to predict the dynamics of ecosystems? *Ecological Modelling* 313: 341–354.
Randle, J.M., M.L. Stroink, and C.H. Nelson. 2014. Addiction and the adaptive cycle: A new focus. *Addiction Research and Theory* 6359: 1–8.
Saint-Béat, B., A.F. Vézina, R. Asmus, H. Asmus, and N. Niquil. 2013. The mean function provides robustness to linear inverse modelling flow estimation in food webs: A comparison of functions derived from statistics and ecological theories. *Ecological Modelling* 258: 53–64.
Salem, P. 2011. The sweet spots in human communication. *Nonlinear Dynamics, Psychology, and Life Sciences* 15: 389–406.
Scharler, U.M. 2012. Ecosystem development during open and closed phases of temporarily open/closed estuaries on the subtropical east coast of South Africa. *Estuarine, Coastal and Shelf Science* 108: 119–131.
Scheffer, M. 2009. *Critical Transitions in Nature and Society*. Princeton University Press, Princeton, NJ.
Scheffer, M., S. Rinaldi, J. Huisman, and F.J. Weissing. 2003. Why plankton communities have no equilibrium: Solutions to the paradox. *Hydrobiologia* 491: 9–18.
Schneider, E., and J.K. Kay. 1994. Life as a manifestation of the second law of thermodynamics. *Mathematical and Computer Modelling* 19: 25–48.
Skene, K.R. 2017. Thermodynamics, ecology and evolutionary biology: A bridge over troubled water or common ground? *Acta Oecologica* 85: 116–125.

Spanbauer, T.L., C.R. Allen, D.G. Angeler, T. Eason, S.C. Fritz, A.S. Garmestani, K.L. Nash, and J.R. Stone. 2014. Prolonged instability prior to a regime shift. *PLoS One* 9: e108936.

Stoy, P., H. Lin, K. Novick, M. Siqueira, and J.-Y. Juang. 2014. The role of vegetation on the ecosystem radiative entropy budget and trends along ecological succession. *Entropy* 16: 3710–3731.

Sundstrom, S.M., D.G. Angeler, C. Barichievy, T. Eason, A.S. Garmestani, L.H. Gunderson, M. Knutson, K.L. Nash, T. Spanbauer, C. Stow, and C.R. Allen. 2018. The distribution and role of functional abundance in cross-scale resilience. *Ecology* 99: 2421–2432.

Sundstrom, S.M., T. Eason, R.J. Nelson, D.G. Angeler, C. Barichievy, A.S. Garmestani, N.A.J. Graham, D. Granholm, L. Gunderson, M. Knutson, K.L. Nash, M. Nyström, T. Spanbauer, C.A. Stow, and C.R. Allen. 2017. Detecting spatial regimes in ecological systems. *Ecology Letters* 20: 19–32.

Tainter, J.A. 1988. *The Collapse of Complex Societies.* Cambridge University Press, Cambridge, UK.

Thapa, R., M. Thoms, and M. Parsons. 2016. An adaptive cycle hypothesis of semi-arid floodplain vegetation productivity in dry and wet resource states. *Ecohydrology* 9: 39–51.

Turchin, P., and S.P. Ellner. 2000. Living on the edge of chaos: Population dynamics of fennoscandian voles. *Ecology* 81: 3099–3116.

Ulanowicz, R.E. 1986. *Growth and Development: Ecosystems Phenomenology.* Springer-Verlag, New York.

Ulanowicz, R.E. 1997. *Ecology, the Ascendent Perspective.* Columbia University Press, New York.

Ulanowicz, R.E., R.D. Holt, and M. Barfield. 2014. Limits on ecosystem trophic complexity: Insights from ecological network analysis. *Ecology Letters* 17: 127–136.

Ulanowicz, R.E., S.E. Jørgensen, and B.D. Fath. 2006. Exergy, information and aggradation: An ecosystems reconciliation. *Ecological Modelling* 198: 520–524.

Upadhyay, R.K. 2009. Dynamics of an ecological model living on the edge of chaos. *Applied Mathematics and Computation* 210: 455–464.

Vallina, S., P. Cermeno, S. Dutkiewicz, M. Loreau, and J. Montoya. 2017. Phytoplankton functional diversity increases ecosystem productivity and stability. *Ecological Modelling* 361: 184–196.

Wagensberg, J., A. Garcia, and R.V. Sole. 1990. Connectivity and information transfer in flow networks: two magic numbers in ecology? *Bulletin of Mathematical Biology* 52: 733–740.

White, E.P., S.K.M. Ernest, and K.M. Thibault. 2004. Trade-offs in community properties through time in a desert rodent community. *American Naturalist* 164: 670–676.

Wiens, J.A. 1989. Spatial scaling in ecology. *Functional Ecology* 3: 385–397.

Wolfram, S. 1984. Universality and complexity in cellular automata. *Physica D* 10: 1–35.

Woodward, G., B. Ebenman, M.C. Emmerson, J.M. Montoya, J.M. Olesen, A. Valido, and P.H. Warren. 2005. Body size in ecological networks. *Trends in Ecology and Evolution* 20: 402–409.

Zorach, A.C., and R.E. Ulanowicz. 2003. Quantifying the complexity of flow networks: How many roles are there? *Complexity* 8: 68–76.

CHAPTER 4

Scales of Coercion: Resilience, Regimes, and Panarchy

David G. Angeler and Craig R. Allen

The Anthropocene is the current geologic epoch in Earth's history and is characterized by human domination of biophysical processes at the planetary scale (Steffen et al. 2018). The ubiquitous influence of humans on planetary processes poses major challenges for the future trajectory of ecological and social-ecological systems at subplanetary scales (Rockström et al. 2009). A primary concern is that these human activities will lead to crossing one or more known, or unknown, critical thresholds, resulting in regime shifts at the local, regional, and global levels (Hughes et al. 2013).

Alternative regimes are characterized by different structures, functions, processes, and feedbacks in a system that result in a new, dynamically stable (fluctuation within a given basin of attraction) state (Holling 1973; Scheffer et al. 1993; Beisner et al. 2003). Alternative regimes are ubiquitous, as indicated by numerous examples across a range of ecological, social, social-ecological, geopolitical, economic, and climatic systems (Miller and Williamson 1988; Biggs et al. 2018; Steffen et al. 2018). Examples include grasslands rapidly becoming encroached by woodland, marine fisheries shifting from cod to lobster dominance, or democracies shifting to authoritarian regimes.

Regimes may be desired or undesired by humanity, reflective of the multiple social values or goals attributed to that regime or services, products, or outputs associated with a particular regime. Undesired regimes provide fewer benefits, such as food or fiber for humans, or they may generate more undesired effects such as extreme weather events, increase in diseases, or geopolitical conflicts (Homer-Dixon 1991; Folke et al. 2004; McMichael et al. 2008). While desired regimes are perpetuated and conserved, much work has focused on transforming undesirable system states into more desired regimes. The need to harness desired system functioning for human welfare leads to an increased examination of alternative system regimes and their management through the lens of resilience (Angeler et al. 2016a). Therefore, ecosystem restoration, rehabilitation, and mitigation (Society for Ecological Restoration [SER] 2004) have all been described through a resilience lens in the context of adaptation, transformation, or stabilization of regimes (Angeler and Allen 2016).

In this chapter, we build on the concepts of coerced resilience and coerced regimes, novel ideas derived from applying resilience theory to regime management (Angeler et al. 2020b). Coerced resilience reflects the types and degree of management actions needed to maintain a regime that has lost the capacity for self-organization and renewal. Such external subsidies may maintain a set of ecosystem services, such as provisioning, associated with a regime. Coercion of a desired regime is the inverse of ecological resilience, ecological resilience being defined as the amount of disturbance needed to shift a system between alternative regimes (Holling 1973; Angeler and Allen 2016). A desired regime with low resilience will require stronger coercion to maintain weak feedbacks of that regime. The trajectory of the current global climate regime, for example, will require substantial societal adaptation and mitigation to avoid a potentially catastrophic regime shift (Steffen et al. 2018). Conversely, a highly resilient degraded regime will also need a high degree of coercion to move the system into a desired regime. Turbid shallow lakes and wetlands can require significant nutrient control, food web manipulation, and technological interventions to mimic the conditions of clearwater lakes (Annadotter et al. 1999; Angeler et al. 2003).

In complex systems of people and nature, coerced resilience clearly has a multiscale connotation. In the following sections we will review the concepts related to resilience coercion and introduce a cross-scale extension of coercion ideas to highlight how management at discrete levels of the system may result in a coerced panarchy. Such a coerced panarchy is discussed theoretically and presented with the current global climate regime as an example.

Concepts

Coerced regimes and *coerced resilience* are two terms recently introduced into resilience literature. Although they are similar, there are fundamental, operationally important differences between the two. In this section we define both terms and distinguish between them.

Coerced resilience

Coerced resilience is a phrase that arose from studies of the resilience of highly managed production ecosystems, such as forestry, agroecosystems, and aquaculture. It refers to resilience that is created as a result of anthropogenic inputs (labor, energy, technology), rather than supplied by the ecological system itself. Such coercion of resilience focuses on and enables the maintenance of high levels of production services from these systems (Rist et al. 2014).

Conceptually, coercion has a main focus on maintaining high production output; that is, it targets only the parts of the ecosystem that are related to desired productivity. Coerced resilience can therefore be regarded as a management form that focuses on specific resilience rather than the broader or general resilience of the ecosystem that supports productivity. Indeed, optimizing outputs from a system is antithetical to managing a broadly resilient system; naturally resilient systems provide (less than optimal) outputs under a variety of conditions, whereas optimizing system outputs focuses on maximizing one or a few outputs under ideal conditions. Thus, coerced resilience is more aligned with "command and control" management (Holling and Meffe 1996), which recognizes that managing parts of the system can in the long run lead to the erosion of system resilience and potentially cause regime shifts (Gunderson 2000). Coercion of this type can reduce resilience of the system under management (Holling and Meffe 1996; Rist et al. 2014).

Coerced regimes

Contrary to coerced resilience, coerced regime management targets the entire system rather than focusing on specific aspects or outputs. Coerced regimes involve management activities that create artificial, non–self-sustaining feedbacks in a system through constant intensive interventions and subsidies that mimic the conditions of a desirable regime (Angeler et al. 2020b). Examples of coerced regimes are omnipresent and occur in diseased human subjects in need of permanent medication for full functioning (Angeler et al. 2018), lake management to mitigate acidification (Angeler et al. 2017), mechanical removal to remove trees from grasslands (Garmestani et al. 2019), or mitigation to preserve the climate (Steffen et al. 2018).

In coerced regimes, because feedbacks are artificial, they mask rather than break the natural feedbacks of undesired system regimes. In practice this means that management often aims but fails to transform degraded regimes to desired regimes (e.g., eutrophic lakes resisting restoration efforts) and therefore only imitates the desired functioning of an untenable prior regime (i.e., the "ghost of a desired regime"). As soon as management is discontinued a system rebounds to the full manifestation of the degraded regime (e.g., a lake returning from clearwater conditions to the degraded aspects of its turbid regime). This notion is well aligned with ecological restoration and resilience theory (Suding et al. 2004; Suding and Gross 2006; Angeler and Allen 2016). Also inherent in coerced regimes is the aim to conserve desirable system regimes that are no longer viable (e.g., the current climate regime) that would flip to an alternative, often undesired regime if management were discontinued (figure 4.1).

Although the concepts of coerced regimes and coerced resilience focus on different aspects—the former imitating systemic conditions of untenable regimes and the latter improving the functioning of a few system variables (e.g., production output)—they are sometimes not mutually exclusive as in some managed ecosystems. For example, lake liming mimics circumneutral conditions within an acidified regime and at the same time targets the enhancement of recreational fisheries (Angeler et al. 2017). Similarly, agroecosystems such as corn, wheat, or soy fields have little or no self-organization, and lost resilience is offset by pesticides and fertilizers, because the management target is enhanced productivity (Rist et al. 2014).

Coercion concepts in resilience science are valuable for envisioning the management challenges ecological, social-ecological, and other complex systems face in a rapidly changing world. Complex systems of people and nature comprise several essential system features, particularly adaptive capacity, thresholds, and scale (Baho et al. 2017; Angeler et al. 2019). These features can be incorporated into coercion concepts to facilitate the envisioning and management of social-ecological challenges. Scale-specific structures and processes inherent in the hierarchical organization of complex systems underpin their resilience because disturbances are scale specific and affect some scales more than others in the system (Nash et al. 2014). Impact-free scales can buffer against disturbances, confer resilience, and maintain the overall system configuration through cross-scale interactions and feedbacks. This resilience is conferred partly by the abundance, redundancy, and diversity of functional traits within and across the scales present in a system (Sundstrom et al. 2018). This suggests that examining coerced resilience and regime concepts through the lens of scale might provide a more detailed picture of potentials and pitfalls of management in an uncertain Anthropocene. In the next section, we discuss coerced regimes from a cross-scale or panarchy perspective.

Panarchy

Panarchy conceptualizes multiscale organization, interconnectedness of scales and dynamic system structure, and the collapse and reorganization inherent

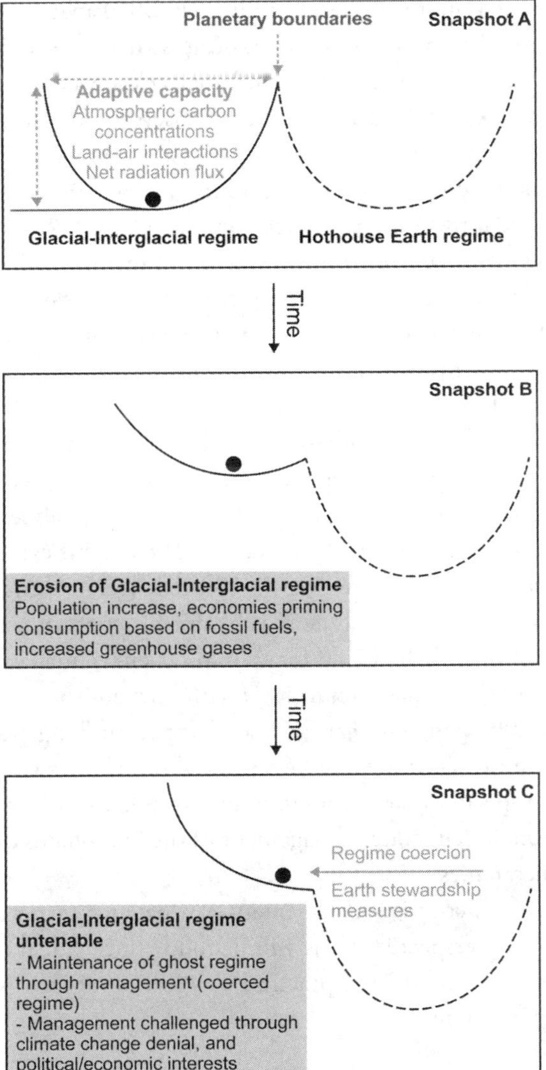

Figure 4.1. Illustration of the global climate system, indicating alternative regimes between a glacial–interglacial climate (left) and hothouse Earth climate (right). Over time resilience has eroded, as indicated by changes in the stability landscape of the glacial–interglacial regime (Snapshot A to Snapshot B to Snapshot C). Coercion through a variety of mechanisms (Earth stewardship measures), symbolized by the arrow that keeps the system out of the current undesired regime, are increasingly needed to maintain the current system from flipping into a hothouse climate regime (Snapshot C). Transformations across societies are needed to remove these coercions to generate a self-organizing regime with increased adaptive capacity.

in adaptive cycles (Gunderson and Holling 2002). Panarchy is a model of complex systems that portrays the cross-scale structure envisioned in hierarchy theory (Allen and Starr 1982) and that acknowledges top-down and bottom-up processes that account for many observed nonlinear dynamics in ecosystems. Panarchy also considers complex system dynamics such as adaptation, conservation, collapse, and reorganization, which are often portrayed as a nested set of adaptive cycles (Gunderson and Holling 2002) (figure 4.2). These adaptive cycles are related to extrinsic (climatic) and intrinsic (e.g., economy, governance) factors that influence, and are influenced by, important interrelated phenomena such as innovation, novelty, and regime changes in social-ecological systems (Allen et al. 2014). The theory can provide a heuristic to conceptualize different aspects of system organization (Allen et al. 2014) or can be formulated into hypotheses for individual premises and empirically tested (Angeler et al. 2015). There have been diverse and successful applications of panarchy in social and ecological systems and analyses (Green et al. 2015; Berkes and Ross 2016; DeWitte et al. 2017), and it can provide quantitative and qualitative underpinning for risk management and vulnerability and risk assessments (Angeler et al. 2016b). In this context, a panarchy may be a particularly suitable heuristic for framing coercive management from a multiscale perspective while accounting for the dynamism at each scale in a system. Specifically, panarchy can be useful for identifying points in adaptive cycles where management can be leveraged to enhance the coercion of a regime within a specific system configuration. In other words, a panarchy can be coerced through deliberate management of distinct phases of the adaptive cycles across hierarchies.

We next discuss coercive management of a panarchy (coerced panarchy) from a conceptual perspective. This will be followed by a real example using the global climate to discuss the practical implication of cross-scale management of dynamic systems.

Coerced Panarchy

Because of the interconnectedness of scales and the dynamic nature of these scale domains, captured by the processes within adaptive cycles, coercive management of a panarchy may have essentially a vertical (cross-scale) and horizontal (within-scale) component. Regarding the vertical component, management interventions at one scale can propagate up or down the entire panarchy. Such cascades are often desirable because not all scales in a panarchy

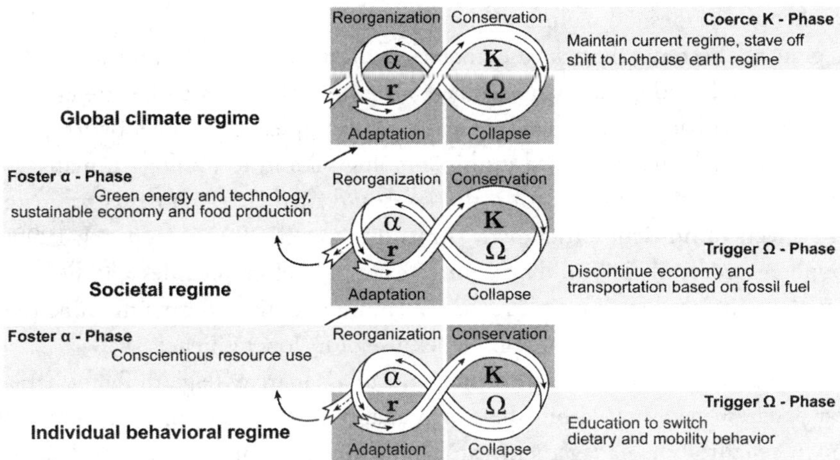

Figure 4.2. Schematic illustration indicating panarchy of change to release the current coercion of the global climate regime. Leverage points are indicated for intervention at other scales. Education and interventions at the individual scale would lead to alternative trajectories of societies. Society regimes characterized by green energy, technology, and economies could be achieved by intervention in different phases of the adaptive cycles at individual and societal scales. Such changes would stave off a transition toward undesired, warmer global regimes.

are amenable to direct interventions or actions, that is, they are not manageable. Imagine a large eutrophic lake with recurring, broad-scale toxic algal blooms. Managing the blooms through direct interventions at the scale at which phytoplankton operates (i.e., fishing of the blooms) is impossible and therefore requires other interventions such as nutrient or top-down food web control. That is, interventions at higher trophic levels (fishing of planktivorous fish or stocking of predatory fish) can induce effects that cascade down the food web and eventually manifest at the trophic level of algal primary producers (Carpenter and Kitchell 1993). Similarly, inducing change at the top level of a panarchy often is impossible and requires a major change induced from below, as demonstrated by the massive revolts during the Arab Spring that lead to the toppling of corrupt authoritarian regimes (Sadiki 2014).

Ideally, management of a panarchy can be facilitated if interventions at a single scale propagate throughout other scales. We speculate that single-scale management of a panarchy might include the need for low coercion of a system in which the resilience of a desired regime is still high or the resilience of a degraded regime low. Building on this line of reasoning, we speculate that single-scale interventions will become insufficient the more eroded the resilience of a desired regime or the more stable the resilience of a degraded regime has become. In such cases management interventions at multiple scales will be needed to strengthen the coercion potential and thus the strength of artificial feedbacks in a panarchy. Again, for example, very resilient eutrophic (degraded) lake regimes need multiple biological and technological management interventions both within the lakes and in their catchments to abate the effects of nutrient enrichment (Annadotter et al. 1999; Madgwick 1999; Søndergaard et al. 2000). The climate example below provides another example of the need for multiple management interventions for maintaining the climate system in its current regime because of its low resilience.

Regarding the horizontal components of a panarchy (e.g., different habitat patches operating within their own adaptive cycles that may or may not be in synchrony), management can exploit the dynamic adaptive cycling of patterns and processes and intervene in the different adaptive cycle phases (from α to Ω) by either freezing, destroying, or boosting the reorganization of these phases. For example, nutrient precipitation in a lake may trigger the Ω phase for phytoplankton and facilitate the growth of submerged plants that contribute to clearwater conditions. Minimizing disturbances to macrophytes, such as sediment resuspension, may consolidate the α phase. As is the case with vertical components of coerced panarchies, redundant interventions within a single phase or across different phases of an adaptive cycle would probably increase the coercion success.

The lake example shows that the management of vertical and horizontal structures in a panarchy can be informed by a coercion analogue of the cross-scale resilience model (Peterson et al. 1998). The cross-scale resilience model posits that the relative resilience of a system is increased with an increasing redundancy of functions within and across scales. An analogous cross-scale coercion model suggests that an increasing redundancy of management interventions within and across scales is needed to artificially maintain a system in a desirable regime with low resilience (i.e., a system facing an imminent regime shift) or to mitigate degraded regimes that are highly resilient.

We next discuss the global climate regime as an example of a coerced panarchy, that is, a system that needs to be coerced into the current desirable climate regime by the implementation of mitigation measures with the aim of preventing it from tipping over into a hothouse Earth regime (Steffen et al. 2018). We particularly highlight how the redundancy of Earth stewardship measures within and across scales in the climate panarchy may influence the coercion potential.

Global Climate Panarchy

Modeling evidence suggests that continuation of past and present human activities will lead to ongoing climate change, as indicated by higher global average temperatures, accelerating rise in sea level, weakened marine jet streams, and shrinking polar ice sheet cover (Intergovernmental Panel on Climate Change [IPCC] 2014). Such changes are expected to occur abruptly and nonlinearly, potentially leading to a new regime in which the climate is substantially and persistently changed. Profound changes in the global climate are already under way and supported by empirical evidence. Examples include regime shifts in forest fire regimes (Westerling et al. 2006), groundwater systems (Figura et al. 2011), arctic marine environments (Kortsch et al. 2012), and other ocean regions in the Northern Hemisphere (Beaugrand et al. 2015).

Steffen et al. (2018) considered the present climate regime as a glacial–interglacial limit cycle with a 100,000-year cycle. This cycle has been self-organizing and maintained during the past 1.2 million years. An alternative regime is described as a hothouse Earth that might have catastrophic consequences for humanity should planet Earth fully shift and stabilize in this regime. Steffen et al. (2018) contended that we need to consider deliberately coercing the Earth system into the glacial–interglacial regime to stave off a regime change to a hothouse Earth (figure 4.1). However, forcing the Earth system artificially into this glacial–interglacial regime would require sustained, massive management efforts and external subsidies by humans because such a regime would no longer be self-organizing. A series of Earth stewardship measures have been identified that could be used to coerce the Earth system to remain in the current glacial–interglacial regime (table 4.1). These measures include a global economy uncoupled from fossil fuels, augmentation of global carbon sinks, technology innovations, new governance agreements, behavioral transformation, and changes in social values.

The global climate provides an example where the resilience of the current glacial–interglacial regime has been increasingly eroded, mainly by the decades-long overexploitation of fossil fuel. Specifically, fossil fuel has become the main subsidy of economic growth and human welfare. The results of externalizing the costs of carbon from the economy has led to increases in the concentrations of carbon dioxide and other greenhouse gases to the atmosphere. The accumulating gases have changed the energy budget of the planet, leading to the effects mentioned above. As Steffen et al. (2018) suggested, the pathological overuse of fossil fuel has exhausted the resilience of the glacial–interglacial

TABLE 4-1. *Examples of partly interdependent management measures that can be mapped onto different phases of the adaptive cycle for coercing the climate panarchy (figure 4.2). Shown are also selected redundancies of these measures and their potential application across scales in the panarchy for increasing their coercive force.*

Measure	Redundancy	Scale
Geoengineering	Solar radiation management. Carbon capture and storage (enhanced weathering, direct air capture; reforestation, afforestation)	Society
Technology	Development of cleaner technology and improved efficiency	Society
Economy	Emission taxes, investments, subsidies, carbon emission trading	Society
Governance	Polycentric; international policy and law	Society
Emission reduction	Low-carbon energy (renewable; biomass, solar, wind, geothermal) and transportation (electric cars, hybrids)	Individual, societies
Urban transformation and building design	Provide and encourage public transport, facilitate walking and bicycling, promote green areas	Individual, societies
Lifestyle and behavior	Dietary shifts to plant-based diet, reduced use of car and air travel, family planning	Individual, societies

regime, and significant management efforts are needed to mitigate the potential devastating consequences of a shift to a hothouse Earth regime. That is, the conditions of the glacial–interglacial regime must be conserved, necessitating the artificial coercion of the entire glacial–interglacial regime. Such coercion currently involves failed attempts to limit inputs of carbon and nitrogen into the atmospheric and oceanic realms of the planet. However, there are indications that the global social–ecological–climatological system is hysteretic. In other words, the path leading to degradation of the system is not the same as the pathway needed to reverse it. That is, management for coercing the current climate regime differs substantially from the human activity that eroded its resilience over time: Simply limiting the use of fossil fuels is unlikely to reverse the warming of the climate.

That the way out of the current delicate situation is extremely challenging is manifested in the complex, insufficient, and ineffective implementation of many Earth stewardship measures in the short and longer term. This may be due to, for example, inflexible policy that is mismatched with environmental challenges (Green et al. 2015), the weak adherence by many countries to international initiatives such as the Paris climate agreement, potential side effects of atmospheric carbon mitigation that can offset potential benefits (e.g., soil carbon sequestration; Powlson et al. 2014), and the speed by which global warming unfolds, manifested in an all-time high of CO_2 emissions at the time this chapter was written (Global Carbon Project 2018). Although Earth stewardship measures (table 4.1) are a minimal requirement to effectively maintain a coerced climate regime trajectory, they allow examination of the complexity of coercion of the global climate from a panarchy perspective. Specifically, examining a coerced climate panarchy may be particularly suitable for envisioning the management challenges (e.g., cross-scale interactions) that are inherent in dynamic wicked problems, such as climate change (Angeler et al. 2016b).

Twidwell et al. (2019) made a clear case that the application of coercive force to management occurs within the bounds of existing policies, which further underscores the need for adaptive law and transformative governance to manage essential structures and functions for maintaining the current Earth system regime (Green et al. 2015; Chaffin et al. 2016; Garmestani et al. 2019). Maintaining structures and functions is critical for the resilience of a system (Holling 1973), especially if management aims to avoid regime changes, such as the shift of the current climate regime to a hothouse Earth regime. In panarchy theory, resilience relates to the adaptive cycling of nested systems through phases of reorganization and collapse. The interconnectedness of hierarchical scales in a panarchy further contributes to system resilience through feedbacks that can stabilize or destabilize system configurations through cross-scale interactions. Accounting for both the adaptive cycling and the interconnectedness of scales provides the potential to envision and implement management under constant system change.

Consider a simplified example of a coerced climate panarchy consisting of three arbitrary scales: the human individual, societies, and the global climate regime (figure 4.2). Each scale shows points in the adaptive cycle at which mitigation measures can be leveraged. At the highest level, the global climate regime, management aims are to freeze the Ω phase to maintain the current glacial–interglacial system and avoid a regime shift and reorganization to a hothouse Earth regime. However, it is clear that at this scale direct management cannot be leveraged to control the Ω phase. Building on interconnectedness as a major tenet of panarchy theory, management interventions at the societal and individual scales can be designed to percolate up to the highest scale in the global climate regime and force the maintenance of the Ω phase and avoid system collapse.

Such coerced panarchy shows that management can target different phases of an adaptive cycle. Such management can be sequential and implement a gale of creative destruction, as described by Austrian economist Josef Schumpeter (2010). In the broadest sense of the climate debate this means that current unsustainable behavior (fossil fuel–based economy) must be destroyed and replaced through the creation of sustainable (green energy) livelihoods at both the societal and individual levels. Such change is necessary because incremental adaptation measures to climate change may be insufficient, necessitating transformation (Kates et al. 2012). In our example, we provide a simplified demonstration of the coercion of the panarchy with Earth stewardship measures that target the α and Ω phases at the individual level. Such measures include changes in current alimentation (e.g., meat-based diet) and transportation habits (e.g., air travel, car). These habits have a high carbon footprint, which should be reduced (inducing the Ω phase). Sustainable livelihoods through the reorganization of their behavior and resource use (reduced consumption, plant-based diet, use of green energy and transportation) resulting from transitioning to a low carbon footprint can then be created and spur the α phase of the cycle (figure 4.2). Similarly, at the societal level, economies, technologies, and production systems that rely on fossil fuels should be replaced (Ω phase) by alternative models that are cleaner and more efficient (α phase) (figure 4.2, table 4.1).

Our model of coerced climate panarchy shows that many Earth stewardship measures, such as those related to green transportation, could be implemented across scales in the panarchy (table 4.1). Specifically, the provision of infrastructure that facilitates green transportation operates mainly at the societal level, whereas the use of such infrastructure depends on the individuals within societies. The model also shows that several categories of Earth stewardship measures are redundant, such as solar radiation management and carbon capture and storage as geo-engineering measures (table 4.1). These examples show that Earth stewardship measures can be redundant within and across scales in the climate panarchy. Within-scale and cross-scale redundancies are critical elements of system resilience (Peterson et al. 1998) and may therefore play a critical role in effectively coercing the current climate in the glacial–interglacial regime. Because the resilience of a system is expected to increase with increasing within-scale and cross-scale redundancy (Allen et al. 2005), we speculate that effective coercion of the glacial–interglacial regime will increase with increased redundancy of climate mitigation measures within and across scales in this particular climate system. Multiple coercion measures are needed in the face of the rapidly eroding resilience of our current climate regime in order to stave off an undesired, catastrophic shift to a hothouse Earth regime.

Conclusions

Unprecedented rates of environmental change in the Anthropocene require management models that guarantee the sustainable development of an increasing human population. The concept of coerced panarchy allows envisioning the leverage of management interventions in a complex system, such as the global climate, to optimize the potential to adapt to or transform in response to societal challenges (Angeler et al. 2020a). Successful long-term stewardship of the Earth will require interdisciplinary efforts involving global partnerships that link researchers, managers, policymakers, and citizens (Chapin et al. 2011). The coerced climate panarchy example identifies the scales and phases of the adaptive cycles that these partnerships could target for the implementation of mitigation measures. This example shows that within-scale and cross-scale aspects will be critical for effectively coercing the climate in the current glacial–interglacial regime. That is, it is likely that the potential for successful coercion increases with an increased redundancy and diversity of Earth stewardship measures within and across scales in the climate regime. Multiple interventions are probably needed to keep the current climate regime in its present low-resilience state. However, although theoretically a series of mitigation measures are available, the implementation of these may face societal rejection (e.g., fewer children) or reluctance to change habits and behavior in the short term. That is, cultural, in addition to economic, political, and technological aspects may potentially limit the application of the full arsenal of Earth stewardship measures and potentially weaken the mitigation of global warming.

A coerced panarchy has the potential to clarify the complexity associated with artificially maintaining desired system conditions. It remains to be seen how effective the coercion of the current climate regime can be in the long term given that atmospheric CO_2 levels continue to increase. Global warming may soon reach levels that exceed our potential for coercion of the climate panarchy even if Earth stewardship measures are implemented at multiple scales.

Acknowledgments

This study was supported by a sabbatical professorship to D.G.A. through the University of Nebraska–Lincoln.

Literature Cited

Allen, C.R., D.G. Angeler, A.S. Garmestani, L.H. Gunderson, and C.S. Holling. 2014. Panarchy: Theory and applications. *Ecosystems* 17: 578–589.

Allen, C.R., L. Gunderson, and A.R. Johnson. 2005. The use of discontinuities and functional groups to assess relative resilience in complex systems. *Ecosystems* 8: 958–966.

Allen, T.F.H., and T.B. Starr. 1982. *Hierarchy: Perspectives for Ecological Complexity.* University of Chicago Press, Chicago.

Angeler, D.G., and C.R. Allen. 2016. Quantifying resilience. *Journal of Applied Ecology* 53: 617–624.

Angeler, D.G., C.R. Allen, C. Barichievy, T. Eason, A.S. Garmestani, N.A.J. Graham, D. Granholm, L. Gunderson, M. Knutson, K.L. Nash, R.J. Nelson, M. Nyström, T.E. Spanbauer, C.A. Stow, and S.M. Sundstrom. 2016a. Management applications of discontinuity theory. *Journal of Applied Ecology* 53: 688–698.

Angeler, D.G., C.R. Allen, and A. Carnaval. 2020a. Convergence science in the Anthropocene: Navigating the known and unknown. *People and Nature* 2: 96–102.

Angeler, D.G., C.R. Allen, A.S. Garmestani, L.H. Gunderson, O. Hjerne, and M. Winder. 2015. Quantifying the adaptive cycle. *PLoS One* 10: e0146053. doi:10.1371/journal.pone.0146053

Angeler, D.G., C.R. Allen, A.S. Garmestani, L.H. Gunderson, and I. Linkov. 2016b. Panarchy use in environmental science for risk and resilience planning. *Environment Systems and Decisions* 36: 225–228.

Angeler, D.G., C.R. Allen, and M.-L. Persson. 2018. Resilience concepts in psychiatry demonstrated with bipolar disorder. *International Journal of Bipolar Disorders* 6: 2.

Angeler, D.G., B.C. Chaffin, S.M. Sundstrom, A. Garmestani, K.L. Pope, D.R. Uden, D. Twidwell, and C.R. Allen. 2020b. Coerced regimes: Management challenges in the Anthropocene. *Ecology & Society* 25: 4.

Angeler, D.G., P. Chow-Fraser, M.A. Hanson, S. Sánchez-Carrillo, and K.D. Zimmer. 2003. Biomanipulation: A useful tool for freshwater wetland mitigation? *Freshwater Biology* 48: 2203–2213.

Angeler, D.G., S. Drakare, R.K. Johnson, S.J. Köhler, and T. Vrede. 2017. Managing ecosystems without prior knowledge: Pathological outcomes of lake liming. *Ecology and Society* 22: 44.

Angeler, D.G., H.B. Fried-Petersen, C.R. Allen, A. Garmestani, D. Twidwell, W. Chuang, V.M. Donovan, T. Eason, C.P. Roberts, S.M. Sundstrom, and C.L. Wonkka. 2019. Adaptive capacity in ecosystems. *Advances in Ecological Research* 60: 1–24.

Annadotter, H., G. Cronberg, R. Aagren, B. Lundstedt, P.Å. Nilsson, and S. Ströbeck. 1999. Multiple techniques for lake restoration. In *The Ecological Bases for Lake and Reservoir Management*, ed. D.M. Harper, B. Brierley, A.J.D. Ferguson, and G. Phillips, 77–85. Developments in Hydrobiology 136. Springer, Dordrecht, The Netherlands.

Baho, D.L., C.R. Allen, A. Garmestani, H.B. Fried-Petersen, S.E. Renes, L.H. Gunderson, and D.G. Angeler. 2017. A quantitative framework for assessing ecological resilience. *Ecology and Society* 22: 17

Beaugrand, G., A. Conversi, S. Chiba, M. Edwards, S. Fonda-Umani, C. Greene, N. Mantua, S.A. Otto, P.C. Reid, M.M. Stachura, L. Stemmann, and H. Sugisaki. 2015. Synchronous marine pelagic regime shifts in the Northern Hemisphere. *Philosophical Transactions of the Royal Society B* 370: 20130272.

Beisner, B.E., D.T. Haydon, and K. Cuddington. 2003. Alternative stable states in ecology. *Frontiers in Ecology and the Environment* 1: 376–382.

Berkes, F., and H. Ross. 2016. Panarchy and community resilience: Sustainability science and policy implications. *Environmental Sciences and Policy* 61: 185–193.

Biggs, R., G. Peterson, and J.C. Rocha. 2018. The regime shifts database: A framework for analyzing regime shifts in social-ecological systems. *Ecology and Society* 23: 9.

Carpenter, S.R., and J.F. Kitchell. 1993. *The Trophic Cascade in Lakes*. Cambridge University Press, Cambridge, UK.

Chaffin, B.C., A.S. Garmestani, L. Gunderson, M. Harm Benson, D.G. Angeler, C.A. Arnold, B. Cosens, R.K. Craig, J.B. Ruhl, and C.R. Allen. 2016. Transformative environmental governance. *Annual Review of Resources and Environment* 41: 399–423.

Chapin, F.S., S.T. Pickett, M.E. Power, R.B. Jackson, D.M. Carter, and C. Duke. 2011. Earth stewardship: A strategy for social–ecological transformation to reverse planetary degradation. *Journal of Environmental Studies and Sciences* 1: 44–53.

DeWitte, S.N., M.H. Kurth, C.R. Allen, and I. Linkov. 2017. Disease epidemics: Lessons for resilience in an increasingly connected world. *Journal of Public Health* 39: 254–257.

Figura, S., D.M. Livingstone, E. Hoehn, and R. Kipfer. 2011. Regime shift in groundwater temperature triggered by the Arctic Oscillation. *Geophysical Research Letters* 38: 1–5.

Folke, C., S. Carpenter, B. Walker, M. Scheffer, T. Elmqvist, L. Gunderson, and C.S. Holling. 2004. Regime shifts, resilience, and biodiversity in ecosystem management. *Annual Reviews in Ecology, Evolution and Systematics* 35: 557–581.

Garmestani, A., J.B. Ruhl, B.C. Chaffin, R.K. Craig, H.F. van Rijswick, D.G. Angeler, C. Folke, L. Gunderson, D. Twidwell, and C.R. Allen. 2019. Untapped capacity for resilience in environmental law. *Proceedings of the National Academy of Sciences* 11: 201906247.

Global Carbon Project 2018. https://www.globalcarbonproject.org/index.htm

Green, O.O., A.S. Garmestani, C.R. Allen, J.B. Ruhl, C.A. Arnold, L.H. Gunderson, N.A.J. Graham, B. Cosens, D.G. Angeler, B.C. Chaffin, and C.S. Holling. 2015. Barriers and bridges to the integration of social-ecological resilience and law. *Frontiers in Ecology and the Environment* 13: 332–337.

Gunderson, L.H. 2000. Resilience in theory and practice. *Annual Review of Ecology and Systematics* 31: 425–439.

Gunderson, L.H., and C.S. Holling. 2002. *Panarchy: Understanding Transformations in Human and Natural Systems*. Island Press, Washington, DC.

Holling, C.S. 1973. Resilience and stability of ecological systems. *Annual Review of Ecology and Systematics* 4: 1–23.

Holling, C.S., and G.K. Meffe. 1996. Command and control and the pathology of natural resource management. *Conservation Biology* 10: 328–337.

Homer-Dixon, T.F. 1991. On the threshold: environmental changes as causes of acute conflict. *International Security* 16: 76–116.
Hughes, T.P., S. Carpenter, J. Rockström, M. Scheffer, and B. Walker. 2013. Multiscale regime shifts and planetary boundaries. *Trends in Ecology & Evolution* 28: 389–395.
Intergovernmental Panel on Climate Change (IPCC). 2014. *Climate Change 2014: Synthesis Report. Contribution of Working Groups I, II and III to the Fifth Assessment Report of the Intergovernmental Panel on Climate Change*, ed. Core Writing Team, R.K. Pachauri and L.A. Meyer. IPCC, Geneva, Switzerland.
Kates, R.W., W.R. Travis, and T.J. Wilbanks. 2012. Transformational adaptation when incremental adaptations to climate change are insufficient. *Proceedings of the National Academy of Sciences of the USA* 109: 7156–7161.
Kortsch, S., R. Primicerio, F. Beuchel, P.E. Renaud, J. Rodrigues, O.J. Lønne, and B. Gulliksen B. 2012. Climate-driven regime shifts in Arctic marine benthos. *Proceedings of the National Academy of Sciences USA* 109: 14052–14057.
Madgwick, F.J. 1999. Restoring nutrient-enriched shallow lakes: Integration of theory and practice in the Norfolk Broads, UK. *Hydrobiologia* 408/409: 1–12.
McMichael, A.J., S. Friel, A. Nyong, and C. Corvalan. 2008. Global environmental change and health: impacts, inequalities, and the health sector. *British Medical Journal* 336: 191–194.
Miller, M.H., and J. Williamson. 1988. The international monetary system: An analysis of alternative regimes. *European Economic Review* 32: 1031–1048.
Nash, K.L, C.R. Allen, D.G. Angeler, C. Barichievy, T. Eason, A.S. Garmestani, N.A.J. Graham, D. Granholm, M. Knutson, R.J. Nelson, M. Nyström, C.A. Stow, and S.M. Sundstrom. 2014. Discontinuities, cross-scale patterns and the organization of ecosystems. *Ecology* 95: 654–667.
Peterson, G., C.R. Allen, and C.S. Holling. 1998. Ecological resilience, biodiversity, and scale. *Ecosystems* 1: 6–18.
Powlson, D. S., C.M. Stirling, M.L. Jat, B.G. Gerard, C.A. Palm, P.A. Sanchez, and K.G. Cassman. 2014. Limited potential of no-till agriculture for climate change mitigation. *Nature Climate Change* 4: 678.
Rist, L., A. Felton, M. Nyström, M. Troell, R.A. Sponseller, J. Bengtsson, H. Österblom, R. Lindborg, P. Tidåker, D.G. Angeler, R. Milestad, and J. Moen. 2014. Applying resilience thinking to production systems. *Ecosphere* 5: 73.
Rockström, J., W. Steffen, K. Noone, A. Persson, F. S. Chapin, E. Lambin, T. M. Lenton, M. Scheffer, C. Folke, H.J. Schellnhuber, B. Nykvist, C.A. de Wit, T. Hughes, S. van der Leeuw, H. Rodhe, S. Sörlin, P.K. Snyder, R. Costanza, U. Svedin, M. Falkenmark, L. Karlberg, R.W. Corell, V.J. Fabry, J. Hansen, B. Walker, D. Liverman, K. Richardson, P. Crutzen, and J. Foley. 2009. Planetary boundaries: Exploring the safe operating space for humanity. *Ecology and Society* 14: 32.
Sadiki, L. 2014. *Routledge Handbook of the Arab Spring*. Routledge, New York.
Scheffer, M., S.H. Hosper, M.L. Meijer, B. Moss, and E. Jeppesen. 1993. Alternative equilibria in shallow lakes. *Trends in Ecology & Evolution* 8: 275–279.
Schumpeter, J.A. 2010. *Capitalism, Socialism, and Democracy.* Routledge, New York.
Society for Ecological Restoration (SER) International Science & Policy Working Group. 2004. *The SER International Primer on Ecological Restoration*. Society for Ecological Restoration International, Washington, DC.

Søndergaard, M., E. Jeppesen, J.P. Jensen, and T. Lauridsen. 2000. Lake restoration in Denmark. *Lakes & Reservoirs: Research & Management* 5: 151–159.

Steffen, W., J. Rockström, K. Richardson, T.M. Lenton, C. Folke, D. Liverman, C.P. Summerhayes, A.D. Barnosky, S.E. Cornell, M. Crucifix, J.F. Donges, I. Fetzer, S.J. Lade, M. Scheffer, R. Winkelmann and H.J. Schellnhuber. 2018. Trajectories of the Earth system in the Anthropocene. *Proceedings of the National Academy of Sciences USA* 115: 8252-8259.

Suding, K.N., and K.L. Gross. 2006. The dynamic nature of ecological systems: Multiple states and restoration trajectories. In *Foundations of Restoration Ecology*, ed. D.A. Falk, 190–209. Island Press, Washington, DC.

Suding, K.N., K.L. Gross, and G.R. Houseman. 2004. Alternative states and positive feedbacks in restoration ecology. *Trends in Ecology & Evolution* 19: 46–53.

Sundstrom, S.M., D.G. Angeler, C. Barichievy, T. Eason, A. Garmestani, L. Gunderson, M. Knutson, K.L. Nash, T. Spanbauer, C. Stow, and C.R. Allen. 2018. The distribution and role of abundance in cross-scale resilience. *Ecology* 99: 2421–2432.

Twidwell, D., C.L. Wonkka, H.H. Wang, W. Grant, C.R. Allen, S.D. Fuhlendorf, A. Garmestani, D. Angeler, C.A. Taylor, U.P. Kreuter, and W.E. Rogers. 2019. Coerced resilience in fire management. *Journal of Environmental Management* 240: 368–373.

Westerling, A.L., H.G. Hidalgo, D.R. Cayan, and T.W. Swetnam. 2006. Warming and earlier spring increase western US forest wildfire activity. *Science* 313: 940–943.

CHAPTER 5

Applications of Spatial Regimes

Craig R. Allen, David G. Angeler, Ahjond Garmestani, Caleb P. Roberts, Shana M. Sundstrom, Dirac Twidwell, and Dan R. Uden

Panarchy theory (Gunderson and Holling 2002) emphasizes the dynamic nature of ecosystems across scales of space and time. The adaptive cycle component of panarchy posits different types of change over time: rapid growth followed by slowing growth of structural aspects of a system. Changes in ecosystem controls and feedbacks generate instabilities that lead to renewal and reorganization within the system. Cross-scale interactions are key in generating instabilities and in how the system reorganizes for future trajectories (Gunderson et al., chapter 1). Such trajectories may be a similar system (regime), or the system may flip into an alternative configuration or regime. Such regime changes are not confined to change in the time domain but often involve changes in the spatial dimensions of the system. This chapter reviews linkages between spatial regimes and panarchy in four parts. We begin with a definition of spatial regimes and how they are bounded. The second section presents methods for identifying spatial regimes. The third section discusses the application of spatial regime concepts to management. We conclude with a discussion of challenges associated with applications of this facet of panarchy theory.

Defining Spatial Regimes

The conceptual linkage between ecological regimes and resilience was introduced by Holling (1973). That was followed by studies on the contribution of spatial attributes to resilience (Cumming 2011; Allen et al. 2016). Resilience is a measure of the amount of disturbance a system can withstand without exceeding a critical threshold and undergoing a regime shift, whereby the system reorganizes into a new regime, rather than buffering or recovering from disturbances and remaining in the same regime (Angeler and Allen 2016). This definition includes another common definition of resilience, that of recovery or bounceback (Allen et al. 2019), but also accounts for nonstationarity of boundaries and critical thresholds that separate regimes. Spatial resilience is the contribution of spatial attributes to the feedbacks that generate resilience in ecosystems and other complex systems and vice versa (Allen et al. 2016).

The ways spatial resilience contributes to our knowledge of system resilience include approaches to identifying leading indicators of regime change, such as critical spatial thresholds (Kéfi et al. 2014); assessing structural and functional effects of spatial components of managed systems in relation to their resilience (Allen et al. 2014; Angeler et al. 2016); determining the role of connectivity, dispersal, and other movements in conferring resilience (Uden et al. 2014); assessing the relevance of network membership for node resilience and the relevance of node composition for network resilience (Keitt et al. 1997; Moore et al. 2015); evaluating relationships between spatial landscape metrics and resilience (Cumming 2011; Uden et al. 2014); and developing approaches for understanding cross-scale interactions in social-ecological systems (Cumming et al. 2015, Sundstrom et al. 2018). These research avenues have been crucial for understanding how different spatial configurations of system attributes contribute to resilience. However, to account for the multiscaled complexity of spatial aspects of resilience, we also need ways to conceptualize and identify the spatial manifestation of the processes and structures that define regimes. We define spatial regimes as *spatially explicit social-ecological systems maintained by feedback mechanisms that exhibit relative homogeneity in structure and function within their geographic boundaries over time* (Allen et al. 2016); these boundaries are likely to move given changing environmental conditions. This definition distinguishes spatial regimes from spatial resilience because its focus makes explicit that the patterns and processes that characterize a given regime are bound by defined spatial extents.

Determination of the spatial extent of regimes is closely related to an ongoing challenge for ecology: the detection of ecological boundaries. The resilience perspective on boundary detection includes the explicit understanding that ecosystems consist of multiple nested scales of pattern and process (chapter 3). This alters boundary detection from something historically done at one level of organization—that of an ecosystem—to the need for boundary detection at multiple scales within a system panarchy. Furthermore, objective methods for identifying the extent of spatial regimes at multiple scales has substantial management implications for tracking the impacts of anthropogenic change on regimes via changes in spatial regimes. For example, in a proof of concept Sundstrom et al. (2017) used homogeneity and discontinuities in avian community structure from an east–west transect of the central United

States to identify spatial regimes. The physical location in space of the avian regime boundaries was compared with expected ecosystem boundaries delineated by U.S. federal ecoregion maps and substantive discrepancies between the location of avian community regimes and ecoregion boundaries; those discrepancies corresponded to land use change along the transects. Roberts et al. (2019) also compared shifts in avian community structure along a north–south transect in the U.S. Great Plains and demonstrated that spatial regimes in avian community structure are shifting north in a manner consistent with climate change. These examples highlight the benefit of a spatial focus as a way to understand and potentially anticipate regime shifts.

Boundary detection

The identification of spatial regimes is urgent given the recent acceleration of anthropogenic climate change, which is expected to shift ecological boundaries. At the largest scale of ecological organization, biomes are a function of climatic and edaphic processes, and their boundaries have long been considered sensitive to climate change (Kent et al. 2006; Danz et al. 2012). Hence, tracking biome boundary shifts is one approach to monitor and evaluate biotic responses to climate change. Dramatic and rapid altitudinal shifts in montane ecological boundaries from climate change (Allen and Breshears 1998; Beckage et al. 2008) have been documented, as well as climate-driven boundary shifts in marine systems (Edwards and Richardson 2004; Grebmeier et al. 2006). Because ecological boundaries typically demarcate the distribution of vegetation and ecosystem type, they provide critical information about the extent and rate of the biological processes shaping the boundary and driving the maintenance of the regime within the boundary (Yarrow and Salthe 2008). This has implications for both environmental management and biological conservation (Kent et al. 2006).

Boundary identification, or the detection of the region separating neighboring ecosystems, has been an active area of research in terrestrial ecology and biogeography and is data intensive and statistically challenging (Kent et al. 2006). The use of remotely sensed vegetation data is less laborious than field work, but the method is poor at distinguishing between physically similar but floristically different vegetation; it may require labor-intensive ground truthing to verify ecological transitions in plant assemblages (Schmidtlein and Sassin 2004). For example, shifts in grassland species composition or structural changes due to land use (i.e., grazing) that cannot be identified using remotely sensed data are nonetheless important to grassland function and the species that inhabit grasslands. Given that typical methods used to detect boundaries are especially weak in ecotones where there is mixed vegetation or unique plant assemblages, the added variability and uncertainty of land use change and climate change heightens the challenge of implementing traditional boundary detection methods (Kent et al. 2006). Boundary detection is further complicated by the multiple spatial extents and resolutions at which different processes and physical patterns are expressed (Strayer et al. 2003) and by the fact that the relationship between abiotic variables such as climate and biotic variables such as vegetation is often nonlinear across boundaries and scales (Danz et al. 2012).

Plant communities may not respond as rapidly as animal communities to direct anthropogenic change and climate change (Pearson 2006), so identifying boundaries in a manner that incorporates both animal and plant communities (and, in the future, social systems) may better represent current biotic and abiotic conditions. In particular, variation in animal population dynamics provides information on the stability of ecosystem mechanisms, processes, and linkages and may serve as an early warning signal of shifting spatial regimes (Cline et al. 2014). However, there is a great deal of uncertainty as to how biotic communities will respond to climate and land use change, because of nonlinear interactions and feedbacks at different scales of organization.

Species' responses to land use and climate change can occur in three primary forms: adaptation, migration, or regional and localized extinction. These three forms interact with each other and affect communities collectively, leading to emergent community dynamics that can be captured with spatial regimes. This complexity becomes evident in the underlying mechanisms associated with these phenomena. Adaptation can occur via behavioral or phenotypic changes (plasticity) or evolution (Gienapp et al. 2008). Neither form of adaptation is expected to be sufficient in the face of rapid global change, which will ultimately lead to extinctions. Response diversity to disturbances appears to be phylogenetically constrained (Helmus et al. 2009), limiting the number of species that are behaviorally plastic. Niche evolution is thought to be rare because of niche conservatism (Martínez-Meyer et al. 2004) and the time-for-speciation effect (Wiens et al. 2010). This leaves migration, or niche tracking, as probably the dominant strategy for coping with climate change (Martínez-Meyer et al. 2004; Huntley et al. 2008), and there is substantial evidence documenting vagile species shifting their ranges to track their climatic niche (Parmesan and Yohe 2003; Tingley et al. 2009).

The complexity of community dynamics is evident from studies of biotic responses to past episodes of climate change over geologic time. Range shifts of populations "varied widely in the timing, magnitude, and direction" such that communities did not migrate as intact units (Williams and Jackson 2007, 476). Long-lived plant species can persist in areas no longer climatically suitable (Dobrowski 2011), which would leave suitable habitat for animal species in areas no longer within their climatic niche. Time lags in vegetation response to climate change can also be indirect (Martin and Maron 2012). Furthermore, there is evidence suggesting that it is more difficult for plant species to migrate than originally thought (Iverson et al. 2004). Finally, pollen records indicate that even successful range shifts can take decades to centuries (Pearman et al. 2008). The ability to shift ranges is further impeded by habitat fragmentation, which has been shown to reduce range shift (Iverson et al. 2004; Thuiller et al. 2008). As a result, range contraction due to a lack of suitable habitat and reduced survivorship within their original range can also be expected (Davis and Shaw 2001; Parmesan 2006). Identifying spatial regimes in vegetation, animal communities, and key structuring processes such as precipitation and temperature provides simple tools for tracking changes in spatial components of resilience over time.

Nonstationary boundaries

Resilience concepts provide a different way to explain how biota may respond to global change. Ecological systems can respond to shifts in external drivers in several ways. Given adequate adaptive capacity, a system can respond to exogenous change through processes that expand the stability domain or basin of attraction, thereby remaining intact in terms of structure and function within set spatial boundaries. If adaptive capacity (*sensu* Angeler et al. 2019) is low, or a disturbance too large and sudden, a disturbance may cause system collapse and reorganization. After disturbance, a system may reorganize around its original processes and structures (Gunderson and Holling 2002), or it may reorganize around a different set of processes and structures that would lead to a regime shift. Such responses in social-ecological systems are well established (see other chapters in this volume), but other responses are possible, involving the spatial nature of ecological and social-ecological systems.

A spatial regime may also move in both space and time (Roberts et al. 2019) as long as self-organizing processes are maintained. Movement of spatial regimes has limits, usually due to edaphic and biogeographical barriers, such as oceans, mountains, topography, soils, or mesoscale climatic conditions. Where a spatial regime encounters a hard boundary, or where self-organizing processes cannot keep up with the speed of the driving processes, regime collapse may occur, and an alternative stable state may emerge in the same location as the previous. Systems such as lakes cannot move, and it is no coincidence that most theory on alternative stable states has emerged from the aquatic sciences and the study of systems that are conveniently bounded. However, most terrestrial systems and aquatic subsystems within oceans do not have such constraints to change (Grady et al. 2019).

Another response to change in spatial regimes is human-generated: transformation (Chaffin et al. 2016). Transformation occurs when a regime is undesirable and human intervention attempts to reduce the resilience of the current undesirable regime, with the goal of establishing a more desirable replacement regime. As global change accelerates and increases the vulnerability of wild lands, agricultural landscapes, and the provision of ecosystem goods and services on which humanity depends, intentional and directed transformation of regimes may become not just desirable but critical.

Boundaries in the context of panarchy

Spatial regimes are predicated on the assumption, with empirical support, that systems strongly self-organize at multiple spatial–temporal scales and are hierarchically structured in a panarchy (Gunderson and Holling 2002; Allen et al. 2014). That is, a focal system, such as a wetland, is embedded in larger systems (a wetland < ecosystem < biome) that are connected with varying intensities of feedbacks within and across spatial and temporal scales. Analyses at different resolutions will identify regimes at different scales, consistent with panarchy theory (Gunderson and Holling 2002). Spatial regimes may help identify the scales most strongly related to processes of different types (e.g., nutrient dynamics within a lake versus climatic factors in a lake landscape). Despite processes occurring over a limited and discrete range of scales (i.e., photosynthesis occurs at a cellular level), processes occurring at different scales are not entirely disconnected from each other (regional climate can affect local nutrient dynamics), and analyses of regimes at multiple scales may provide insight into appropriate scales for management and policy interventions (Garmestani and Allen 2014). Similarly, panarchies and spatial regimes that are geographically distant are nonetheless probably linked through telecoupling (Carlson et al. 2018). The assessment of scale in ecological systems is important because disturbances, including those resulting from environmental change, affect ecosystems in scale-specific ways (Nash et al. 2014a). An understanding of scale-specific processes provides managers with a realistic assessment of vulnerabilities and the resilience of ecosystems to environmental change. Explicit consideration of intact and altered scale domains in analyses of global change impacts provides opportunities to tailor more specific management actions. For instance, management might be better matched to target the reinforcement of critical ecological functions at scales that are lower in stress to buffer against the potential loss of functions at affected scales, thereby fostering ecosystem resilience (Angeler et al. 2013).

In practice, the selection of scales in ecological research is often arbitrary. A widespread approach used to understand the effects of structure at various scales on populations, communities, and their habitat use is to draw buffers of various sizes around a focal area and collect metrics from these increasingly large buffers (Angeler et al. 2016). Similarly, important scales may be defined by delineating local habitat and regional scales in metacommunity ecology (Leibold et al. 2004) or distinguishing between reach, catchment, and inter-catchment scales by aquatic ecologists (Allan and Castillo 2007). Thus, the selection of scale in such studies is generally subjective (Wheatley and Johnson 2009) and typically reflects the analyst's view about which connections are important to the study at hand and which can be ignored. However, rather than arbitrarily assigning scales or a priori assuming what the most critical scales are, we can objectively identify scales of analysis by finding thresholds or discontinuities in the degree of homogeneity in system structure and composition across space. Such homogeneity is probably driven by positive interactions between biotic and abiotic species and processes. Ignoring the interaction between space and time when searching for patterns indicating early warnings and regime shifts can lead to ecological misinterpretations of underlying structure of state variables. Using frameworks that do not require or use a priori knowledge of regimes but instead focus on transitions that are detectable and measurable, it is possible to identify spatial risks or vulnerabilities to transitions and then concentrate management activities where they are most needed and will be most effective.

Identifying spatial regimes across multiple scales spanning individual land parcels to biomes may be an advancement for the implementation of resilience in practice. Resilience concepts have often been greatly simplified in their application. Resilience-based management frameworks and inclusion into policy are therefore most often based on expert opinion (e.g., current state-and-transition models; Twidwell et al. 2013), abstract concepts such as deep knowledge, or reductionist univariate metrics of richness or ecotonal diversity.

Quantifying spatial thresholds provides a more robust and theoretically aligned approach than has been available previously. The inclusion of spatial boundary detection into resilience management may resolve debates that currently foster indecision and inaction. For example, it is difficult to determine when the best decision is to conserve ecosystem services derived from legacies of past evolutionary feedbacks or to acknowledge when managing for past regimes is unlikely, a phenomenon where spatial regimes are mobilizing at broad scales and neighboring regimes are expanding into new territories (Uden et al. 2019; Angeler et al. 2021).

Methods to Identify Spatial Regimes

Most research on regime shifts has been temporally focused and retrospective: how a system in a particular place changed over time. Some researchers (e.g., Brock and Carpenter 2012) have proposed indicators of impending regime shifts. Some of the statistical indicators of regime shifts are assumed to represent generic system properties that behave in similar and predictable ways across system types (Dakos et al. 2008, 2011), and therefore detailed mechanistic knowledge is not necessary for their use. However, most such metrics are univariate and derived from simulation studies or from systems with fixed boundaries, such as lakes. This precludes the biological reality of interacting, multivariate systems. In a univariate statistical analysis, variables selected may or may not represent system behavior; the critical variables driving system transitions are typically unknown. Brock and Carpenter (2012) cited this lack of knowledge as a fundamental problem in leading indicators research.

Spatial regime detection is possible without imposing arbitrary, taxonomic classification of a single dominant trait at ecosystem levels—such as the delineation of vegetation into grasslands, forests, wetlands, or agricultural lands—which has been a common practice in the mapping and monitoring of life on Earth because of the difficulty of detecting spatial mobility of complex ecological patterns (Roberts et al. 2018). Revisiting our preconceptions about ecological assembly and using a spatial regime approach may reconfigure our views on the organization of nature and how to coexist at the boundaries of sudden regime change.

In the case of spatial regimes that move in space over time, traditional metrics provide feedback only when the spatial transition moves over a monitoring point, and this is often too late to effect change. Other methods are therefore needed that can detect a moving regime shift.

Fisher information

Fisher information (FI) is a statistical approach that involves and assesses observable variables that characterize unknown factors of a system's condition (Fisher 1922). FI merges multiple variables into an index to identify regimes and regime shifts and has been used to assess temporal change in terrestrial (Mayer et al. 2007) and aquatic systems (Eason et al. 2016). FI shows promise for identifying spatial regimes, because it can be used to evaluate a variety of data, with no minimum or maximum number of variables needed (Sundstrom et al. 2017). In order to further strengthen the FI index, Spearman rank order correlation has been used in addition to FI to identify which variables or groups of variables are critical for shaping the FI index (Eason and Cabezas 2012). Using multiple quantitative approaches to assess a complex system reduces some of the weaknesses of univariate (and some multivariate) quantitative approaches that are dependent on significant assumptions about the dynamics of complex systems.

Discontinuity analysis

Discontinuity analysis provides a tool to objectively identify cross-scale structure in complex systems. The approach is appealing because it can use a number of proxy variables (such as body mass) to delineate breaks in system attributes across scales. Scales may be identified as follows. First, discontinuities can be assessed by collecting size or mass data from a census of organisms from an ecosystem and analyzing the distribution of body sizes or mass for gaps in the distribution (Angeler et al. 2016). The recommendation is to use several methods to identify discontinuities in rank-ordered size distributions and assess for convergence in results (Stow et al 2007). The most robust methods are a Bayesian classification and regression tree (Chipman et al. 1998) and a recent update to a kernel density estimation approach, called the Discontinuity Detector (Barichievy et al. 2018), and cluster analysis can also be used to confirm results. These methods can be adjusted to account for differences in the life history traits of organisms (e.g., species with determinate vs. indeterminate growth; Nash et al. 2014b) or abundances (Sundstrom et al. 2018).

Discontinuity theory has promising application for assessing spatial resilience. In landscape ecology, Urban et al. (1987) described how components in a hierarchical system are separated by breaks into defined landscape units. A landscape is decomposed into fundamental units, that is, spatially discrete patches whose internal structure or function is significantly different from that of its surroundings (Burnett and Blaschke 2003). This spatial representation of scale-specific patterns and processes is explicitly recognized in the hierarchical patch dynamics paradigm (Wu and Loucks 1995). The discontinuous structure of pattern and process has been empirically demonstrated in landscapes (Gillson 2004), animal movement patterns in response to resource distribution (Fauchald and Tveraa 2006), and scale-specific distribution of functional feeding groups of invertebrates in a stream network (Göthe et al. 2014) or lake system (Angeler et al. 2015).

Although discontinuity analyses have been effective for identifying the number of dominant scales present in animal communities, there has been little application in other taxonomic groups with modular growth and those that lack body mass or size data, such as plants. Kerkhoff and Enquist (2007) have shown an elegant approach to evaluate scaling relationships (based on power laws) in forest ecosystems by using tree density and diameter in forests. Therefore, metrics such as density and diameter may also be amenable to analysis via the discontinuity approach to identify scaling patterns. Alternatively, spatial modeling allows the application of discontinuity theory to these groups (Angeler et al. 2016). Many of these methods can be used to assess discontinuities in the abiotic environment and therefore have research potential in the hydrological, ecotoxicological, and geochemical sciences to address management issues related to water quality and pollution.

Spatial covariance

The utility of spatial covariance for regime shift imaging and tracking (Uden et al. 2019) is based on two systemic properties. The first is that any system exists in one regime at a time (Holling 1973; Folke et al. 2010), and the second is that transitions from one ecological regime to another exhibit spatial order at one or more scales of organization (Allen et al. 2016; Roberts et al. 2019). Given that alternative ecological regimes cannot occupy the same space at the same scale in time, the simultaneous existence of multiple alternative regimes means that every regime is neighbored in space by alternative regimes. When computed in a moving window algorithm with spatially continuous data (e.g., percentage cover of plant functional groups), strongly negative spatial covariance provides a geographic transition signal of regime boundaries that can be tracked through time (Uden et al. 2019). Strongly negative spatial covariance between two entities highlights geographic transition zones characterized by spatial irregularities (Norberg and Cumming 2008) in the two entities' relational organization (i.e., tendency to separate from one another in space). Spatial covariance provides a measure of the degree of coexistence between two entities. Simply put, spatial covariance acts as an edge detection technique for spatial regimes and, importantly, one closely aligned with resilience theory.

Wombling

Wombling (Womble 1951), was developed to avoid subjective, discrete classification schemes of ecological systems by estimating the probability of a given location being a spatial boundary between ecological entities (Barbujani and Sokal 1990). Wombling has strong potential as a spatial regime detection method because it can incorporate two spatial dimensions, providing the ability to detect spatial boundaries in open, complex systems, and it can use univariate or multivariate data (Kent et al. 2013; Diniz-Filho et al. 2016). Thus, if repeated over time and with suitable data for detecting regimes (e.g., biotic community composition), wombling can probabilistically identify spatial boundaries between regimes and track their movement over time. Wombling has been implemented for a wide range of spatial analysis purposes, including quantifying the spatial patterns and vulnerabilities to disease and invasive species spread (Ma and Carlin 2007; Fitzpatrick et al. 2010), locating landscape barriers of gene flow and spatially distinct genotypes (Diniz-Filho et al. 2016), and identifying spatial boundaries of bird and butterfly communities across ecotonal gradients (Kent et al. 2013).

Management Applications

Environmental management priorities are typically established without quantitative information on how to cope with potential thresholds in social-ecological systems. This lack of information often limits successful intervention actions or knowledge of the extent to which past regimes have shifted. The quantification of spatial regime boundaries serves as an indicator of environmental change that provides a foundation for threshold-based management in an era of nonstationarity. Many of the failed management efforts of the twenty-first century are tied to pathologies surrounding past legacies of equilibrium conditions of an idealized climax regime. Quantitative delineation of spatial regimes provides opportunities to move past static mapping of idealized potential vegetation and animal distribution envelopes. Geographic information system analysis falls into this category and often represents a modern form of cartography representing snapshots of more dynamic behavior in nature. Integration of resilience into spatial classification of composition, structure, process, and function has yet to occur and is needed to more fully appreciate complexity associated with emergence, persistence, collapse, and reorganization that underscore the resilience of ecological regimes.

Managing spatial regimes

The concept of spatial regimes can be a useful application for the management of social-ecological systems. Analyses to date have shown directional changes in spatial regimes consistent with theorized responses to global change (Roberts et al. 2019). Spatial regime shifts in the Northern Hemisphere occur more rapidly with increasing latitude, which is consistent with arctic amplification (Roberts et al. 2019). Therefore, mapping spatial regimes, identifying spatial regime boundaries, and determining their rate of movement allows assessment of the vulnerability of particular sites. Sites further from moving boundaries are less likely to experience regime shifts, given a lack of hard boundaries that would prevent movement.

Protected areas far from regime boundaries, those close to boundaries, and those that have recently been overtaken by a moving spatial regime should be managed differently. Roberts et al. (2019) showed that protected grassland areas in Kansas (Fort Riley) have recently had a southern spatial regime move past them. This regime boundary seems to separate Great Plains grasslands (Fort Riley) from a moving front of woody species, especially eastern red cedar (*Juniperus virginiana*), which has recently moved across Kansas and is now spreading in Nebraska. Propagule pressure (cedar seeds) on Fort Riley from the now surrounding cedar regime means that maintaining the previous grassland regime requires ever more vigilant and intensive management, exacerbated by the loss of spatial connectivity preventing spatially contagious processes such as fire from occurring on small and isolated protected areas. Remnant patches such as this are "ghosts of regimes past" (Angeler et al. 2020).

Spatial regimes display a spatial hierarchical organization, a panarchy. A given regime identified at a focal scale (such as the southern Great Plains grasslands; Roberts et al. 2019) will contain multiple spatial regimes at smaller scales, such as shortgrass prairie, tallgrass prairie, woodland, and wetland patches. The scale of analysis chosen will affect regime identification. Nonstationary regimes in space over time drive a suite of changes, including in ecosystem services, agriculture, and plant and animal distributions. For example, the regime (a red cedar woodland regime replacing the grassland regime) that has moved from the mid-Great Plains to southern Nebraska in the last 50 years has included Lone Star ticks, increasing the vulnerability of human populations to a suite of tickborne diseases.

In some cases, management objectives and actions continue to perpetuate relic spatial regimes that are increasingly vulnerable to larger scales in a panarchy. For example, some marine reserves are managed under largely preservationist management schemes, as a way of buffering against unwanted regime shifts (Hughes et al. 2015). Even so, adaptive frameworks for managing coral reef systems are not currently able to offset accelerating global change, such as the ongoing bleaching of the Great Barrier Reef. In the Anthropocene, the utility of geographically fixed reserves in the face of rapidly accelerating environmental change is debatable. A key question is whether reserves serve as sources of adaptive and transformative capacity (e.g., genetic diversity, breeding grounds) for larger-scale ecosystems or if they are relics of the past that no longer serve their intended purpose because of shifting baselines (e.g., temperature, ocean acidification). Shifting spatial regimes as a result of global change may necessitate the shifting of the location of reserves as well, which we recognize may be an unpalatable management option.

Identification of complex multiscale regimes across broad landscapes

There is similarity in the concept of spatial regimes, when applied broadly, to plate tectonics—a biological tectonics of sorts. Geological phenomena such as subversion, induction, and catastrophic change can occur with biotic systems, when spatial regimes move in response to global change. The degree to which these phenomena occur in the context of spatial regimes is testable. In many ways, managing spatial regimes is more straightforward than managing individual species. Land masses consist of multiple spatial regimes that respond to drivers of change and to each other, which are further hierarchically organized in a similar manner to the way that biogeographers conceptualize biomes, ecosystems, and habitats. Following from resilience and panarchy theory, each regime will have its own characteristic structure and driving processes and its own characteristic scaling structure.

Management Challenges

Application of spatial regimes is an emerging frontier in understanding cross-scale environmental changes. However, use of spatial regime shifts to understand and manage such changes is currently limited by data constraints, the predictive power of spatial regimes, and outstanding theoretical questions concerning spatial regimes. Each is discussed in turn.

Data constraints

Appropriate monitoring is needed for assessing the speed and direction of spatial change and the width of areas of transience (the region analogous to an ecotone, which exists between adjacent spatial regimes). Difficulties in acquiring and analyzing data of sufficient resolution and extent in both space and time have been an obstacle to early warning and regime shift detection. In lieu of such data and analyses, qualitative assessment, expert opinions, and fixed-scale models (e.g., state-and-transition models) have been used. Fortunately, monitoring data meeting these requirements are increasingly available. For example, data from the U.S. Geological Survey Breeding Bird Surveys meet these requirements, although gaps exist in both space and time. The Rangeland Analysis Platform (Jones et al. 2018) is another promising dataset, featuring vegetation functional group percentage cover as a continuous raster (30 × 30 m) for most of the western United States from 1984 onward.

Predictive power

One major criticism of traditional early warning indicators is that they have low predictive power and tend toward false positives (Hastings and Wysham 2010; Burthe et al. 2016), and spatial regimes are not immune to this same criticism (Clements and Ozgul 2018). The logic behind early warning indicator application has long been that, given the possibility of hysteresis after regime shift, false positives are less harmful than false negatives. The same logic applies to spatial regimes that are moving, shifting, and contracting across geographies. This means simple inductive predictions derived from historic and current spatial regime boundary movement trajectories and speeds can motivate continental-scale natural resource planning (Roberts et al. 2019). But because false positives also incur costs to society (assuming the false positives are acted on), exploring the ability of spatial regimes to make accurate out-of-sample predictions is a nontrivial pursuit.

Outstanding questions

Traditionally, resilience theory has conceptualized regime shifts as fundamental changes in the processes and structures of social-ecological systems. To operationalize this notion, resilience-based management has advocated that establishing the characteristic scale-specific structure of a system is a prerequisite to determining whether a regime shift has occurred. Thus, the traditional steps in detecting or predicting a regime shift are to determine system processes and structures (e.g., ecosystem services provided, historic feedbacks or species composition) and determine whether these processes or structures have been or are in danger of being altered. But the spatial regime concept subverts these steps: Implementing spatial regimes requires only identifying spatial boundaries between regimes. No a priori assessment of historic baselines is needed. Although this concept makes spatial regimes more abstract and amorphous than the traditional approach with (arbitrarily) agreed-upon regimes, it reveals critical quantitative and philosophical questions: How can spatial regimes be tracked if they are not fixed, and what is the role of historic baselines in spatial regime detection and tracking? These questions have been obstacles for applying alternative state and resilience theory since their inception, but spatial regimes offer a quantitative, objective framework for answering and moving past them.

Theory posits spatial regimes should manifest hierarchically at discrete, discontinuous scales. Studies have hinted at the hierarchical nature of spatial regimes (e.g., Angeler et al. 2013; Roberts et al. 2019), but to date these propositions have not been rigorously tested. A closely related question—to what degree are spatial regimes nested—is also outstanding. Hierarchy, scaling properties, and nestedness all have profound implications for ecological resilience and vulnerability to regime shifts, so understanding the scaling of spatial regimes is critical for applying them in management. In a practical sense, because there are infinite choices of scales, if spatial regimes do manifest at discrete, discontinuous, and hierarchical scales, the identification of those scales should facilitate spatial regime tracking and the associated derivation of management applications more tractable. Finally, there is uncertainty regarding what hard boundaries might exist to movement, how much substitution is possible between system elements before the system represents an alternative regime, and when and why a regime might collapse and reorganize instead of move, in the absence of hard boundaries.

Conclusions

Under global change, spatial regimes may resist change, adapt via changes in processes or biota, move in space over time, or collapse and reorganize. Little is currently known about how spatial regimes respond to particular forcings, but there is evidence that movement is possible and that tracking spatial regime movement in space allows early warning of regime shifts for any given location over time. Consideration of the spatial dimensions of resilience and regime shifts has great potential for creating planning horizons that provide sufficient time for transformation or adaptation, as opposed to leading indicators of regime shifts that are based solely on temporal data, which provide warning only after it is too late to affect action (Biggs et al. 2009). The underlying relevant concepts and theories of complex adaptive systems, resilience, spatial resilience, and panarchy provide the foundation for a rich set of hypotheses and interpretations, even if their heuristic simplifications will need to be benchmarked against the complexities that characterize nature. Although the ideas we present might be considered Clementian (Clements 1916), they are not; the organization of complex adaptive systems accounts for interchangeability of individual elements (e.g., species) as conditions change; a regime is dependent on process–structure relationships rather than invariance of components. Spatial regimes, like resilience and panarchy from which it emerged, encompass the multiscaled nature of complex systems but explicitly include spatial and temporal components, which have not previously been linked. This chapter, and the methods described herein, helps avoid the problem of arbitrary human scale imposition while accounting for nonstationary and multiscale processes in dynamic landscapes.

Acknowledgments

This material is based on work supported by the National Science Foundation under grant nos. DGE-1735362 and 1920938. Any opinions, findings, and conclusions or recommendations expressed in this material are those of the authors and do not necessarily reflect the views of the National Science

Foundation. The findings and conclusions in this manuscript have not been formally disseminated by the U.S. Environmental Protection Agency and should not be construed to represent any agency determination or policy. Any use of trade names is for descriptive purposes only and does not imply endorsement by the U.S. government.

Literature Cited

Allan, J.D., and M.M. Castillo. 2007. *Stream Ecology: Structure and Function of Running Waters*. Springer, New York.

Allen, C.D., and D.D. Breshears. 1998. Drought-induced shift of a forest woodland ecotone: Rapid landscape response to climate variation. *Proceedings of the National Academy of Sciences* 95: 14839–14842.

Allen, C.R., D.G. Angeler, B. Chaffin, D. Twidwell, and A. Garmestani. 2019. Resilience reconciled. *Nature Sustainability* 2: 898–900.

Allen, C.R., D.G. Angeler, G.S. Cumming, C. Folke, D. Twidwell, and D.R. Uden. 2016. Quantifying spatial resilience. *Journal of Applied Ecology* 53: 625–635.

Allen, C.R., D.G. Angeler, A.S. Garmestani, L.H. Gunderson, and C.S. Holling. 2014. Panarchy: Theory and application. *Ecosystems* 17: 578–589.

Angeler, D.G., and C.R. Allen. 2016. Quantifying resilience. *Journal of Applied Ecology* 53: 617–624.

Angeler, D.G., C.R. Allen, C. Barichievy, T. Eason, A.S. Garmestani, N.A.J. Graham, D. Granholm, L. Gunderson, M. Knutson, K.L. Nash, R.J. Nelson, M. Nyström, T. Spanbauer, C.A. Stow, and S.M. Sundstrom. 2016. Management applications of discontinuity theory. *Journal of Applied Ecology* 53: 688–698.

Angeler, D.G., C.R. Allen, and R.K. Johnson. 2013. Measuring the relative resilience of subarctic lakes to global change: Redundancies of functions within and across temporal scales. *Journal of Applied Ecology* 50: 572–584.

Angeler, D.G., C.R. Allen, D.R. Uden, and R.K. Johnson. 2015. Spatial scaling patterns and functional redundancies in a changing boreal lake landscape. *Ecosystems* 18: 889–902.

Angeler, D.G., B. Chaffin, S. Sundstrom, A. Garmestani, K. Pope, D. Uden, D. Twidwell, and C.R. Allen. 2020. Coerced regimes: Management challenges in the Anthropocene. *Ecology and Society* 25(1): 4. https://doi.org/10.5751/ES-11286-250104

Angeler, D.G., H.B. Fried-Petersen, C.R. Allen, A. Garmestani, and D. Twidwell. 2019. Adaptive capacity in ecosystems. *Resilience in Complex Socioecological Systems* 60: 1.

Angeler, D.G., C. Roberts, D. Twidwell, and C.R. Allen. 2021. Low contribution of rare species to resilience and adaptive capacity in novel spatial regimes arising from biome shifts caused by global change. *bioRxiv*. https://doi.org/10.1101/2020.01.29.924639

Barbujani, G., and R.R. Sokal. 1990. Zones of sharp genetic change in Europe are also linguistic boundaries. *Proceedings of the National Academy of Sciences* 87: 1816–1819.

Barichievy, C., D.G. Angeler, T. Eason, A.S. Garmestani, K.L. Nash, C.A. Stow, S. Sundstrom, and C.R. Allen. 2018. A method to detect discontinuities in census data. *Ecology and Evolution* 2018: 1–10.

Beckage, B., B. Osborne, D.G. Gavin, C. Pucko, T. Siccama, and T. Perkins. 2008. A rapid upward shift of a forest ecotone during 40 years of warming in the Green Mountains of Vermont. *Proceedings of the National Academy of Sciences* 105: 4197–4202.

Biggs, R., S.R. Carpenter, and W.A. Brock. 2009. Turning back from the brink: Detecting an impending regime shift in time to avert it. *Proceeding of the National Academy of Sciences* 106: 826–831.

Brock, W.A., and S.R. Carpenter. 2012. Early warnings of regime shift when the ecosystem structure is unknown. *PLoS One* 7: e45586.

Burnett, C., and T. Blaschke. 2003. A multi-scale segmentation/object relationship modelling methodology for landscape analysis. *Ecological Modelling* 168: 233–249.

Burthe, S.J., P.A. Henrys, E.B. Mackay, B.M. Spears, R. Campbell, L. Carvalho, B. Dudley, I.D.M. Gunn, D.G. Johns, S.C. Maberly, L. May, M.A. Newell, S. Wanless, I.J. Winfield, S.J. Thackeray, and F. Daunt. 2016. Do early warning indicators consistently predict nonlinear change in long-term ecological data? *Journal of Applied Ecology* 53: 666–676.

Carlson, A.K., J.G. Zaehringer, R.D. Garrett, R.F.B. Silva, P. Furumo, A.N. Raya Rey, A. Torres, M.G. Chung, Y. Li, and J. Liu. 2018. Toward rigorous telecoupling causal attribution: a systematic review and typology. *Sustainability* 10: 4426.

Chaffin, B.C., A.S. Garmestani, L.H. Gunderson, M.H. Benson, D.A. Angeler, C Arnold, B. Cosens, R.K. Craig, J.B. Ruhl, and C.R. Allen. 2016. Transformative environmental governance. *Annual Review of Environment and Resources* 41: 399–423.

Chipman, H.A., E. George, and R.E. McCulloch. 1998. Bayesian CART model search. *Journal of the American Statistical Association* 93: 935–948.

Clements, C.F., and A. Ozgul. 2018. Indicators of transitions in biological systems. *Ecology Letters* 21: 905–919.

Clements, F.E. 1916. *Plant Succession: An Analysis of the Development of Vegetation*. Carnegie Institution of Washington, Washington, DC.

Cline, T.J., D. Seekell, S. Carpenter, M. Pace, J. Hodgson, J. Kitchell, and B.C. Weidel. 2014. Early warnings of regime shifts: Evaluation of spatial indicators from a whole-ecosystem experiment. *Ecosphere* 5(8): 102.

Cumming, G.S. 2011. *Spatial Resilience in Social-Ecological Systems*. Springer, Dordrecht, The Netherlands.

Cumming, G.S., C.R. Allen, N.C. Ban, D. Biggs, H.C. Biggs, D.H.M. Cumming, A. De Vos, G. Epstein, M. Etienne, K. Maciejewski, R. Mathevet, C. Moore, M. Nenadovic, and M. Schoon. 2015. Understanding protected area resilience: A multi-scale, social-ecological approach. *Ecological Applications* 25: 299–319.

Dakos, V., S. Kefi, M. Rietkerk, E.H. van Nes, and M. Scheffer. 2011. Slowing down in spatially patterned ecosystems at the brink of collapse. *American Naturalist* 177: E153–E166.

Dakos, V., M. Scheffer, E.H. van Nes, V. Brovkin, V. Petoukhov, and H. Held. 2008. Slowing down as an early warning signal for abrupt climate change. *Proceedings of the National Academy of Sciences* 105: 14308–14312.

Danz, N.P., L.E. Frelich, P.B. Reich, and G.J. Niemi. 2012. Do vegetation boundaries display smooth or abrupt spatial transitions along environmental gradients? Evidence from the prairie–forest biome boundary of historic Minnesota, USA. *Journal of Vegetation Science* 24: 1129–1140.

Davis, M.B., and R.G. Shaw. 2001. Range shifts and adaptive responses to quaternary climate change. *Science* 292: 673–680.

Diniz-Filho, J.A.F., T.N. Soares, and M.P. de Campos Telles. 2016. Geographically weighted regression as a generalized wombling to detect barriers to gene flow. *Genetica* 144: 425–433.

Dobrowski, S. Z. 2011. A climatic basis for microrefugia: The influence of terrain on climate. *Global Change Biology* 17: 1022–1035.

Eason, T.H., and Cabezas. 2012. Evaluating the sustainability of a regional system using Fisher information in the San Luis Basin, Colorado. *Journal of Environmental Management* 94: 41–49.

Eason, T., A.S. Garmestani, C.A. Stow, M. Alvarez-Cobelas, C. Rojo, and H. Cabezas. 2016. Managing for resilience: an information theory-based approach to assessing ecosystems. *Journal of Applied Ecology* 53: 656–665.

Edwards, M., and A.J. Richardson. 2004. Impact of climate change on marine pelagic phenology and trophic mismatch. *Nature* 430: 881–884.

Fauchald, P., and T. Tveraa. 2006. Hierarchical patch dynamics and animal movement pattern. *Oecologia* 149: 383–395.

Fisher, R.A. 1922. On the mathematical foundations of theoretical statistics. *Philosophical Transactions of the Royal Society A* 222: 309–368.

Fitzpatrick, M.C., E.L. Preisser, A. Porter, J. Elkinton, L.A. Waller, B.P. Carlin, and A.M. Ellison. 2010. Ecological boundary detection using Bayesian areal wombling. *Ecology* 91: 3448–3455.

Folke, C., S.R. Carpenter, B. Walker, M. Scheffer, T. Chapin, and J. Rockström. 2010. Resilience thinking: Integrating resilience, adaptability and transformability. *Ecology and Society* 15(4): 20. http://www.ecologyandsociety.org/vol15/iss4/art20/

Garmestani, A.S., and C.R. Allen. 2014. *Social-Ecological Resilience and Law.* Columbia University Press, New York.

Gienapp, P., C. Teplitsky, J. Alho, J. Mills, and J. Merila. 2008 Climate change and evolution: Disentangling environmental and genetic responses. *Molecular Ecology* 17: 167–178.

Gillson, L. 2004. Evidence of hierarchical patch dynamics in an east African savanna? *Landscape Ecology* 19: 883–889.

Göthe, E., L. Sandin, C.R. Allen, and D.G. Angeler. 2014. Quantifying spatial scaling patterns and their local and regional correlates in headwater streams: Implications for resilience. *Ecology and Society* 19(3): 15. http://dx.doi.org/10.5751/ES-06750-190315

Grady, J.M., B.S. Maitner, A.S. Winter, K. Kaschner, D.P. Tittensor, S. Record, F.A. Smith, A.M. Wilson, A.I. Dell, P.L. Zarnetske, H.J. Wearing, B. Alfaro, and J.H. Brown. 2019. Metabolic asymmetry and the global diversity of marine predators. *Science* 363: 4220.

Grebmeier, J., J. Overland, S. Moore, E. Farley, E. Carmack, L.W. Cooper, K.E. Frey, J.H. Helle, F.A. McLaughlin, and S.L. McNutt. 2006. Major ecosystem shift in the Northern Bering Sea. *Science* 311: 1461–1464.

Gunderson, L.H., and C.S. Holling (Eds.). 2002. *Panarchy: Understanding Transformations in Human and Natural Systems.* Island Press, Washington, DC.

Hastings, A., and D.B. Wysham. 2010. Regime shifts in ecological systems can occur with no warning. *Ecology Letters* 13: 464–472.

Helmus, M.R., M.J. Paterson, N.D. Yan, and B.W. Keller. 2009. Communities contain closely related species during ecosystem disturbance. *Ecology Letters* 13: 162–174.

Holling, C.S. 1973. Resilience and stability of ecological systems. *Annual Review of Ecology and Systematics* 4: 1–23.

Hughes, T.P., J. Day, and J. Brodie. 2015. Securing the future of the Great Barrier Reef. *Nature Climate Change* 5: 508–511.

Huntley, B., Y.C. Collingham, S.G. Willis, and R.E. Green. 2008. Potential impacts of climatic change on European breeding birds. *PLoS One* 3(1): e1439. https://doi.org/10.1371/journal.pone.0001439

Iverson, L.R., M.W. Schwartz, and A.M. Prasad. 2004. How fast and far might tree species migrate in the eastern United States due to climate change? *Global Ecology and Biogeography* 13: 209–219.

Jones, M.O., B.W. Allred, D.E. Naugle, J.D. Maestas, P. Donnelly, L.J. Metz, J. Karl, R. Smith, B. Bestelmeyer, C. Boyd, J.D. Kerby, and J.D. McIver. 2018. Innovation in rangeland monitoring: annual, 30 m, plant functional type percent cover maps for US rangelands, 1984–2017. *Ecosphere* 9: e02430.

Kéfi, S., V. Guttal, W.A. Brock, S.R. Carpenter, A.M. Ellison, V.N. Livina, D. A. Seekell, M. Scheffer, E.H. van Nes, and V. Dakos. 2014. Early warning signals of ecological transitions: Methods for spatial patterns. *PLoS One* 9: e92097.

Keitt, T.H., D.L. Urban, and B.T. Milne. 1997. Detecting critical scales in fragmented landscapes. *Conservation Ecology* 1: 4.

Kent, R., O. Levanoni, E. Banker, G. Pe'er, and S. Kark. 2013. Comparing the response of birds and butterflies to vegetation-based mountain ecotones using boundary detection approaches. *PloS One* 8(3): e58229.

Kent, M., R.A. Moyeed, C.L. Reid, R. Pakeman, and R. Weaver. 2006. Geostatistics, spatial rate of change analysis and boundary detection in plant ecology and biogeography. *Progress in Physical Geography* 30: 201–231.

Kerkhoff, A.J., and B.J. Enquist. 2007. The implications of scaling approaches for understanding resilience and reorganization in ecosystems. *BioScience* 57: 489–499.

Leibold, M.A., M. Holyoak, N. Mouquet, P. Amarasekare, J.M. Chase, M.F. Hoopes, R.D. Holt, J.B. Shurin, R. Law, D. Tilman, M. Loreau, and A. Gonzalez. 2004. The metacommunity concept: A framework for multi-scale community ecology. *Ecology Letters* 7: 601–613.

Ma, H., and B.P. Carlin. 2007. Bayesian multivariate areal wombling for multiple disease boundary analysis. *Bayesian Analysis* 2: 281–302.

Martin, T.E., and J.L. Maron. 2012. Climate impacts on bird and plant communities from altered animal-plant interactions. *Nature Climate Change* 2: 195–200.

Martinez-Meyer, E., A.T. Peterson, and W. Hargrove. 2004. Ecological niches as stable distributional constraints on mammal species, with implications for Pleistocene extinctions and climate change projections for biodiversity. *Global Ecology and Biogeography* 13: 305–315.

Mayer, A.L., C.W. Pawlowski, B.D. Fath, and H. Cabezas. 2007. Applications of Fisher information to the management of sustainable environmental systems. In *Exploratory Data Analysis Using Fisher Information*, ed. B.R. Frieden and R.A. Gatenby, 217–243. Springer-Verlag, London.

Moore, C., J. Grewar, and G.S. Cumming. 2015. Quantifying network resilience: Comparison before and after perturbation shows strengths and limitations of network metrics. *Journal of Applied Ecology* 53: 636–645.

Nash, K.L., C.R. Allen, D. Angeler, C. Barichievy, A.S. Garmestani, N.A.J. Graham, D. Granholm, M. Knutson, J. Nelson, M. Nyström, S. Riley, C.A. Stow, and S.M. Sundstrom. 2014a. Discontinuities, cross-scale patterns and the organization of ecosystems. *Ecology* 95: 654–667.

Nash, K.L., C.R. Allen, C. Barichievy, M. Nyström, S. Sundstrom, and N.A.J. Graham. 2014b. Habitat structure and body size distributions: Cross-ecosystem comparison for taxa with determinate and indeterminate growth. *Oikos* 123: 971–983.

Norberg, J., and G.S. Cumming. 2008. *Complexity Theory for a Sustainable Future*. Columbia University Press, New York.

Parmesan, C. 2006. Ecological and evolutionary responses to recent climate change. *Annual Review of Ecology, Evolution, and Systematics* 37: 637–669.

Parmesan, C., and G. Yohe. 2003. A globally coherent fingerprint of climate change impacts across natural systems. *Nature* 421: 37–42.

Pearman, P.B., C.F. Randin, O. Broennimann, P. Vittoz, W.O. van der Knaap, R. Engler, G.L. Lay, N.E. Zimmermann, and A. Guisan. 2008. Prediction of plant species distributions across six millennia. *Ecology Letters* 11: 357–369.

Pearson, R.G. 2006. Climate change and the migration capacity of species. *Trends in Ecology and Evolution* 21: 111–113.

Roberts, C.P., C.R. Allen, D.G. Angeler, and D. Twidwell. 2019. Shifting avian spatial regimes in a changing climate. *Nature Climate Change* 9: 562–568.

Roberts, C.P., J.L. Burnett, V.M. Donovan, C. Wonkka, C.H. Bielski, C.R. Allen, A. Garmestani, D.G. Angeler, T. Eason, S. Sundstrom, and H.E. Birge. 2018. Early warnings for state transitions. *Rangeland Ecology & Management* 71: 659–670.

Schmidtlein, S., and J. Sassin. 2004. Mapping of continuous floristic gradients in grasslands using hyperspectral imagery. *Remote Sensing of Environment* 92: 126–138.

Stow, C.A., C.R. Allen, and A.S. Garmestani. 2007. Evaluating discontinuities in complex systems: toward quantitative measures of resilience. *Ecology and Society* 12(1): 26. http://www.ecologyandsociety.org/vol12/iss1/art26/

Strayer, D.L., M.E. Power, W.F. Fagan, S.T.A. Pickett, and J. Belnap. 2003. A classification of ecological boundaries. *BioScience* 53: 723–729.

Sundstrom, S.M., D.G. Angeler, C. Barichievy, T. Eason, A. Garmestani, L. Gunderson, M. Knutson, K.L. Nash, T. Spanbauer, C. Stow, and C.R. Allen. 2018. The distribution and role of functional abundance in cross-scale resilience. *Ecology* 99: 2421–2432.

Sundstrom, S., T. Eason, R.J. Nelson, D.G. Angeler, C.R. Allen, C. Barichievy, A.S. Garmestani, N.A.J. Graham, D. Granholm, L. Gunderson, M. Knutson, K.L. Nash, M. Nystrom T. Spanbauer, and C.A. Stow. 2017. Detecting spatial regimes in ecosystems. *Ecology Letters* 20: 19–32.

Thuiller, W., C. Albert, M.B. Araújo, P.M. Berry, M. Cabeza, A. Guisan, T. Hickler, G.F. Midgley, J. Paterson, F.M. Schurr, M.T. Sykes, and N.E. Zimmermann. 2008. Predicting global change impacts on plant species' distributions: Future challenges. *Perspectives in Plant Ecology, Evolution and Systematics* 9: 137–152.

Tingley, M.W., W.B. Monahan, S.R. Beissinger, and C. Moritz. 2009. Birds track their Grinnellian niche through a century of climate change. *Proceedings of the National Academy of Sciences* 106(suppl): 19637–19643.

Twidwell, D., S.D. Fuhlendorf, C.A. Taylor Jr., and W.E. Rogers. 2013. Refining thresholds in coupled fire–vegetation models to improve management of encroaching woody plants in grasslands. *Journal of Applied Ecology* 50: 603–613.

Uden, D.R., M.L. Hellman, D.G. Angeler, and C.R. Allen. 2014. The role of reserves and anthropogenic habitat for functional connectivity and resilience of ephemeral wetlands. *Ecological Applications* 24: 1569–1582.

Uden, D.R., D. Twidwell, C.R. Allen, M.O. Jones, D.E. Naugle, J.D. Maestas, and B.W. Allred. 2019. Spatial imaging and screening for regime shifts. *Frontiers in Ecology and Evolution* 7: 407.

Urban, D.L., R.V. O'Neill, and H.H. Shugart Jr. 1987. Landscape ecology. *BioScience* 37: 119–127.

Wheatley, M., and C.J. Johnson. 2009. Factors limiting our understanding of ecological scale. *Ecological Complexity* 6: 150–159.

Wiens, J.J., D.D. Ackerly, A.P. Allen, B.L. Anacker, L.B. Buckley, H.V. Cornell, E.I. Damschen, T.J. Davies, J.A. Grytnes, S.P. Harrison, B.A. Hawkins, R.D. Holt, C.M. McCain, and P.R. Stephens. 2010. Niche conservatism as an emerging principle in ecology and conservation biology. *Ecology Letters* 13: 1310–1324.

Williams, J.W., and S.T. Jackson. 2007. Novel climates, no-analog communities, and ecological surprises. *Frontiers in Ecology and the Environment* 5: 475–482.

Womble, W.H. 1951. Differential systematics. *Science* 114: 315–322.

Wu, J., and O.L. Loucks. 1995. From balance of nature to hierarchical patch dynamics: A paradigm shift in ecology. *Quarterly Review of Biology* 70: 439–466.

Yarrow, M.M., and S.N. Salthe. 2008. Ecological boundaries in the context of hierarchy theory. *Biosystems* 92: 233–244.

Chapter 6

An Engineering Perspective on Managing for Resilience and Panarchy

Ian Pumo, Margaret Kurth, Stephanie Galaitsi, and Igor Linkov

The five costliest natural disasters in the United States occurred in the first two decades of the 21st century: hurricanes Katrina (2005), Harvey (2017), Maria (2017), and Irma (2017) and superstorm Sandy (2012). Other notable disasters include a drought and heatwave in 2012 and wildfires in the western United States in 2004, 2013, 2017, 2018, 2020. In 2019, a novel coronavirus (COVID-19) spread in a few months to become a global pandemic. The collective responses to these disasters highlight the need for cross-scale, cross-sector planning and management.

Decision makers and policymakers acknowledge that complex systems of people, ecosystems, and technology cannot be universally robust or resilient to disturbances, such as natural disasters. Disruptions that test system resilience will occur and hence must be addressed by the organizations that operate and manage these complex systems. However, different conceptualizations of resilience (i.e., what constitutes resilience for different systems and what configurations of system components contribute to or erode resilience) can lead to difficulties for managers. Identifying the type of resilience in question is necessary because different resilience conceptualizations and definitions have implications for how social-ecological systems (SESs) are managed and governed (Allen et al. 2019). Perhaps unsurprisingly, then, a universal framework for moving the concept into practice remains elusive (Linkov and Trump 2019; Pimm et al. 2019).

Since the word *resilience* began appearing in different scholarly literatures in the early 1970s, different definitions have been derived and applied by different disciplines. More recently, Allen et al. (2019) summarized three different conceptualizations of resilience, describing resilience as a process, a rate, or an emergent property of systems. All three conceptualizations are contextualized by or applied to a system subjected to an external stress, shock, or disturbance. Resilience as a process has been applied in areas of human psychology, human community studies in the social sciences, and disaster response management organizations (Allen et al. 2019). Resilience as a rate emerged from engineering and the physical sciences to describe how quickly a system returned to equilibrium (Holling 1996). Resilience as an emergent property arose from the interplay between the life sciences and research on complex systems and characterizes the amount of change an SES can absorb before shifting to a new configuration with different processes and structures (Holling 1973).

Engineering and disaster management practitioners describe resilience in the context of external shocks and disturbances to a geographically defined system (Galaitsi et al. 2021). The National Research Council (2012) defined resilience as a system's ability to prepare for, absorb, recover from, and adapt to shocks and stresses over time (PARA). The PARA framework was developed in part after Hurricane Katrina, as practitioners hoped to mitigate impacts of future disasters across a range of human, social, economic, and environmental sectors. Robustness and resistance dictate how much shock is absorbed by a system (Grafton et al. 2019), and rapidity and resourcefulness dictate the recovery trajectory (Bruneau et al. 2003). Other authors refer to repairability, loose coupling, functional redundancy, and modularity, among other considerations as key ingredients of the process of resilience (Jackson 2016). Because many resilience frameworks and definitions exist (Allen et al. 2019), attempts to apply *resilience* without context leads to inconsistent and sometimes conflicting operationalization schemes. Ambiguous terms can make quantifying resilience difficult, especially in policymaking applications (Allen et al. 2019; Linkov and Trump 2019).

All *resilience* definitions share the common threads of defining a system (such as an individual human, the city of New Orleans, or the lower Mississippi drainage basin); the system's structure (i.e., how elements are related); and the system's ability to continue to produce a function by persisting, recovering, or adapting despite disruptions (Connelly et al. 2017). Social-ecological resilience (resilience as an emergent property of SESs) diverges from other conceptions of resilience by explicitly acknowledging and accounting for the possibility of multiple configurations (e.g., oligotrophic vs. eutrophic states of a lake) of SESs (Holling 1973; Allen et al. 2019). Managing resilience in an SES is dependent on designation of geographic system boundaries, relevant timescales, human capacity for change, and human preferences for an array of system outputs (Carpenter et al. 2001; Cutter 2016; Meerow et al. 2016).

As a general proposition, we propose that resilience management is sometimes too narrowly defined in terms of scope, timescales, and assumedly inflexible beneficiaries. Much effort has been spent attempting to resolve inconsistent or apparently incompatible definitions of *resilience*. We propose that convergence on a single definition is unnecessary and may be counterproductive: Resilience management for diverse systems will require similarly diverse approaches and strategies that could be founded on different conceptualizations of resilience. This chapter explores facets of resilience through the lens of panarchy to frame alternative options for practitioners who play key roles in the design, construction, operation, and maintenance of social, ecological, and technological resource systems.

Resilience and Panarchy

Panarchy theory emerged from resilience (Holling 1973) and is a framework that acknowledges that a dynamic system can be defined at distinct scale domains but that aspects of the system's behavior and nonlinear dynamics (such as resilience, however defined) can be attributed to cross-scale linkages (Gunderson and Holling 2002). Panarchy theory gives resilience managers a framework for identifying and assessing elements and connections that influence resilience at multiple spatial and temporal scales. Once identified, these elements and connections can be used to assess resilience, develop predictive models, and provide decision support for managing resilience (Garmestani and Benson 2013). Managers can benefit from contextualizing their resilience efforts in the dynamics emphasized by panarchy—individual adaptive cycles, cross-scale interactions, and the potential for sudden nonlinear system behavior—because these dynamics are not always obvious and can be difficult or impossible to predict. These interactions can produce surprising changes in system components, organization, and function. This has additional implications for resilience managers, because they need both to attempt to account for scale issues and to acknowledge the extents and limits of plans and predictive models. We describe three general resilience types, related to both Allen et al. (2019) and Davidson et al. (2016), and use the panarchy framework to help managers influence system resilience to suit their objectives.

Aspects of resilience

Many studies, approaches, and applications have been made of ecological, social-ecological, urban, disaster, and community resilience (Holling 1973; Gunderson and Holling 2002; Meerow et al. 2016; Connelly et al. 2017;

Paton and Johnston 2017; Fox-Lent and Linkov 2018; Zemba et al. 2019; Hynes et al. 2020). These studies share core elements that can be grouped into three general types or aspects of postdisturbance resilience: elasticity, or the ability to bounce back after a disturbance; the ability to adapt; and the ability to transform (figure 6.1). Engineering resilience is both the capacity of a system to recover to its previous state without reorganizing and a measure of the rate of recovery; social-ecological resilience includes engineering resilience but is more focused on system dynamics when the ability to recover is exceeded and the system is nonstationary (Allen et al. 2019). Social-ecological resilience is characterized by the capacity to absorb disturbance without changing system structures and processes (through adaptation, in part), or, when the system is undesirable, human agency is used to change the system to a desired state, transformation (Chaffin et al. 2016). Again, we emphasize that the choices for defining resilience and deciding which aspects of resilience to focus on define frameworks for thinking about resilience management, thus shaping how resilience managers operationalize resilience and then seek to influence it.

We advocate using the panarchy framework for resilience management so that dynamism, system states, and impacts across scales will be explicitly recognized in decisions about management objectives. Acknowledging the cyclic, multiscale nature of systems, as described by panarchy, may prompt stakeholders to consider whether their goal is to keep their system in its current configuration (i.e., regime or state) and attempt to foster adaptation or to plan for or initiate transformation to a new regime as necessary (Chaffin et al. 2016). The chosen goal defines system success and failure, thus driving the analyst's focus and recommendations.

Consciously or not, managers operate with a resilience perspective that establishes goals and acceptable measures; panarchy clarifies the approach because it helps explicitly account for those measures and goals across multiple scales (Angeler et al. 2016, 2018). Selecting the aspects of resilience focuses managers' analysis and management efforts on a specific part of the adaptive cycle.

Gunderson and Holling's (2002) notions of different types of system change each generate different types of social learning: incremental, sporadic, and transformative. Traditionally, human-managed physical systems have prioritized stability over the natural variation that SESs often exhibit. This prioritization limits the acceptable types of system change and resulting types of social learning. For example, oxbow lakes are common reminders of sporadic changes in river geography, but such changes cannot be accommodated in

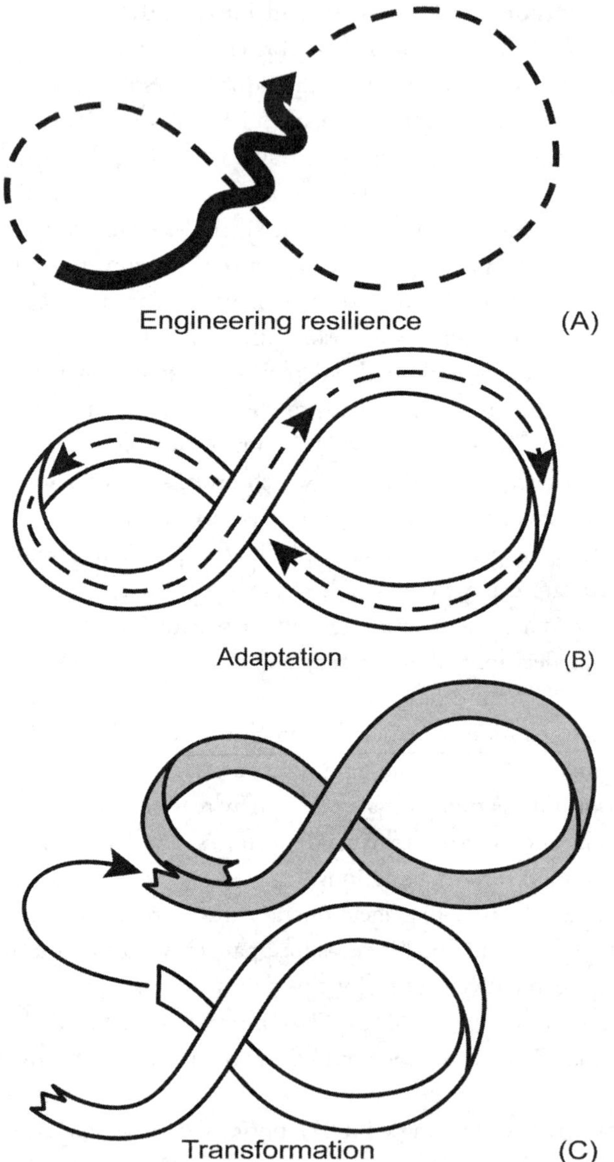

Figure 6.1. Engineering resilience, adaptation, and transformation in context of adaptive cycles. (A) Engineering resilience returns a system to design operating conditions after perturbations. (B) System adaptation occurs when disturbances lead to actions that increase flexibility, while retaining predisturbance processes and structures. (C) Transformation involves an alteration of structures and processes to a more desired system.

regions of high infrastructural density because investments in infrastructure must be protected, creating a tension in SESs.

Planners often seek the stability of engineering resilience, yet as our systems change more rapidly with global change and upscaling of anthropogenic effects, especially climate change, stability becomes a false promise. All systems interact within their broader context, and resisting change clearly has limits: A lighthouse cannot continue operating when its foundation erodes into the sea. Focusing on engineering resilience may provide stability for a time but at the cost of not exploring other approaches that could better accommodate broader scale changes. This is especially important when cross-scale interactions and tipping points play a role in system function. Planners may better meet their goals if they are able to account for resilience through adaptive cycles across scales of time and space and to consider potential reorganization into a new configuration (i.e., transformation).

Thinking in terms of cycles and multiple scales, as panarchy dictates, nudges resilience managers to consider cross-scale interactions and subsequent tipping points or system surprises. Although it may be hard or impossible to predict when tipping points will occur (Pimm et al. 2019; but see Carpenter et al. 2011; Spanbauer et al. 2016), considering their possibility and preparing for them is essential to managing resilience. A manager may see tipping points or surprises as an impending failure, something to manage through, or an opportunity for transformation, and that view will affect how they plan for these surprises. For example, if an unanticipated event would lead to catastrophic failure, then the system must be designed to be fail safe. On the other hand, if managers want to maximize chances for transformation, then they might seek opportunities to increase random or unplanned impacts across scales. At any rate, simply ignoring the possibility of collapse or allowing it without any concern for the impacts is negligent and poor policy (DeWitte et al. 2016), so potential collapse should be considered in terms of the adaptive cycle. In the following section we discuss resilience through an engineering lens.

Engineering resilience

In this section we explore several examples of resilience management, pointing out how perspective influences the scope and direction of analysis and how panarchy can be used to enrich our understanding of resilience. From the engineering perspective, resilience is displayed when a system experiences disruption but recovers in a way that preserves its organization and critical functions.

The resilience manager aims to maintain the system's status quo (equilibrium state) with the minimum loss of function after a disruption and the minimum recovery time (Davidson et al. 2016).

Transportation networks are a common engineered system: They are designed and operated to maintain vehicle flow, subject to disruptions in the network. In urban settings, this flow is critical to sustaining economic activity and output. Kurth et al. (2020b) considered how shutting down 1–5 percent of road networks in several cities affected each city's gross domestic product (GDP). Transportation resilience, or lack thereof, is expressed by how much the network disruption caused vehicle delays and resultant impacts on economic activity. Networks that maintain travel times despite increasing disruption are deemed more resilient than those that do not. Implicit in this analysis is the assumption that the networks are largely fixed and cannot be changed.

However, the engineering resilience approach can still consider and reveal the influences of cross-scale interactions. Kurth et al. (2020b) found significantly different relationships between disruptions and impacts. While some cities experienced linear increases in impact on GDP costs, others showed rapid exponential failure. For example, closing 1–5 percent of the road network in Los Angeles led to a decline in GDP of 1–9 percent, but similar closure rates in San Francisco led to much greater declines of 9–51 percent (Kurth et al. 2020b). These dramatic and nonlinear impacts are a reminder that cross-scale interactions are omnipresent and that the engineering resilience approach could be fundamentally improved with more explicit consideration of these interactions. Further work building on this study could explore what interactions caused the disparate responses or could identify how other socioeconomic systems (entertainment events, planned road closures, media) influence disruptions and readiness.

It is worth noting that cross-scale interactions can lead to positive surprises in fixed systems. For example, neither Los Angeles's Carmageddon in 2011 nor Seattle's viaduct removal in 2018–2019 resulted in traffic jams, presumably because significant public awareness of these major road closures led citizens to avoid travel entirely. What is particularly interesting about these cases is that drivers did not pick another route, which would have led to traffic jams elsewhere. Instead, drivers made major changes to their daily practices, and we briefly saw what a transformed transportation system might look like in Los Angeles and Seattle.

Although it can be useful in the right context, the engineering resilience approach is limited because it is relegated to a single part of the adaptive cycle and focuses only on return time (figure 6.1A). In this circumstance, prioritizing engineering resilience can make systems rigid or brittle because they are designed to minimize or reflexively accommodate damage rather than expand the system's function (Linkov and Trump 2019). Increasing the resilience of existing built infrastructure without reconfiguring how critical functions occur will take the form of either one-for-one exchange of components (e.g., upgrade HVAC), addition of components to increase resistance to disturbance (e.g., floodproofing), or external measures to improve the function of existing components (e.g., sensors to improve damage detection). Over time it becomes likelier that these reforms will be insufficient to manage the inevitably changing nature of disruptions that a system might face.

Adaptation

Systems adapt when they are flexible enough to navigate a disturbance or to go through an adaptive cycle without changing to a different set of processes and structures. One way a manager attempts to provide this flexibility is by influencing reorganization. Adaptive systems are systems that have greater apparent malleability or leeway but still tend to remain fundamentally unchanged after reorganization (figure 6.1B). One example of adaptation within an engineering focus is government support of mutual aid as an important component of disaster response. Because first responders are limited in resources, survivors provide much of the initial response immediately after disasters (Drury et al. 2019). How survivors do and do not support one another is influenced largely by whether emergency services are trained to control or to empower them (Drury et al. 2019). In this case, policymakers must try to maintain the two critical functions of protecting life and property and preserving government stability during and after disasters. An adaptive perspective treats survivors as essential participants who must be empowered and supported in disaster response and recovery. In panarchy terms, leaders allow a temporary collapse of control but influence the reorganization phase such that the system will return to its previous state of conservation. The government provides some flexibility in how the critical function is provided by giving survivors the tools and latitude to support the response themselves. By planning to relinquish some control and proactively establishing the terms of reorganization during and after the disaster, policymakers act to ensure that the parallel critical function of government stability returns easily once the emergency passes and the system is reestablished.

Transformation

In the context of resilience (*sensu* Holling 1973), transformation involves a transition away from the status quo of system functionality by changing system processes and structures with human agency (Chaffin et al. 2016). This generally means purposely exceeding the resilience of a system and then guiding the reorganization process to foster a more desirable system state (figure 6.1C). One example of transformation is the transition from gray to green infrastructure in water and sewer systems in the United States. In this case, the critical function was to improve water quality in U.S. cities by transforming the structures (water and sewer infrastructure) and processes (hydrologic cycle) in urban water systems. Transformation has occurred at small spatial scales in many cities in the type of water infrastructure used (i.e., gray to green), with potential to scale up via panarchy (Green et al. 2015). In addition, the transition from gray to green infrastructure can generate many other ecosystem services, including green space, improved aesthetics, urban agriculture, wildlife corridors, and habitat for beneficial arthropods (e.g., pollinators) (Herrmann et al. 2016a).

Although systems can flip with (transformation) or without (regime shift) human agency, the possibility that some recognizable functions can survive these changes in configuration has critical positive and negative implications for resilience managers. When systems are repeatedly marked by undesirable outcomes or structures, system designers and resilience managers might attempt to position the system for transformation in order to replace those structures (see Chaffin et al. 2016). Counterintuitively, resilience managers hoping to flip a system must search for ways to make the system both less robust and resistant in terms of engineering resilience and less flexible in terms of social-ecological resilience.

The ongoing COVID-19 global pandemic may present in retrospect another example of transformation. As roads emptied across the world, some governments altered infrastructure use. In Paris, perhaps the most dramatic example, roads that previously carried heavy car traffic are now pedestrianized for non–car users, with the government offering €400 subsidies for families to buy cargo bikes to transport children (Hidalgo 2020), all part of an effort to change the use of road infrastructure. It remains to be seen what existing systems will survive the disruptions of COVID-19 and what new systems will emerge.

An adaptive cycle heuristic can be used to depict patterns of resilience, adaptation, and transformation (figure 6.1). An adaptive cycle is only one part of a panarchy. The broader adaptive capacity (Angeler et al. 2019; Garmestani et al. 2019) of a system includes new formulations, which should incorporate cross-scale interactions at various scales. The full spectrum of panarchy is fundamentally at odds with the way current society frames development. Planners channel investments into long-term infrastructure that re-creates itself over time within engineered systems such as cities, electricity grids, and coastal communities. Such systems can recover and even adapt but might not tolerate transformation, especially at large scales that require abandoning infrastructure before the investment has paid off. Yet environmental change data increasingly indicate that the assumption of contextual stationarity is no longer reliable (Craig 2010). Panarchy provides an avenue for planners to explore their systems and system functionality more broadly and across scales, to determine management options that work best for the changes anticipated in the future. Resilience can mean many things, and planners should have a holistic understanding of the benefits and limitations of each definition for their system objectives.

The different conceptions and focal aspects of resilience have critical implications for how systems are managed. This has significant implications, because any designed structure or system, from infrastructure to governments to global economies, can be considered from alternative perspectives. For example, a pipe in isolation is best assessed from an engineering resilience perspective. However, once that pipe becomes part of an SES, the pipe becomes subject to the social-ecological resilience of the SES (Herrmann et al. 2016b; Allen et al. 2019).

Resilience in the U.S. Army Corps of Engineers

The U.S. Army Corps of Engineers (USACE) provides engineering solutions for many of the toughest challenges faced by the United States, including the design and construction of systems that reduce disaster risk in a cost-efficient manner. USACE missions often entail management of SESs. For example, the agency manages inland flood risk with levees, dams, and reservoir operations. Similarly, a wide range of coastal engineering infrastructure aims to protect people and assets from waves and storm surge along the coastlines. Navigability of waterways depends on dredging operations to ensure sufficiently deep and wide channels, as well as the operation of locks and dams to maintain reliable water levels for safe passage. These USACE missions have been authorized for nearly two centuries, and undoubtedly, implementers have improved and honed management through scientific advances and pragmatic considerations, as well as being pushed and pulled by outside forces.

As of 2015, the agency is required to consider resilience in all aspects of its mission (USACE and Bostick 2017). The agency's resilience worldview is derived from the U.S. National Academy of Sciences definition of disaster resilience: the ability to prepare and plan for, absorb, recover from, and more successfully adapt to adverse events (National Research Council 2012). Presidential executive orders direct the USACE to optimize infrastructure performance in the face of natural disasters (executive order no. 13653, 2013). In turn, the USACE (2020) has developed internal guidelines to implement resilience principles. In general, the agency is accustomed to controlling systems through engineered solutions that maintain the design objectives of that system. For example, the Everglades drainage basin water management system in Florida was designed and built to control flooding in support of urban development and agricultural production. Conservation of nature was not the original intent of the Everglades water system, but in recent years consideration has been made to protect biodiversity. In this worldview, resilience was sought through strategies that reduce the risk of disruptions or damage to human-made infrastructure, help recover their functions rapidly after disruptions or damage, and support adaptation to reduce future disruptions. The mounting externalities and unintended consequences over the years provide evidence that designing to maintain the status quo might not always be desirable; furthermore, climate change and its attendant increase in disruptions could make it an impossible goal.

To date, many resilience enhancements considered by the USACE have tended to be implicitly understood—as opposed to explicitly measured—and manifest almost exclusively as damage avoidance, that is, the resilience of the built environment (resilience as a process; Allen et al. 2019). In part, this is because of aspects of current practice that limit the capacity of the agency to explicitly consider and assess resilience (project timelines and budgets are prohibitive) or to have a role in projects after construction (nearly all projects are passed to local sponsors after construction). In other words, the agency's current operational processes limit the scope of resilience considerations to engineering resilience (i.e., damage avoidance, and even then an incomplete version because the agency cannot manage many of its projects after construction). However, there is pressure on the agency to reimagine how projects can be resilient and support resilient communities. This pressure comes from constituents, the U.S. Congress, and agency employees. A primary factor driving interest in resilience is the need for forward-looking strategies to help society cope with uncertain futures including the impacts of a changing climate (Linkov et al. 2014; Rosati et al. 2015).

In many respects, the USACE project portfolio is well positioned to manage resilience for uncertain climate futures, including by realigning natural and built environments. However, adherence to the narrow definition of engineering resilience may prevent the agency from gaining benefits of full-scale implementation of resilience-based solutions and addressing emerging threats including climate change. Panarchy may help further operationalize resilience: Considering adaptive cycles and potential cross-scale interactions would require the agency and its partners to specify reframed and forward-looking objectives. These objectives should match the temporal and geographic scope and scale of prospective interventions. To benefit from the panarchy framing for resilience work, USACE decision makers and stakeholders could ask, Resilience of what? Resilience for whom? Resilience when and where? What functions does the system, asset, or program deliver that society, the economy, or the environment depends on and therefore need to be persistent?

Nature-based resilience management

One strategy for managing resilience in USACE projects relies on the use of natural and nature-based features (NNBFs). USACE is currently considering how NNBFs can be used in projects for coastal storm risk management. Despite many benefits of NNBFs (Bridges et al. 2015; Kurth et al. 2020a), congressional requirements (Government Accountability Office 2019), and growing demand from coastal communities, NNBFs remain rare (Congressional Research Service 2020). A panarchy focus can provide insight into why USACE resilience efforts fall short and how to realize the promise of NNBFs: The existing physical, institutional, and organizational systems may need to undergo some reorganization and learning, orchestrated at different scales, to accomplish better outcomes than current resilience management (Garmestani and Benson 2013).

Traditional flood risk management uses the engineering resilience perspective: The engineering objective is to minimize the magnitude of current and future probabilistic flood damage to property (USACE 2017). In this framing, the critical functions of the protected community persist because they are never interrupted. In an oversimplified version of this world, USACE operates in a narrow band that protects a specific vulnerable system from a specific hazard.

Recovery and adaptation are simple in this worldview: If protective infrastructure is damaged, it is repaired back to the predamage state; in cases where adaptation has been planned, protective infrastructure is retrofitted. From the USACE perspective, adaptation arises from human management, not natural system change. Nature is to be controlled. Most importantly, very little is changeable about the system, so alternative project design options are limited to those that do not require system reconfiguration or trade-offs. As long as stressors stay within the design parameters, such infrastructure can provide broad protections. NNBF design alternatives generally do not fit into the engineering resilience perspective because a one-for-one exchange of conventional coastal armoring to NNBFs is unlikely to maintain the status quo (i.e., damage cannot be fully avoided, and so an NNBF is not deployed). This narrow focus causes the USACE to miss out on long-term and broad scope contributions of NNBFs, which can contribute more to resilience on longer timelines.

The resilience approach that is dominated by risk management and bouncing back has been deemed successful for decades, but it is not universally viable for the future, a fact that is evident from the increasing losses suffered with every new year of natural disasters (National Oceanic and Atmospheric Administration 2020). Vulnerability of coastal settlements is growing much faster than engineering resilience solutions can keep pace for a variety of reasons (Burkett and Davidson 2012). The characteristics of hazards are changing so that past designs are not suited to future events (Cheng and AghaKouchak 2014), and designs that offer protection from future hazards are sometimes undesirable or cost prohibitive (Bostick et al. 2018). Unfortunately, intense human activity paired with coastal armoring has stripped away the natural defenses afforded by near-coast marine features, which we have starved of sediment and stressed beyond their capacity (Burkett and Davidson 2012). Furthermore, the single-objective focus on avoiding short-term economic damages disregards other benefits (e.g., nonmonetary value of ecosystem goods and services) and biases solutions toward protection of high-value property. This tendency to compound social inequities (Chuang et al. 2019) and to neglect nonmonetary benefits calls into question whether federal policy reflects broad societal values. Considering resilience *of what* and *for whom* might reveal misalignment of goals and current practice.

One of the traps of engineering resilience in the context of the USACE is that cross-scale interactions are discounted or ignored. A social-ecological resilience approach could value and harness these interactions instead but would require expanding beyond boundaries of single USACE missions and authorities (Garmestani and Benson 2013). As an example, sediment is a common thread throughout USACE work. USACE projects often interrupt sediment transport dynamics, intentionally or consequently. In general, sediment has not always been viewed as a resource but instead as a waste product. Many coastal systems are starved of sediment because channelized rivers deposit less sediment, and USACE dredging operations typically move sediment to landside sites or offshore disposal cells (i.e., out of the local sediment system). Although the problem has long been recognized, beneficial reuse of sediment to renourish sediment-deficient areas is not common practice, because of narrow perspectives as well as real constraints on cost, various logistical challenges, and stakeholder opposition (Bailey et al. 2009). But system management that invokes social-ecological resilience to look beyond damage avoidance could require managers to foster sediment dynamics so that both natural and engineered coastal features remain stable or become more robust. The hope of NNBFs is that they have some capacity to recover or be more easily recovered after erosive events compared with non-NNBF counterparts and to adapt or be adapted to changing environmental factors (e.g., wetland migration that keeps pace with sea level change) (Cunniff and Schwartz 2015).

The USACE is better positioned to implement NNBFs in support of the navigation mission than the flood risk management missions because of the peculiarities of each mission's business case. Flood risk management requires that the benefits of construction (i.e., damages avoided) exceed the costs. Uncertainty about the level of flood risk protection provided by NNBFs challenges benefit accounting so greatly that those benefits have simply been excluded from economic justifications (CRS 2020). On the other hand, the navigation mission requires only that the mission is executed cost effectively. Here, using dredged sediment to build NNBFs can save money, because placing sediment in nearby systems and at sites that do not have to be acquired and prepared often reduces or eliminates disposal costs (USEPA and USACE 2007). Furthermore, material reuse in NNBFs can lead to quantifiable ecosystem benefits and therefore helps navigation projects meet requirements for mitigation when they have ecologically damaging side-effects or are coordinated with the ecosystem restoration mission (USEPA and USACE 2007).

Resilience management and governance

New resilience management strategies, including greater consideration of NNBFs, can include USACE efforts to build multiagency and multijurisdiction relationships beyond its governance boundaries (Garmestani and Benson 2013). When the USACE moves outside its usual narrow band of space between hazard and vulnerable societal assets, the entities that operate in the larger domain need to be involved, especially to realize a greater set of benefits and to reduce the effective costs of such a change. For example, operating in the natural environment involves navigating regulatory requirements, which can pose technical, administrative, and cost barriers; newly created NNBFs may take the place of other regulated habitat types, and there may be new requirements to conduct monitoring. Aligning agency objectives can expand the benefits that projects produce and increase access to funding, by way of other parties interested in those benefits. The latter outcome is crucial for NNBFs and for the USACE because current practice is generally not compatible with the lifecycle cost profile of NNBFs, including the need to monitor performance and perform adaptive management such as physically intervening to realign NNBF structures and intended functions.

Similarly, USACE designs can be given more flexibility and effectiveness if they are paired with local action. Sometimes worthy actions are identified by USACE studies but cannot be implemented because they are not economically justified, are not deemed a federal interest, or fall outside USACE authorized activities. For example, the recoverability and adaptability of local critical functions sometimes depend on public and private components and capabilities that are beyond the USACE's influence. The resilience of communities adjacent to USACE projects can influence what these projects can achieve. Resilience research often highlights infrastructure alongside multiple other dimensions that have a role in resilience outcomes (Koliou et al. 2018). However, communities do not necessarily build these capacities if they do not know that residual risk and risk of failure exist for all infrastructure. With better knowledge about risks and future outlooks, communities may elect to take on an approach based on social-ecological resilience. How the exchange between communities, the USACE, and other influential parties proceeds will vary. In the most extreme cases, communities may relocate or elect not to rebuild after a disaster. Other community decisions may be driven by the local economy and the role of natural resources and beauty. Others still could be dictated by land use preferences and transportation choices. The extent to which communities rely on adaptive versus transformative governance will depend on their vision and some balance of deliberate and unbridled change (see Chaffin et al. 2016).

Adopting an adaptive or transformative approach is challenging given both the institutional commitment to the status quo and its beneficiaries' expectations for it. The existing framework of immediate objectives makes finding points of intervention in the current system challenging. For example, the most straightforward way to reduce vulnerability to coastal flooding is by retreating from flood-prone areas, but this strategy is rare, and the population in vulnerable coastal areas is actually growing. Major infrastructure removal seems drastic, but so is building new infrastructure (e.g., gray vs. green water infrastructure) that will exist for the next 100 years (Herrmann et al. 2016a; Herrmann et al. 2016b). Managing resilience must be matched to political and social feasibilities; whereas some areas are suited to and ready for transformation, other areas may need more incremental approaches to build an evidence base and to overcome hesitancy. If the probability of some flooding is recognized and accepted, sacrificial dunes or berms that give a limited sense of protection could be preferable to hard structures that have some chance of failing catastrophically. Embracing change would allow us to reconsider what we truly value and shape our systems in that direction rather than working endlessly to keep them in their current configuration. Panarchy offers other perspectives on how resilience can be managed, which will lead to new management objectives and scoping that can provide avenues for intervention in light of rapidly accelerating environmental change.

Literature Cited

Allen, C.R., D.G. Angeler, B.C. Chaffin, D. Twidwell, and A. Garmestani. 2019. Resilience reconciled. *Nature Sustainability* 2: 898–900.

Angeler, D.G., C.R. Allen, A.S. Garmestani, L.H. Gunderson, and I. Linkov. 2016. Panarchy use in environmental science for risk and resilience planning. *Environment Systems and Decisions* 36: 225–228.

Angeler, D.G., C.R. Allen, A. Garmestani, K.L. Pope, D. Twidwell, and M. Bundschuh. 2018. Resilience in environmental risk and impact assessment: Concepts and measurement. *Bulletin of Environmental Contamination and Toxicology* 101: 1–6.

Angeler, D.G., H. Peterson, C.R. Allen, A. Garmestani, D. Twidwell, W. Chuang, V.M. Donovan, T. Eason, C.P. Roberts, S.M. Sundstrom, and C.L. Wonkka. 2019. Adaptive capacity in ecosystems. *Advances in Ecological Research* 60: 1–24.

Bailey, S.E., T.J. Estes, P.R. Schroeder, T.E. Myers, J.D. Rosati, T.L. Welp, L.T. Lee, W.V. Gwin, and D.E. Averett. 2009. *Sustainable Confined Disposal Facilities for Long-Term Management of Dredged Material.* DOER Technical Notes Collection. ERDC TN-DOER-D10. U.S. Army Engineer Research and Development Center, Vicksburg, MS.

Bostick, T.P., E.B. Connelly, J.H. Lambert, and I. Linkov. 2018. Resilience science, policy and investment for civil infrastructure. *Reliability Engineering & System Safety* 175: 19–23.

Bridges, T.S., P.W. Wagner, K.A. Burks-Copes, M.E. Bates, Z.A. Collier, C.J. Fischenich, C.D. Piercy, E.J. Russo, D.J. Shafer, B.C. Suedel, J.Z. Gailani, and J.D. Rosati. 2015. *Use of Natural and Nature-Based Features (NNBF) for Coastal Resilience*. ERDC SR-15-1. US Army, Engineer Research and Development Center, Vicksburg. https://http://el.erdc.usace.army.mil/

Bruneau, M., S.E. Chang, R.T. Eguchi, G.C. Lee, T.D. O'Rourke, A.M. Reinhorn, M. Shinozuka, K. Tierney, W. A. Wallace, and D. von Winterfeldt. 2003. A framework to quantitatively assess and enhance the seismic resilience of communities. *Earthquake Spectra* 19: 733–752.

Burkett, V.R., and M.A. Davidson. 2012. Coastal impacts, adaptation and vulnerability: A technical input to the 2012 National Climate Assessment. *Cooperative Report to the 2013 National Climate Assessment.*

Carpenter, S., B. Walker, J.M. Anderies, and N. Abel. 2001. From metaphor to measurement: Resilience of what to what? *Ecosystems* 4: 765–781.

Carpenter, S.R., J.J. Cole, M.L. Pace, R. Batt, W.A. Brock, J. Coloso, J.R. Hodgson, J.F. Kitchell, D.A. Seekell, L. Smith, and B. Weidel. 2011. Early warning of regime shifts: A whole-ecosystem experiment. *Science* 332: 1079–1082.

Chaffin, B.C., A.S. Garmestani, L.H. Gunderson, M.H. Benson, D.G. Angeler, C.A. Arnold, B. Cosens, R.K. Craig, J.B. Ruhl, and C.R. Allen. 2016. Transformative environmental governance. *Annual Review of Environment and Resources* 41: 399–423.

Cheng, L., and A. AghaKouchak. 2014. Nonstationary precipitation intensity–duration–frequency curves for infrastructure design in a changing climate. *Scientific Reports* 4. https://doi.org/10.1038/srep07093

Chuang, W., T. Eason, A. Garmestani, and C.P. Roberts. 2019. Impact of Hurricane Katrina on the coastal systems of southern Louisiana. *Frontiers in Environmental Science* 7: 68. https://doi.org/10.3389/fenvs.2019.00068

Congressional Research Service. 2020. *Flood Risk Reduction from Natural and Nature-Based Features: Army Corps of Engineers Authorities*. Congressional Research Service, Washington, DC. https://crsreports.congress.gov/product/pdf/R/R46328

Connelly, E.B., C.R. Allen, K. Hatfield, J.M. Palma-Oliveira, D.D. Woods, and I. Linkov. 2017. Features of resilience. *Environment Systems and Decisions* 37: 46–50.

Craig, R.K. 2010. Stationarity is dead—Long live transformation: Five principles for climate change adaptation law. *Harvard Environmental Law Review* 34: 9–75.

Cunniff, S., and A. Schwartz. 2015. *Performance of Natural Infrastructure and Nature-based Measures as Coastal Risk Reduction Features*. Environmental Defense Fund, New York. https://www.edf.org/sites/default/files/summary_ni_literature_compilation_0.pdf

Cutter, S.L. 2016. Resilience to what? Resilience for whom? *Geographical Journal* 182: 110–113.

Davidson, J.L., C. Jacobson, A. Lyth, A. Dedekorkut-Howes, C.L. Baldwin, J.C. Ellison, N.J. Holbrook, M.J. Howes, S. Serrao-Neumann, L. Singh-Peterson, and T.F. Smith. 2016. Interrogating resilience: Toward a typology to improve its operationalization. *Ecology and Society* 21: 27. https://doi.org/10.5751/ES-08450-210227

DeWitte, S.N., M.H. Kurth, C.R. Allen, and I. Linkov. 2016. Disease epidemics: Lessons for resilience in an increasingly connected world. *Journal of Public Health* 39: 254–257.

Drury, J., H. Carter, C. Cocking, E. Ntontis, S.T. Guven, and R. Amlôt. 2019. Facilitating collective psychosocial resilience in the public in emergencies: Twelve recommendations based on the social identity approach. *Frontiers in Public Health* 7. https://doi.org/10.3389/fpubh.2019.00141

Fox-Lent, C., and I. Linkov. 2018. Resilience matrix for comprehensive urban resilience planning. In *Resilience-Oriented Urban Planning*, ed. Y. Yamagata and A. Sharifi, 29–47. Springer International Publishing, Cham, Switzerland. https://doi.org/10.1007/978-3-319-75798-8_2

Galaitsi, S.E., B.D. Trump, J.M. Keisler, and I. Linkov. 2021. The need to reconcile concepts that characterize systems facing threats. *Risk Analysis* 41: 3–15.

Garmestani, A.S., and M.H. Benson. 2013. A framework for resilience-based governance of social-ecological systems. *Ecology and Society* 18 (1): 9. http://www.ecologyandsociety.org/vol18/iss1/art9/

Garmestani, A., J.B. Ruhl, B.C. Chaffin, R.K. Craig, H.F.M.W. van Rijswick, D.G. Angeler, C. Folke, L. Gunderson, D. Twidwell, and C.R. Allen. 2019. Untapped capacity for resilience in environmental law. *Proceedings of the National Academy of Sciences* 116: 9899–19904.

Government Accountability Office. 2019. *Army Corps of Engineers Consideration of Project Costs and Benefits in Using Natural Coastal Infrastructure and Associated Challenges.* GAO-19-319. Government Accountability Office, Washington, DC.

Grafton, R.Q., L. Doyen, C. Béné, E. Borgomeo, K. Brooks, L. Chu, G.S. Cumming, J. Dixon, S. Dovers, D. Garrick, and A. Helfgott. 2019. Realizing resilience for decision-making. *Nature Sustainability* 2: 907–913.

Green, O.O., A.S. Garmestani, C.R. Allen, L.H. Gunderson, J.B. Ruhl, C.A. Arnold, N.A.J. Graham, B. Cosens, D.G. Angeler, B.C. Chaffin, and C.S. Holling. 2015. Barriers and bridges to the integration of social-ecological resilience and law. *Frontiers in Ecology and the Environment* 13: 332–337.

Gunderson, L.H., and C.S. Holling. 2002. *Panarchy: Understanding Transformations in Human and Natural Systems.* Island Press, Washington, DC.

Herrmann, D.L. K. Schwarz, W.D. Shuster, A. Berland, B.C. Chaffin, A.S. Garmestani, and M.E. Hopton. 2016a. Ecology for the shrinking city. *BioScience* 66: 965–973.

Herrmann, D.L., W.D. Shuster, A.L. Mayer, and A.S. Garmestani. 2016b. Sustainability for shrinking cities. *Sustainability* 8(9): 911. https://doi.org/10.3390/su8090911

Hidalgo, D. 2020. More bicycles, slower speeds, a more livable city: Paris mayor Anne Hidalgo plans an ambitious second term. *The City Fix* (blog), July 15. https://thecityfix.com/blog/bicycles-slower-speeds-livable-city-paris-mayor-anne-hidalgo-plans-ambitious-second-term-dario-hidalgo/

Holling, C.S. 1973. Resilience and stability of ecological systems. *Annual Review of Ecology, Evolution and Systematics* 4: 1–23.

Holling, C.S. 1996. Engineering resilience vs. ecological resilience. In *Engineering Within Ecological Constraints*, ed. Peter C. Schulze, 31–44. National Academy Press, Washington, DC.

Hynes, W., B.D. Trump, P. Love, and I. Linkov, I. 2020. Bouncing forward: A resilience approach to dealing with COVID-19 and future systemic shocks. *Environment, Systems, Decisions* 40: 174–184.

Jackson, S. 2016. *Principles for Resilient Design: A Guide for Understanding and Implementation in IRGC Resource Guide on Resilience.* EPFL International Risk Governance Center, Lausanne, Switzerland.

Koliou, M., J.W. van de Lindt, T.P. McAllister, B.R. Ellingwood, M. Dillard, and H. Cutler 2018. State of the research in community resilience: Progress and challenges. *Sustainable and Resilient Infrastructure* 5: 1–21.

Kurth, M.H., R. Ali, T.S. Bridges, B.C. Suedel, and I. Linkov. 2020a. Evaluating resilience co-benefits of engineering with nature projects. *Frontiers in Ecology and Evolution* 8: 149. doi:10.3389/fevo.2020.00149

Kurth, M., W. Kozlowski, A. Ganin, A. Mersky., B. Leung, M. Kitsak, and I. Linkov. 2020b. Lack of resilience in transportation networks: Economic implications. *Transportation Research Part D: Transportation and Environment* 86: 102419.

Linkov, I., T. Bridges, F. Creutzig, J. Decker, C. Fox-Lent, W. Kröger, J.H. Lambert, A. Levermann, B. Montreuil, J. Nathwani, R. Nyer, O. Renn, B. Scharte, A. Scheffler, M. Schreursand, and T. Thiel-Clemen. 2014. Changing the resilience paradigm. *Nature Climate Change* 4: 407–409.

Linkov, I., and B.D. Trump. 2019. *The Science and Practice of Resilience.* Springer, Amsterdam.

Meerow, S., J.P. Newell, and M. Stults. 2016. Defining urban resilience: A review. *Landscape and Urban Planning* 147: 38–49.

National Oceanic and Atmospheric Administration. 2020. *U.S. Billion-Dollar Weather and Climate Disasters.* National Centers for Environmental Information, Asheville, NC. https://www.ncdc.noaa.gov/billions/

National Research Council, Committee on Increasing National Resilience to Hazards and Disasters, Committee on Science, Engineering, and Public Policy, Policy and Global Affairs, and Policy and Global Affairs. 2012. *Disaster Resilience: A National Imperative.* National Academies Press, Washington, DC. https://doi.org/10.17226/13457

Paton, D., and D. Johnston. 2017. *Disaster Resilience: An Integrated Approach.* Charles C Thomas Publisher, Springfield, IL.

Pimm, S. L., I. Donohue, J. M. Montoya, and M. Loreau. 2019. Measuring resilience is essential to understand it. *Nature Sustainability* 2: 895–897.

Rosati, J.D., K.F. Touzinsky, and W.J. Lillycrop. 2015. Quantifying coastal system resilience for the US Army Corps of Engineers. *Environment, Systems, and Decisions* 35: 196–208.

Spanbauer, T.L., C.R. Allen, D.G. Angeler, T. Eason, S.C. Fritz, A.S. Garmestani, K.L. Nash, J.R. Stone, C.A. Stow, and S.M. Sundstrom. 2016. Body size distributions signal a regime shift in a lake ecosystem. *Proceedings of the Royal Society B* 283: 20160249.

U.S. Army Corps of Engineers (USACE). 2017. *Engineer Regulation 1105-2-101 Risk Assessment for Flood Risk Management Studies.* USACE, Washington DC. https://www.publications.usace.army.mil/Portals/76/Publications/EngineerRegulations/ER_1105-2-101.pdf

U.S. Army Corps of Engineers (USACE). 2020. *ECB 2020-06 Implementation of Resilience Principles.* USACE, Washington DC.

U.S. Army Corps of Engineers (USACE) and Thomas Bostick. 2017. *EP 1100-1-2 Resilience Initiative Roadmap. Engineering Pamphlet,* October 2017. USACE, Washington DC. https://www.publications.usace.army.mil/Portals/76/Publications/EngineerPamphlets/EP_1100-1-2.pdf?ver=2017-11-02-082317-943

U.S. Environmental Protection Agency (USEPA) and U.S. Army Corps of Engineers (USACE). 2007. *Identifying, Planning, and Financing Beneficial Use Projects Using Dredged Material. Beneficial Use Planning Manual.* U.S. EPA, EPA842-B-07-001, USACE, Washington, DC.

Zemba, V., E.M. Wells, M.D. Wood, B.D. Trump, B. Boyle, S. Blue, C. Cato, and I. Linkov. 2019. Defining, measuring, and enhancing resilience for small groups. *Safety Science* 120: 603–616.

CHAPTER 7

Mapping Panarchy to Improve Visualization of Complex Environmental Change

Dirac Twidwell, Daniel R. Uden, Caleb P. Roberts, Brady W. Allred, Matthew O. Jones, David E. Naugle, and Craig R. Allen

Panarchy has served as a mental model that describes changes in complex, dynamic systems of people and nature over time (Gunderson and Holling 2002). A next step for advancing panarchy theory is to empirically map complex system dynamics both across space and through time. Mapping panarchy provides opportunities to ground the concept, make it more tractable, and facilitate exploration of the spatial interdependencies of resilience, regime shifts, and cross-scale change. In this chapter, we apply data from environmental monitoring and spatial imaging technologies to map and visualize aspects of panarchy.

One reason for the lack of progress in applying panarchy has been a sparsity of data with sufficient resolution and extent for representing spatiotemporal change within and across scales of organization in real-world systems and the computing capacity to leverage such data. Recent innovations in data and technology have converged to make this application possible. One such convergence involves the linking of landcover data and geospatial cloud computing to detect regime shifts in ecosystems (Uden et al. 2019). Regime shift imaging and screening is enabled by a data-based innovation within the Rangeland Analysis Platform (RAP; Jones et al. 2018). The RAP is used to map annual cover estimates of plant functional groups for the entire western United States at 30-meter resolution and uses geospatial cloud computing from Google Earth Engine (Gorelick et al. 2017).

Uden et al. (2019) proposed spatial covariance as a resilience-based metric capable of screening for spatially explicit state transitions between different basic plant functional groups. Broadly, covariance is a measure of how much two random variables change together (Le and Zidek 2006; Houlahan et al. 2018) and is capable of highlighting correlations in data that may be smoothed over or hidden by averaging operations (Frasinski et al. 1989). Accordingly, spatial covariance highlights spatial relationships (i.e., spatial correlations) between variables (Wagner 2003). When computed in a moving window (i.e., neighborhood) algorithm, spatial covariance produces a continuous surface that quantifies the degree to which two plant functional groups coexist in geographic space (Uden et al. 2019). Positive spatial covariance indicates the tendency of functional groups to spatially increase and decrease together, spatial covariance of zero indicates a lack of spatial relationship between the functional groups, and negative spatial covariance indicates the tendency of the functional groups to spatially separate from one another (i.e., segregate, exclude one another). Because a system can exist in only one regime at a time (Folke et al. 2010), strong negative spatial covariance is useful for delineating geographic boundaries between neighboring spatial regimes at different moving window sizes. Tracking spatial regime boundaries, ecotones, over time allows identification of boundaries that are persistent and nonstationary (i.e., moving), as opposed to those that exhibit transient or stationary behavior (Uden et al. 2019). From a practical standpoint, screening and imaging regime shifts may provide earlier warnings for proactive management of undesirable transitions, before the manifestation of their effects (i.e., signs and symptoms), compared with other ecological metrics that are diagnostic.

Ecological transitions or regime shifts are related to system resilience, defined as the amount of disturbance needed to flip the system from one regime to another (Gunderson 2000). Resilience varies over time within systems and correlates with four phases of ecosystem development and change (Gunderson and Holling 2002). Four sequential phases of ecosystem change (exploitation, conservation, collapse, and reorganization) create an adaptive cycle (figure 7.1). Adaptive cycles (each at a particular spatiotemporal domain) are linked across a hierarchy of ecological organization to form a panarchy. At each domain of scale, the forward loop in the adaptive cycle is consistent with ecological succession and emphasizes periods of growth and conservation (figure 7.1). A critical transition in the adaptive cycle occurs as the system enters the back loop, which describes a period of collapse and reorganization. In theory, this transition from the forward to the back loop can manifest as a spatially explicit signal of a transition between alternative, neighboring spatial regimes. Imaging and screening for the leading edge of regime shifts captures this critical transition (figure 7.1), highlighting signals of spatial patterns (i.e., boundaries) where one regime ends, or collapses in geographic space, and another regime begins (Uden et al. 2019). This leading edge can then be tracked over time to provide cues into the directional movement of regime shifts, provide

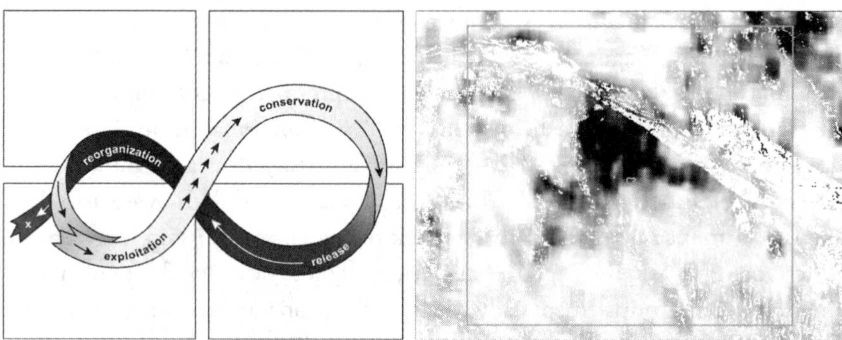

Figure 7.1. Mapping panarchy draws from the theory of the adaptive cycle, recent advances in spatial resilience theory, and next-generation technologies developed for rangeland monitoring. The classic depiction of the adaptive cycle is shown here. Shading on the adaptive cycle correspond to transitions, or boundaries, between neighboring spatial regimes and emphasizes a critical transition from the forward loop to the backward loop in the adaptive cycle. This critical spatial transition in the adaptive cycle occurs as one regime breaks down and reorganizes into another spatial regime. This process is shown here through recent advances in the spatial imaging and screening for alternative grass to woody spatial regimes that lead to undesired changes in ecosystem services in Great Plains grasslands (mapping product shown here for the Loess Canyons, Nebraska, ecoregion). Imaging has been optimized (following Uden et al. 2019) to reflect ecoregion-scale boundaries between grass to woody regimes and involves an analytical approach capable of capturing cross-scale discontinuities in the system that describe the spatial order of multiple independent but nested adaptive cycles in a panarchy.

more advanced warning of ecological change, and identify heightened vulnerability to intact ecosystems before regime shifts manifest (Roberts et al. 2019; chapter 5, this volume).

We focus this chapter on a well-studied regime shift that occurs in grassland ecosystems through woody plant encroachment. Such transitions occur on multiple continents (Stevens et al. 2017) and are a core example of ecological regime shifts resulting from factors operating across scales (Walker and Abel 2002). Woody plant invasion is regarded as one of the biggest threats to the sustainability of the Great Plains biome (Engle et al. 2008). Indeed, hundreds of millions of dollars have been spent—with little evidence of large-scale success—in an effort to combat the woody plant encroachment problem (Twidwell et al. 2013a). We use a panarchy framework to envision the complexities of grass–woody transitions in the Great Plains biome and in doing so provide an example of how panarchy can be applied to complicated social-ecological issues that involve ecological regime shifts. Analytically, we compute spatial covariance between percentage cover of herbaceous plants (forbs and grasses, but we will use the shorthand of grasses) and percentage cover of trees across moving windows of different sizes. Spatial covariance is calculated in multiple years from data collected between 2000 and 2018.

The remainder of this chapter covers five key topics related to panarchy. First, panarchy depicts scales of hierarchical organization and allows delineation of regime boundaries at multiple scales. Such depictions are key to visualizing complex system dynamics. Second, panarchy emphasizes the potential for collapse and reorganization to manifest at any domain of scale in systems with a clear spatial component. This helps us avoid tendencies to limit or bias our perception of a single or narrow range of scales that are assumed relevant for observation and interpretation. Third, the adaptive cycle inherently differentiates between alternative pathways of renewal versus transitions to alternative states after disturbance. Fourth, panarchy underscores the cross-scale connectedness of systems and the potential for regime shifts to occur not just at a preexisting scale domain and to emerge at new scales in the system. We end this chapter with an eye to the future, discussing next-generation capabilities and the potential to further our understanding of complex environmental change through additional explorations at the intersections of panarchy theory, data, and spatial imaging technologies.

Tracking Dynamic Discontinuities through Space and Time

Spatial boundaries of ecosystems represent emergent structures (and functions) that are compartmentalized at discrete scale domains (Garmestani et al. 2009). We map these spatial breaks between functional plant types across a range of scales within the Great Plains. Such breaks are key transition points that occur between forward and back loops in a panarchy (figure 7. 2). In this example, a series of maps were created that depict Gunderson and Holling's (2002) classic representation of panarchy as hierarchically nested adaptive cycles, each representing discrete domains of scale within the system. These maps represent the spatial pattern of the forward loop, the back loop, and the transition between them that is unique to a particular scale in the system. We transpose this color pattern onto the graphic illustration of the adaptive cycle in order to indicate which scales in this system are conserved and which exhibit critical spatial transitions between competing grass–woody alternative states.

Mapping and visualizing multiple scales in a panarchy has the potential to resolve problems created by observers who view systems at different scale

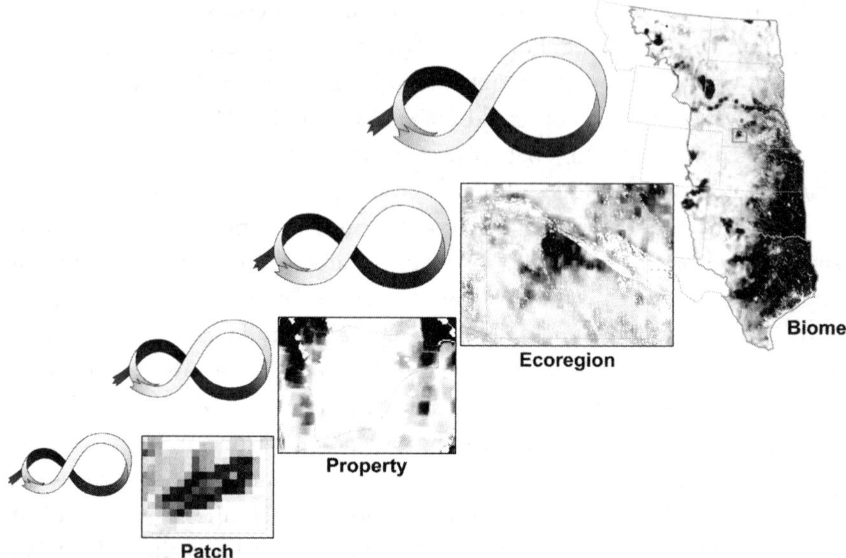

Figure 7.2. Mapping grass to woody spatial transitions as a panarchy to assist decision making and planning at multiple domains of scale. Shown here are maps of grass (white)–woody (black) transitions at scales ranging from the patch (windbreak) to the biome (Great Plains temperate grassland). Four scales are shown here and are organized as a series of adaptive cycles in a panarchy. Each map depicts the spatial context of individual adaptive cycles, and images are customized to reflect boundaries in grass–woody transitions at that particular domain of scale (protocols described in Uden et al. 2019).

domains. Such cross-scale analyses may resolve debates that arise between individuals and groups who perceive change at distinct spatial and temporal scales. For example, geographic boundaries within the United States were established at the township, county, state, and national levels, thereby representing discontinuities in administrative jurisdiction. Yet ecological processes do not conform to such hierarchies; indeed, no single appropriate scale exists to study ecological dynamics (Levin 1992). A scale-limited view of landscape-level vegetation change may lead to biases of potential structure–function relationships and the types of interventions needed to manage such change (Wu and Loucks 1995). For example, a private property owner might assess vegetation change within his property boundaries, whereas the director of a state government resource agency holds a statewide perspective. Both of these views tend to exclude bottom-up and top-down processes that influence ecological responses.

Panarchy avoids this tendency and emphasizes the potential for collapse and reorganization to manifest at any scale (Allen et al. 2014). Collapse and

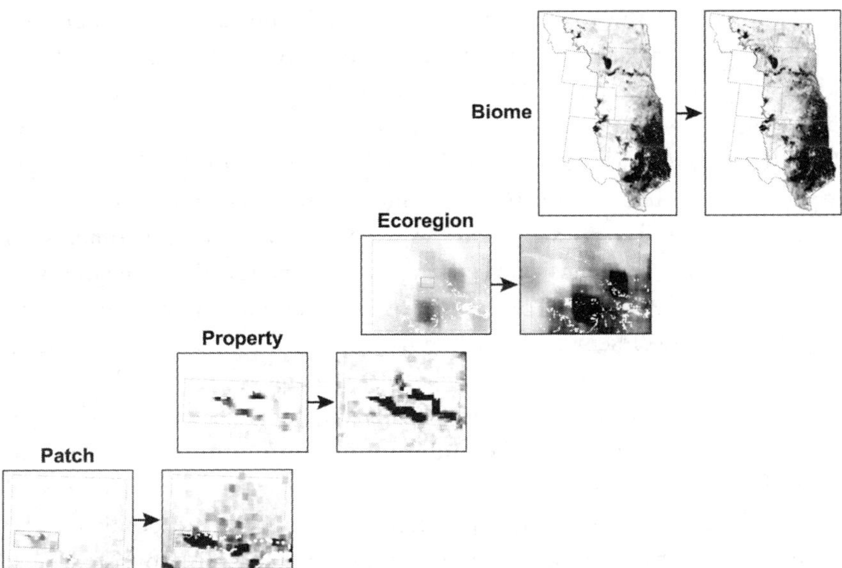

Figure 7.3. Imaging system collapse and reorganization as a panarchy. Shown here are real examples of multiple scales in the system transitioning from grassland to woody plant dominance. Collapse of grassland (lighter areas) and reorganization toward woody plant dominance (darker areas) occurs more rapidly at the patch scale, whereas collapse of the Great Plains temperate grassland biome is much slower. This is a central proposition of panarchy theory, and these mapping products visualize the premise that collapse can occur at any range of scales in the system. Images are optimized to reveal the leading edge of grass–woody transitions at each domain of scale from 2000 to 2018 (following Uden et al. 2019).

reorganization can manifest spatially as directional change (chapter 5, this volume) or other abrupt change in the location of spatial boundaries separating alternative spatial regimes over time. In the Great Plains, grass–woody transitions are occurring at all ranges of scales (figure 7.3). As expected in a panarchy, finer scales of organization (e.g., patch) are transitioning more rapidly compared with larger scales (e.g., biome).

Distinguishing Renewal from Transition

As systems move through an adaptive cycle, a critical juncture occurs during the reorganization phase, when the system either enters a state of renewal or transitions to an alternative state (figure 7.1). Renewal leads to restoration of structures and functions similar to the state that occurred before they entered the back loop of the adaptive cycle. It should be noted that this does not imply a return to identical reference or starting conditions (Holling 2004), which is often assumed in studies that test for differences in plant community composition and structures before and after a disturbance (e.g., Schimmel and Granstrom 1996; Van Leeuwen et al. 2010; Donovan et al. 2020). Changes in the relative abundance of species within a plant association can occur after a disturbance, and the system can still remain in the same functional state (Briske et al. 2005), which is consistent with the perspective of engineering resilience (Allen et al. 2019). Areas that undergo a transition to an alternative vegetation state represent a shift in dominance from one basic functional group to another, with corresponding changes in structure–function relationships (Fath et al. 2015). Examples of such state changes include erosion, when perennial grass cover disappears, becoming bare ground; woody encroachment, a shift from a herbaceous state to one dominated by woody plants, or annual invasions, when sites dominated by perennial grasses shift to dominance by annual plants. These transitions result in multistate dynamics consistent with ecological resilience theory. Novel, nonrandom, and discontinuous structures emerge after transitions to alternative regimes, at least based on a particular scale of observation.

Such discontinuities can be visualized to aid our interpretation of complex dynamics. The emergence of more orderly phenomena can be tracked after the more disorderly phases of destruction and reorganization. For example, concerns over the potential for wildfire-induced erosion and the destabilization of vegetation over broad areas of the Great Plains have led to decades of rehabilitation policies and programs (Wright 1979; Arterburn et al. 2018). One would therefore expect to observe a major departure toward bare ground dominance and less perennial vegetation. Instead, analyses of vegetation response in Great Plains grasslands demonstrate a high degree of similarity to pre-wildfire conditions (Donovan et al. 2020), consistent with the notion of renewal in the adaptive cycle. Grassland still dominates, and the degree of departure therefore represents small changes in the grand scheme of vegetation dynamics and change. Alternatively, the same wildfire occurring in non-resprouting *Juniperus* woodland can result in collapse of the woody-dominated state, if fires are consistent with conditions needed to operate above *Juniperus* mortality thresholds (Twidwell et al. 2009, 2013b). Grasslands reemerge to dominance after the wildfire (Twidwell et al. 2020), representing a major departure from previous structure–function relationships. To document these changes, we monitored the vegetation before and after two well-known wildfires, one in

2012 and one in 2001, that occurred in the central Great Plains and screened for transitions of alternative grass–woody states and mapped changes in their distributions. We present the results for the 2012 wildfire first.

In 2012, the Region 24 Complex wildfire occurred where the Sandhills prairie ecoregion meets the easternmost distribution of ponderosa pine with a high-density juniper encroached understory. These can be thought of as neighboring regimes, that is, a grass regime adjacent to a conifer (i.e., woody) regime. The 2012 wildfire occurred during a record drought, fostering fears that the wildfire would result in large-scale destabilization of perennial grassland vegetation and subsequent erosion. The implication here is that the resilience of the grass state to wildfire was low and would readily collapse, similar to the expectations for the conifer regime. Yet screening for this transition showed rapid renewal of the grassland-dominated areas (figure 7.4). Field inventories found that aboveground grassland biomass reached levels equivalent to biomass in unburned areas in less than two years (Arterburn et al. 2018). After the same fire, the ponderosa pine and juniper-dominated corridor collapsed and underwent a transition to a novel regime, characterized as a mixed grass–deciduous plant community (Roberts et al. 2018). This example illustrates a case where renewal occurred in the grassland regime, whereas a transition to an alternative vegetation type occurred in the adjacent conifer regime.

The 2001 Gothenburg wildfire resulted in a clear signal of destruction and reorganization in a juniper-dominated landscape. After this wildfire, grassland rapidly reemerged to dominance (figure 7.4), and the establishment of stabilizing feedbacks would have resulted in continued persistence of the grassland state at a large scale. However, heterogeneity within the fire perimeter allowed some woody patches to escape fire damage and the corresponding spatial legacies of the previous woody-dominant state to serve as internal propagule sources that facilitated reestablishment of the woody-dominated state over time. In addition, external propagule sources provided additional repositories for the reestablishment of the woody-dominant state at the wildfire perimeter. By 2018, the system was again transitioning toward woody dominance (figure 7.4), representing the absence of stabilizing grassland feedbacks that would have fostered long-term persistence of the grassland state.

Visualizing these wildfires and considering the spatial context of approximately twenty years of grass–woody transitions reinforces two major theoretical precepts of panarchy. First, transitions become difficult to compare based on differences, or mismatches, in the time frame at which complex systems reorganize after disturbance and perturbation (Allen and Hoekstra 1990). The

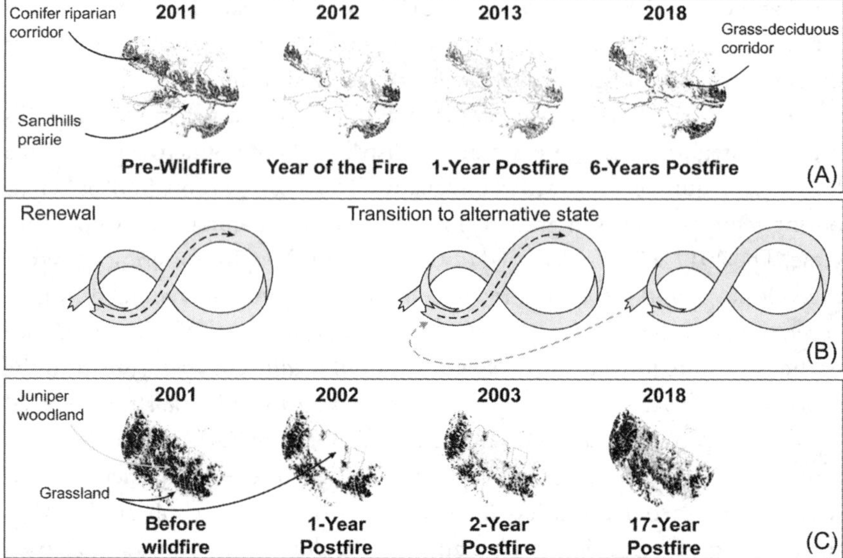

Figure 7.4. Time series maps showing different trajectories following wildfires in two different regions in Nebraska. (A) In 2012, the Region 24 wildfire complex burned in a Sandhills prairie (white), which bordered a riparian corridor dominated by ponderosa pine with an eastern redcedar understory (dark gray). Before the wildfire (2011), both regimes were in the K phase of the adaptive cycle. Afterward, rapid renewal was observed in the Sandhills prairie community, whereas the conifer riparian corridor transitioned to a novel grass–deciduous regime, as indicated in the maps for 2012, 2013, and 2018. (B) The adaptive cycle as a mental model showing the contrasting trajectories that occurred in this system. The Sandhills prairie underwent renewal, still in the prairie regime. However, the conifer corridor transitioned from a woody regime to an alternative ecological regime of prairie. (C) As a more complex example, the 2001 Gothenburg wildfire of Nebraska occurred when a grassland landscape was undergoing slow collapse as a result of years of juniper encroachment. This can be interpreted as a wildfire occurring when the grassland-dominated system was in the back loop of the adaptive cycle. After the wildfire, rapid renewal of the grassland regime was observed for the first few years (white regions), but the absence of stabilizing grassland feedbacks in the system set the stage for history to repeat itself; subsequently, the system has been undergoing a second phase of release, reorganization, and increasing transition toward woody plant dominance (increase in size of dark areas in the maps from 2002 through 2018).

time frames at which renewal or transition to an alternative vegetation type occurs can vary widely for vegetation based on the life form, the scale of observation, and interactions between the abiotic and biotic environment. For example, the potential for forests to recover after stand-replacing fires may take decades or longer, whereas renewal in grasslands to the same type of fire can occur rapidly (weeks or years) (Roberts et al. 2020). We interpreted the recovery of cedar woodland after a wildfire (figure 7.4) to be a temporal sequence that cycled between alternative states of woody, then grassland, and back to woody again. Such dynamic change could be viewed as simply the nature of renewal in this particular system and a reflection that the preconceived identities we assign to fit the norm (e.g., alternative woody and grass states) do not convey the types of complexities inherent in other ecosystems (e.g., easternmost distribution of ponderosa pine in the Great Plains; Donovan et al. 2019; Roberts et al. 2020). Although distinguishing one type of regime shift from another is important for continued academic exploration and debate, we demonstrate how the ability to detect and map the boundaries separating ecological regimes allows this type of complexity to be tracked over time. Such an approach has the potential to foster greater inference on how complex systems function, outside of ideological norms, and ultimately to lead to more informed decisions on how and where to prioritize management resources in a challenging and perplexing system.

Panarchy also means that in many real-world systems we should expect complex and novel structures to be the rule, not the exception, after disturbance-induced collapse (Gunderson and Holling 2002). These complexities are partially the result of how top-down and bottom-up structures and processes are operating relative to the scale of disturbance. In the Gothenburg wildfire example, woody dominance at the scale above that of the wildfire fostered more rapid recovery to cedar woodland, whereas the same configuration of alternative grass–woody regimes would have responded differently if grassland had dominated the matrix instead of cedar woodland. This cross-scale connectedness is a critical feature of the panarchy and forces one to expect greater complexity and a lack of "true" experimental replication from more reductionist approaches (Allen et al. 2014). New spatial technologies with greater computational power are making it increasingly possible to consider cross-scale connectedness as part of how systems recover or change after disturbance. The spatial boundaries separating neighboring regimes and how those boundaries change as a result of both internal and external features of the system (figure 7.3) are much more intuitive for management and planning when visualized and mapped in this way.

Tracking Regime Changes from Smaller to Larger Scales

Panarchy was one of the early concepts to differentiate between regime shifts that occur within a fixed-scale domain and those that move from smaller to larger scales, or scale up, and enter a novel level of hierarchical organization. This type of regime shift represents a type of cross-scale change (Allen et al. 2014). We refer these spatial regimes as transcending past scaling norms. The concept of spatial regimes brings explicit attention to the scaling properties of a given regime and the understanding that all regimes are bounded in geographic space (Roberts et al. 2019). To date, exploration of the scaling up of regimes has been difficult, largely because of limitations in monitoring data. Documentation of cross-scale change necessitates monitoring data that are cross-scale and of sufficient spatial and temporal grain and extent for capturing change over multiple areas and time spans.

Imaging and tracking shifts in the boundaries of an ecological regime can illustrate these concepts in a manner that is easy to understand, for both scientists and the general public. For example, woody plants have been planted throughout the Great Plains as shelterbelts, effectively introducing localized scales of novelty into a large-scale intact grassland matrix. This "improvement" practice has continued for nearly a century, without careful consideration of whether woody patches would expand across space (scale up) to become woody landscapes (Donovan et al. 2018). This pattern of scaling up can take multiple forms (figure 7.5). In one example, scaling up occurred through regime expansion, or swelling. We consider this one of the simplest representations of how a regime can move from a more localized scale to a broader one—in this case, a jump from the patch to the landscape level. In contrast, the second example represents a more diffuse process of scaling up that is still nonrandom but more loosely coupled from the spatial boundaries of the woody regime itself. This makes the scaling-up process less predictable and more difficult to target, with interventions aimed at preventing further transcendence of scale in the system.

The degree to which regimes can scale up in the Anthropocene should be carefully considered as people continue to introduce novelty into ecological systems at unprecedented rates. Looking to the future, it is no longer a question of whether the introduction of woody patches into grasslands is expanding but

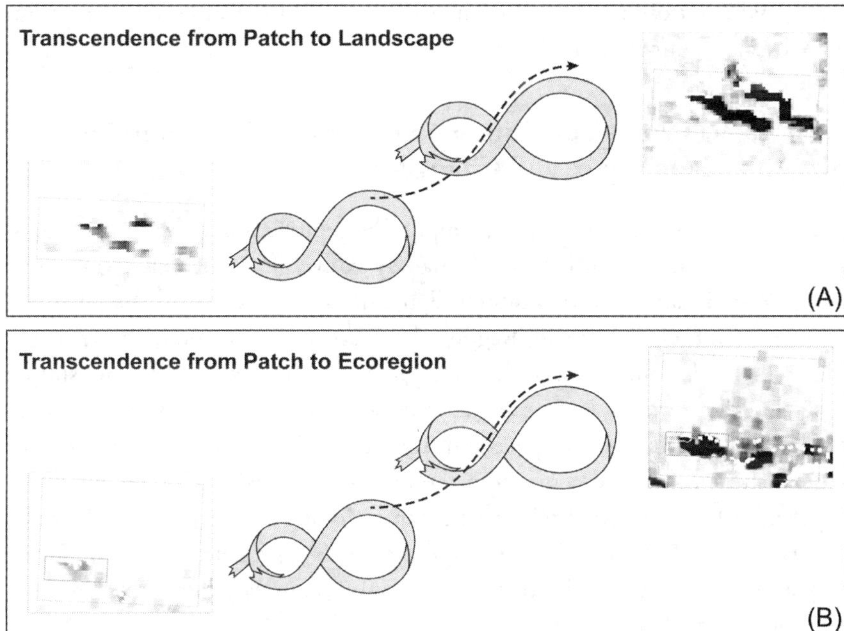

Figure 7.5. Images of cross-scale change and the scaling up of a regime shift. The panarchy concept differentiates two types of regime shifts that enter into a new level of hierarchical organization. Such scaling up can generate two different patterns. (A) Regime expansion through spatially contagious transition with a strong neighbor effect. The dark areas depict a windbreak that expands (lower left to upper right) due to woody plant encroachment into a grassland landscape. (B) Regime transition showing a spread through a more diffuse process, as indicated by encroachment (of trees into prairies) at an ecoregion scale. These examples show the complexity involved in cross-scale regime change. Even with this simple example, a purposeful woody planting can transcend scales in multiple ways, leading to unintended changes evident at larger spatial scales.

rather whether there are any limits to the scaling up of these woody patches and whether the entire Great Plains biome is vulnerable to collapse (*sensu* Engle et al. 2008; Twidwell et al. 2013c). The simple examples of scaling up presented here have important implications for other systems. As one regime scales up, it does so at the expense of the integrity and scaling properties of the neighboring regime that previously filled that geographic space. This trade-off is grounded in the fact that all ecological regimes are finite in geographic space. In other words, where one spatial regime ends, another regime begins (Roberts et al. 2019). For this reason, regime shifts that transcend scale are likely to be some of the most consequential to suites of ecosystem services (Birge et al.

2016), which arose from long-term conservation of self-organized structures across scales in a panarchy.

Next-Generation Capabilities: Tracking Multiple Complex Signals of Persistence and Change

We foresee a future where spatiotemporal vegetation monitoring platforms, geospatial cloud computing, and panarchy converge to further advance our abilities to visualize complex system dynamics that gave rise to the panarchy concept as a useful heuristic. Regime shifts can occur at one or more of the hierarchical, discontinuous scales of a panarchy, with another regime simply replacing the previous regime at the same scale. These regime shifts are readily detected and tracked in space and time (figure 7.3; Uden et al. 2019). But regime shifts can also emerge from the novelty churned out between discontinuous scale breaks (figure 7.5; Allen and Holling 2010; Allen et al. 2014; Sundstrom et al. 2018). These emergent regimes represent both an opportunity and a challenge to the future of mapping panarchy and understanding environmental change.

Higher-resolution vegetation data would allow better detection of the emergence of regime shifts at small scales. After the emergence of regime shift signals, continued tracking of their movement within and across scales could reveal characteristic temporal signatures associated with specific forms of transitions (e.g., scaling up of trees vs. invasive annual grasses). In addition to detecting emergence at small scales, continued screening for change at additional scales—from properties, to ecoregions, to biomes—can be used to identify locations experiencing regime shifts at broad but not local scales and to prioritize management actions for them accordingly. Collectively, such advances can allow even more proactive screening and earlier warning of undesirable shifts in vegetation regimes, before the onset of signs and symptoms.

However, sifting through every possible scale to search for emerging regime shifts is impractical when the data grain and extent range from a baseball diamond to a continent and can lead to data dredging pitfalls (Zuur et al. 2010). One approach to tackling this challenge is to analytically identify functional spatial scales in systems over time (Dray et al. 2012), and many scale detection techniques (e.g., discontinuity analysis, distance-based Moran's eigenvector mapping, wavelet analysis) exist (Angeler et al. 2015, 2016; James et al. 2010). Monitoring changes in the number or characteristics of functional scales would indicate an emerging regime shift (Allen et al. 2014), and mapping could be used to pinpoint the location and extent of such change. Tracking the emergence of regime shifts in this way would also bring the screening of regime shifts and mapping of panarchy into the realm of machine learning, where changes in the number and characteristics of functional scales could be inputs into an algorithm that continuously tracks a panarchy in space over time.

Mapping spatial transitions, as described here, and mapping spatial regimes (chapter 5) highlight the frontier represented broadly under the concept of spatial resilience (Cumming 2011; Allen et al. 2016). Spatial resilience considers explicitly the spatial attributes that confer resilience, including the potential for cross-scale dynamic behavior to occur across an ecological hierarchy. Many maps and mapping schemes reflect the hierarchical order of nature, and attention is increasing on the need to depict hierarchical structures as complex and dynamic at any range of scales—consistent with panarchy theory. As tools and technologies continue to develop rapidly, more objective mapping of not only ecological systems but complex social-ecological systems is making it easier to visualize complex, dynamic behavior across scales. Combining such maps with approaches that map vulnerability to regime shifts (Uden et al. 2019) holds enormous promise for the application of resilience and panarchy theory.

Literature Cited

Allen, C.R., D.G. Angeler, B.C. Chaffin, D. Twidwell, and A.S. Garmestani. 2019. Resilience reconciled. *Nature Sustainability* 2: 898–900.

Allen, C.R., D.G. Angeler, C. Folke, D. Twidwell, D. Uden, and G. Cumming. 2016. Quantifying spatial resilience. *Journal of Applied Ecology* 53: 625–635.

Allen, C.R., D.G. Angeler, A.S. Garmestani, L.H. Gunderson, and C.S. Holling. 2014. Panarchy: Theory and application. *Ecosystems* 17: 578–589.

Allen, T.F., and T.W. Hoekstra. 1990. The confusion between scale-defined levels and conventional levels of organization in ecology. *Journal of Vegetation Science* 1: 5–12.

Allen, C.R., and C.S. Holling 2010. Novelty, adaptive capacity, and resilience. *Ecology and Society* 15: 24. http://www.ecologyandsociety.org/vol15/iss3/art24/

Angeler, D.G., C.R. Allen, C. Barichievy, T. Eason, A.S. Garmestani, N.A.J. Graham, D. Granholm, L. Gunderson, M. Knutson, K.L. Nash, R.J. Nelson, M. Nyström, T. Spanbauer, C.A. Stow, and S.M. Sundstrom. 2016. Management applications of discontinuity theory. *Journal of Applied Ecology* 53: 688–698.

Angeler, D.G., C.R. Allen, D.R. Uden, and R.K. Johnson. 2015. Spatial patterns and functional redundancies in a changing boreal lake landscape. *Ecosystems* 18: 889–902.

Arterburn, J.R., D. Twidwell, W.H. Schacht, C.L. Wonkka, and D.A. Wedin. 2018. Resilience of Sandhills grassland to wildfire during drought. *Rangeland Ecology & Management* 71: 53–57.

Birge, H., K. Pope, A. Garmestani, and C.R. Allen. 2016. Adaptive management for ecosystem services. *Journal of Environmental Management* 183: 343–352.

Briske, D.D., S.D. Fuhlendorf, and F.E. Smeins. 2005. State-and-transition models, thresholds, and rangeland health: A synthesis of ecological concepts and perspectives. *Rangeland Ecology and Management* 58: 1–10.

Cumming, G.S. 2011. Spatial resilience: Integrating landscape ecology, resilience, and sustainability. *Landscape Ecology* 26: 899–909.

Donovan, V.M., J.L. Burnett, C.H. Bielski, H.E. Birgé, R. Bevans, D. Twidwell, and C.R. Allen. 2018. Social–ecological landscape patterns predict woody encroachment from native tree plantings in a temperate grassland. *Ecology and Evolution* 8: 9624–9632.

Donovan, V.M., C.P. Roberts, C.L. Wonkka, D.A. Wedin, and D. Twidwell. 2019. Ponderosa pine regeneration, wildland fuels management, and habitat conservation: Identifying trade-offs following wildfire. *Forests* 10: 286.

Donovan, V.M., D. Twidwell, D.R. Uden, T. Tadesse, B.D. Wardlow, C.H. Bielski, M.O. Jones, B.W. Allred, D.E. Naugle, and C.R. Allen. 2020. Resilience to large, "catastrophic" wildfires in North America's grassland biome. *Earth's Future* 8: e2020EF001487.

Dray, S., R. Pélissier, P. Couteron, M.J. Fortin, P. Legendre, P.R. Peres-Neto, E. Bellier, R. Bivand, F.G. Blanchet, M. De Cáceres, A.-B. Dufour, E. Heegaard, T. Jombart, F. Munoz, J. Oksanen, J. Thioulouse, and H. H. Wagner. 2012. Community ecology in the age of multivariate multiscale spatial analysis. *Ecological Monographs* 82: 257–275.

Engle, D.M., B.R. Coppedge, and S.D. Fuhlendorf. 2008. From the dust bowl to the green glacier: Human activity and environmental change in Great Plains grasslands. In *Western North American Juniperus Communities*, 253–271. Springer, New York.

Fath, B.D., C.A. Dean, and H. Katzmair. 2015. Navigating the adaptive cycle: An approach to managing the resilience of social systems. *Ecology and Society* 20(2): 24. http://dx.doi.org/10.5751/ES-07467-200224

Folke, C., S.R. Carpenter, B. Walker, M. Scheffer, T. Chapin, and J. Rockström. 2010. Resilience thinking: Integrating resilience, adaptability, and transformability. *Ecology and Society* 15(4): 20. http://www.ecologyandsociety.org/vol15/iss4/art20/

Frasinski, L.J., K. Codling, and P.A. Hatherly. 1989. Covariance mapping: A correlation method applied to multiphoton multiple ionization. *Science* 246: 1029–1031.

Garmestani, A.S., C.R. Allen, and L. Gunderson. 2009. Panarchy: Discontinuities reveal similarities in the dynamic system structure of ecological and social systems. *Ecology and Society* 14(1): 15. http://www.ecologyandsociety.org/vol14/iss1/art15/

Gorelick, N., M. Hancher, M. Dixon, S. Illyushchenko, D. Thau, and R. Moore. 2017. Google Earth Engine: Planetary-scale geospatial analysis for everyone. *Remote Sensing of Environment* 202: 18–27.

Gunderson, L.H. 2000. Ecological resilience—in theory and application. *Annual Review of Ecology and Systematics* 31: 425–439.

Gunderson, L.H., and C.S. Holling. 2002. *Panarchy: Understanding Transformations in Human and Natural Systems.* Island Press, Washington, DC.

Holling, C.S. 2004. From complex regions to complex worlds. *Ecology and Society* 9(1): 11. http://www.ecologyandsociety.org/vol9/iss1/art11/

Houlahan, J.E., D.J. Currie, K. Cottenie, G.S. Cumming, C.S. Findlay, S.D. Fuhlendorf, P. Legendre, E.H. Mudavin, D. Noble, R. Russell, R.D. Stevens, T.J. Willis, and S.M. Wondzell. 2018. Negative relationships between species richness and temporal variability are common but weak in natural systems. *Ecology* 99: 2592–2604.

James, P.M., R.A. Fleming, and M.J. Fortin. 2010. Identifying significant scale-specific spatial boundaries using wavelets and null models: Spruce budworm defoliation in Ontario, Canada as a case study. *Landscape Ecology* 25: 873–887.

Jones, M.O., B.W. Allred, D.E. Naugle, J.D. Maestas, P. Donnelly, L.J. Metz, J. Karl, R. Smith, B. Bestelmeyer, C. Boyd, J.D. Kerby, and J.D. McIver. 2018. Innovation in rangeland monitoring: Annual, 30 m, plant functional type percent cover maps for US rangelands, 1984–2017. *Ecosphere* 9: e02430.

Le, N.D., and J.V. Zidek. 2006. *Statistical Analysis of Environmental Space–Time Processes*. Springer, New York.

Levin, S.A. 1992. The problem of pattern and scale in ecology: The Robert H. MacArthur award lecture. *Ecology* 73: 1943–1967.

Roberts, C.P., C.R. Allen, D.G. Angeler, and D. Twidwell. 2019. Shifting avian spatial regimes in a changing climate. *Nature Climate Change* 9: 562–566.

Roberts, C.P., V.M. Donovan, S.M. Nodskov, E.B. Keele, C.R. Allen, D.A. Wedin, and D. Twidwell, 2020. Fire legacies, heterogeneity, and the importance of mixed-severity fire in ponderosa pine savannas. *Forest Ecology and Management* 459: 117853.

Roberts, C.P., D. Twidwell, J.L. Burnett, V.M. Donovan, C.L. Wonkka, C.L. Bielski, A.S. Garmestani, D.G. Angeler, T. Eason, B.W. Allred, and M.O. Jones. 2018. Early warnings for state transitions. *Rangeland Ecology and Management* 71: 659–670.

Schimmel, J., and A. Granstrom. (1996). Fire severity and vegetation response in the boreal Swedish forest. *Ecology* 77: 1436–1450.

Stevens, N., C.E. Lehmann, B.P. Murphy, and G. Durigan. 2017. Savanna woody encroachment is widespread across three continents. *Global Change Biology* 23: 235–244.

Sundstrom, S.M., D.G. Angeler, C. Barichievy, T. Eason, A.S. Garmestani, L.H. Gunderson, M. Knutson, K.L. Nash, T. Spanbauer, C. Stow, and C.R. Allen. 2018. The distribution and role of functional abundance in cross-scale resilience. *Ecology* 99: 2421–2432.

Twidwell, D., B.W. Allred, and S.D. Fuhlendorf. 2013a. National-scale assessment of ecological content in the world's largest land management framework. *Ecosphere* 4: 1–27.

Twidwell, D., C.H. Bielski, R. Scholtz, and S.D. Fuhlendorf. 2020. Advancing fire ecology in 21st century rangelands. *Rangeland Ecology and Management*. https://doi.org/10.1016/j.rama.2020.01.008

Twidwell, D., S.D. Fuhlendorf, D.M. Engle, and C.A. Taylor Jr. 2009. Surface fuel sampling strategies: Linking fuel measurements and fire effects. *Rangeland Ecology and Management 62:* 223–229.

Twidwell, D., S.D. Fuhlendorf, C.A. Taylor Jr., and W.E. Rogers. 2013b. Refining thresholds in coupled fire–vegetation models to improve management of encroaching woody plants in grasslands. *Journal of Applied Ecology* 50: 603–613.

Twidwell, D., W.E. Rogers, S.D. Fuhlendorf, C.L. Wonkka, D.M. Engle, J.R. Weir, U.P. Kreuter, and C.A. Taylor Jr. 2013c. The rising Great Plains fire campaign: Citizens' response to woody plant encroachment. *Frontiers in Ecology and the Environment* 11: e64–e71.

Uden, D.R., D. Twidwell, C.R. Allen, M.O. Jones, D.E. Naugle, J.D. Maestas, and B.W. Allred. 2019. Spatial imaging and screening for regime shifts. *Frontiers in Ecology and Evolution* 7: 407.

Van Leeuwen, W.J., G.M. Casady, D.G. Neary, S. Bautista, J.A. Alloza, Y. Carmel, L. Wittenberg, D. Malkinson, and B.J. Orr. 2010. Monitoring post-wildfire vegetation response with remotely sensed time-series data in Spain, USA and Israel. *International Journal of Wildland Fire* 19: 75–93.

Wagner, H. 2003. Spatial covariance in plant communities: Integrating ordination, geostatistics, and variance testing. *Ecology* 84: 1045–1057.

Walker, B.H., and N.A. Abel. 2002. Resilient rangelands: Adaptation in complex systems. In *Panarchy: Understanding Transformations in Human and Natural Systems*, ed. L. Gunderson and C.S. Holling, 293–314. Island Press, Washington, DC.

Wright, H.A. 1979. *Fire Ecology and Prescribed Burning in the Great Plains: A Research Review*. U.S. Forest Service Technical Report, Intermountain Forest and Range Experiment Station.

Wu, J., and O.L. Loucks. 1995. From balance of nature to hierarchical patch dynamics: A paradigm shift in ecology. *Quarterly Review of Biology* 70: 439–466.

Zuur, A.F., E.N. Ieno, and C.S. Elphick. 2010. A protocol for data exploration to avoid common statistical problems. *Methods in Ecology and Evolution* 1: 3–14.

PART III

DIFFUSION OF PANARCHY CONCEPTS

CHAPTER 8

Capacities for Navigating Large-Scale Sustainability Transformations: Exploring the Revolt and Remembrance Mechanisms for Shaping Collapse and Renewal in Social-Ecological Systems

Per Olsson, Carl Folke, and Michele-Lee Moore

The combination of the Decade for Action declared by the United Nations to rapidly achieve the Sustainable Development Goals with the current COVID-19 crisis has led to urgent calls for transformations that can accelerate change toward a more equitable and sustainable world. However, the demand for these kinds of transformations requires a focus on understanding large-scale systemic change, that is, changes at the social-ecological scales that can match the Anthropocene challenges (Olsson et al. 2017). The various approaches in the field of transformations to sustainability hold promise for analyzing and navigating these changes (Leach et al. 2010; Olsson et al. 2014; Loorbach et al. 2017; Patterson et al. 2017; Blythe et al. 2018; Köhler et al. 2019; Leichenko and O'Brien 2019). However, much of our understanding of transformations is derived from studies of local or small-scale transformations (Olsson et al. 2017). Fewer studies focus on alternative approaches that can have impact at larger scales (Waddell et al. 2015; Olsson et al. 2017).

Panarchy theory provides a conceptual basis for cross-scale interactions rather than one that assumes linearity and scalability (Gunderson and Holling 2002). This is particularly important for understanding system change and how new ideas, initiatives, and movements can have an impact beyond the local and national levels and influence larger scales (Olsson et al. 2017). Several disciplines have a long history of studying and exploring such interactions. For example, the literature on transition theory focuses on niche–regime interactions and how new, more sustainable configurations of sociotechnological systems emerge and take hold (Rotmans et al. 2001; Smith 2007; Geels 2010; Loorbach 2010; Avelino and Wittmayer 2016; Frantzeskaki et al. 2017;

Mylan et al. 2019). Similarly, resilience scholars have used panarchy to generate insights on cross-scale interactions and systems reconfiguration as part of transformations in social-ecological systems (Gunderson and Holling 2002; Garmestani and Benson 2013; Allen et al. 2014; Chaffin et al. 2016). This chapter builds on this work to further explore these interactions for large-scale transformative change.

We begin with a definition of transformations in the context of panarchy. We then use panarchy thinking to develop a model that can help increase the understanding of complex interactions across scales and generate new insight on transformations toward sustainable and equitable futures. In the same way as for the panarchy model, the concepts of revolt and remembrance are central to the model we develop and present and are used to explore cross-scale interactions. We apply the model to transformations in natural resource governance regimes in Chile, Uzbekistan, and South Africa. Finally, we acknowledge the need to go beyond niche and regime levels for understanding the large-scale transformations that are necessary for successfully addressing Anthropocene risks and challenges at the global level. Therefore, we use the panarchy model to explore the case of the emergence of neoliberalism and a market-based economy and the shift to a new global economic system. We use the insights from this case to discuss the implications for people–planet futures and conclude by discussing how these insights may elucidate how transformations are telecoupled across the world and how to increase the speed, magnitude, and direction of transformations to sustainability that are urgently needed to deal with the Anthropocene challenges and risks.

Transformations and Panarchy

Resilience theory defines transformation as a specific type of change, distinct from the types of change that support persistence or adaptation (Walker et al. 2004; Folke et al. 2010). Transformation entails changing the system conditions that created the problems in the first place (Westley et al. 2006) and shifting from one basin of attraction to another one. Transformation is the shifting of a system through human agency from one regime to a new regime characterized by alternative sets of processes and structures (Olsson et al. 2010; Chaffin et al. 2016). This means fundamentally changing self-reinforcing feedbacks that keep the system on a certain trajectory, and in the context of resilience and sustainability, this involves reconfigurations of the social-ecological system (Olsson et al. 2008; Chapin et al. 2010). Transformations toward sustainable and equitable futures are multilevel and multiphase processes (Loorbach et al. 2008; Olsson et al. 2010). Moore and Milkoreit (2020, 4) defined transformations as "shifts in the way authority, power, and resources are structured and flow in a particular social system, the practices and processes that reflect and reproduce those structures, the norms, values, and beliefs that underpin those structures and processes, and the way that all of these are connected to ecological systems across multiple scales." This means that transformations entail "a change that restructures, reconnects, and remakes the meanings of relationships between people and between people and the ecosystems in which they are embedded" (Moore and Milkoreit 2020, 4).

Panarchy theory addresses cross-scale dynamics of transformation from at least two perspectives, as described in the following paragraphs. First, panarchy theory highlights that transformational changes do not just occur simultaneously, nor in some synchronous, sequential fashion across scales. Social-ecological systems are dynamic, changing through adaptive cycles, represented by the now widely recognized panarchy heuristic (figure 8.1; Gunderson and Holling 2002). Scholarship from other fields has contributed to our understanding of dynamics at particular scales. Schot and Geels (2008) and Smith and Raven (2012) described niches as protected, identifiable spaces within a system where novelty can be introduced and experimental approaches can be nurtured. Such niches can emerge, develop, and collapse for any number of reasons, such as having insufficient resources or the loss of key champions that are important for upholding and developing the niche (Olsson et al. 2007; Westley et al. 2013; Moore et al. 2018). Similarly, governance regimes, institutions, and policies can go through cycles of abrupt change (Herrfahrdt-Pähle et al. 2020). Repetto (2006) applied the notion of punctuated equilibrium to describe policy changes over time. Kingdon (1984) found that problems, policies, and politics interact to create a window of opportunity, which is a prerequisite for institutional change. Although often considered as a set of slower variables, values and norms can also go through cycles of change (Nyborg et al. 2016). But these dynamics alone are insufficient for understanding transformation.

The second aspect of change that panarchy theory highlights describes how dynamic levels and scales present in a given panarchy interact and shape one another. Gunderson and Holling (2002) portrayed panarchy with smaller levels nested within large ones. The smaller levels were understood to contain fast dynamics, with rapid activities such as invention, experimentation, and testing occurring, whereas the larger levels are depicted as slower, stabilizing,

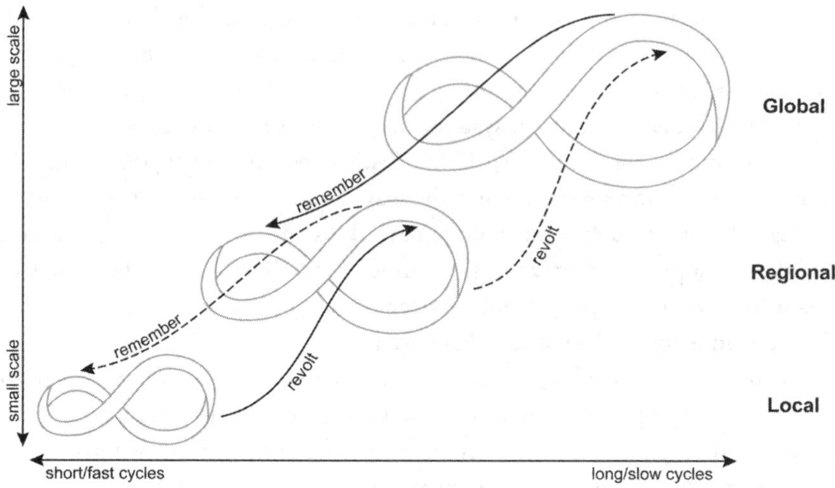

Figure 8.1. Panarchy of interconnected adaptive cycles (local, regional, and global) at different spatial and temporal scales.

and an accumulation of memory of system dynamics. Holling et al. (2002, 75) explained, "There are potentially multiple connections between phases of the adaptive cycle at one level and phases at another level." Two specific types of connection are referred to as "revolt" and "remember." We argue that it is these revolt and remembrance connections across nested, dynamic scales that can illuminate stronger understandings of the cross-scale interactive aspects of transformations.

Remembrance provides context and experience for reorganization after collapse or periods of instability. Such processes have been described as ecological memory (Nyström and Folke 2001; Allen et al. 2016), social or institutional memory (Folke et al. 2003), and social-ecological memory (Barthel et al. 2010; Whyte 2018). The significance of the role of memory in the back loop of the adaptive cycle for renewal after disturbance and release has been emphasized (Berkes and Folke 2002) as well as the recombination of diverse social-ecological memories for novelty and social innovation (Gunderson and Holling 2002).

However, remembrance or system memory can also be a hindrance to disruptive change and renewal, if it increases system inertia and blocks transformative change. For example, institutional structures that are embedded in certain values can perpetuate unsustainable and inequitable outcomes (Moore et al. 2014; Scoones et al. 2020). These patterns can be perpetuated, even if the system is disturbed, because of strong self-reinforcing feedbacks that are difficult to break. Revolts can arise from interactions between radical niches, each of which holds different sets of values and positions that actively challenge, break down, and lead to changes in the status quo. But the mere existence of radical niches does not always lead to disruptive or systemic, transformative change. Such niches can be eradicated or coopted by the status quo system if the remembrance mechanisms are too strong (Herrfahrdt-Pähle et al. 2020).

In contrast, revolt may cause undesired systemic change when there is too little or no remembrance, which occurs when the social-ecological system is too open or too weak to deal with the revolt, be it an extreme event, a pandemic, or an exploding market. But generally, revolt seldom takes place in a vacuum, and remembrance can be navigated to nurture the conditions that will move a revolt up the panarchy and cause a transformation (Olsson et al. 2014). This gets at the core of resilience thinking, how periods of rapid change interplay with periods of slow change, that is, the complex dynamics of the interplay of fast and slow variables (Folke et al. 2010). In other words, the dynamics between revolt and remembrance of a panarchy are central for understanding and guiding transformative change.

Furthermore, we propose that the new conditions of the Anthropocene of widespread connectivity and fast speed at larger levels of the panarchy also provide a new context for understanding the dynamics of revolt and remembrance across levels and scales. The current situation is therefore not as hierarchical as portrayed in the original panarchy model, and the large scale might not be slow. More specifically, there may be situations where the revolt is caused by rapid fluxes of information at larger levels causing cascading effects down to smaller levels and altering deeper and slower remembrance in local places. We interpret such situations as inverted panarchies of revolt and remembrance dynamics.

We propose that transformation involves not just a single panarchy but multiple panarchies, which we refer to here as parallel panarchies. Parallel structures appear in the work of Avelino (2017), who used transitions theory to explore power dynamics in multilevel systems. Avelino described a set of niches at the micro level (shown as white squares in figure 8.2) that are embedded in, fit with, and support the operation of the current regime and values that are the basis for that regime. Other niches (black squares in figure 8.2) oppose and challenge the status quo of the extant dominant system and outline a radically different set of practices, rules, and roles based on different values and would require other institutional regimes (Avelino 2017). Both

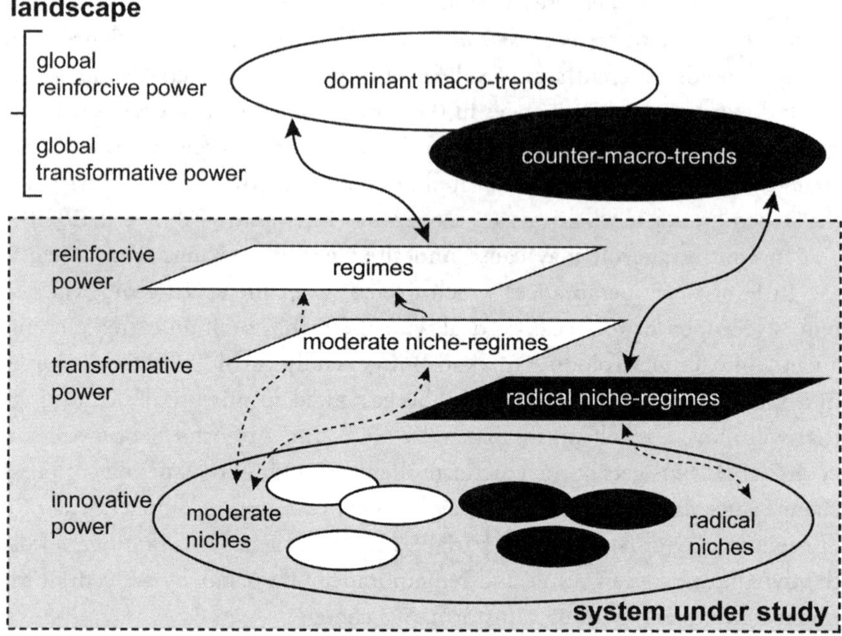

Figure 8.2. Power transition framework indicating connections between larger-scale system (top) and smaller system (system under study). The black and white cross-scale interactions can be viewed as parallel panarchies. The white one is the dominating, status quo system and the black the emerging parallel one. Arrows represent synergetic power dynamics, including exchanges between innovative, transformative, and reinforcing powers. Reproduced with permission © John Wiley and Sons.

niche groups can exist at the same time, in parallel, and can play key parts in transformations.

Building on Avelino's work (2017), we propose that transformations do not merely involve the collapse of one panarchical system replaced by a new system. Instead, the radically new and the dominant old systems coexist for a period of time, and it is the complex, dynamic interactions between these that will shape the transformation (Moore 2017; Geels 2018). This proposal challenges the notion that transformations occur in a linear fashion; that is, an innovation must be developed within a niche, and then that must be transferred at higher scales to affect a larger number of people. Evidence shows this is seldom the case for transformations (Moore et al. 2015; Chaffin et al. 2016; Preiser et al. 2018). Nor is the reverse necessarily true; that is, a higher scale sets a certain policy or law, which sets in motion a transformation that spreads across other scales. Instead, evidence shows how dynamics across scales, both vertically and horizontally, are continuously shaping the transformation process in a complex interplay (Geels 2018; Garmestani et al. 2020; Herrfahrdt-Pähle et al. 2020).

In the context of panarchy theory, we therefore propose that it is useful to think in coexisting, parallel panarchies exhibiting complex dynamics of the fast and the slow and where the revolt and remembrance mechanisms operate in new ways. In the following section, we build on the work by Herrfahrdt-Pähle et al. (2020) to focus on, and exemplify, how radical niches can lead to disruptive, large-scale, systemic change and the establishment of new dynamic, multilevel systems or panarchies.

Case Studies

Three case studies are presented to demonstrate how panarchies change over time: fishery governance in Chile and water governance in South Africa and Uzbekistan. Through these examples of systemic transformation we will illustrate the interplay of revolt and remembrance in parallel panarchies.

Chile

During the Pinochet dictatorial regime in the 1970s, Chile adopted an open-access fishery governance regime as part of the neoliberal and market-based economic system that was emerging globally (Gelcich et al. 2010). The open-access fisheries, with few rules and regulations, led to overexploitation of marine resources such as loco (*Concholepas concholepas*) stocks. Because the loco functions as a keystone species, population declines led to ecological collapse and regime (Gelcich et al. 2010).

The decline in loco stocks and degradation of coastal ecosystems triggered collaboration between artisanal fishers and scientists to start developing alternative approaches to manage and govern fisheries in three different coastal areas of Chile (Castilla 1994). These local experiments constituted radical niches or new approaches for managing and governing marine resources and configurations of the social-ecological system. These practices were the basis of subsequent approaches to fisheries described as adaptive co-management and governance (Gelcich et al. 2010).

Pinochet's regime began to crumble in the late 1980s, and through elections a new government was established in 1989. The new political regime wanted to stop overexploitation of marine resources and saw the need to replace the open-access governance with new ways of governing natural resources. Knowledge from the collaborative experiments played a key role in formulating the new fishery policies (Gelcich et al. 2010). A national confederation of fishers that had been dormant for seventeen years during the Pinochet era became a bridging organization that was central in moving the experience from local experiments to the national policy process. This network provided the political leverage that was needed to move through the back loop of the adaptive cycle, through the reorganization phase, and create the new legislative system, including the Fishery and Aquaculture Law (FAL) of 1991. The new practices that had been developed in the informal collaborations and experiments at the niche level needed rules and regulations at the regime level that supported such practices, and the FAL was the first step in that process (Gelcich et al. 2010). The FAL paved the way for an inclusive negotiating process to create a new management system, where marine resources were managed through Territorial User Rights for Fisheries (TURF), which was operationalized in 1997.

A revolt mechanism of the panarchy in the Chile case was the capacity of the reemerging network to carry knowledge from the micro level to the meso level. The values changed from one of extraction that supported the top-down, open-access regime to one of stewardship with values that supported a more bottom-up, ecosystem-based regime, in turn supported by a democratic system.

Importantly, although the FAL and TURF was decided on quite early after the political regime shift, it took seven years of debates between fishers, managers, and politicians to negotiate how marine tenure was going to be allocated to fishers (Gelcich et al. 2010). During these years, through a process of institutional bricolage, key individuals in the Undersecretary of Fishery, jointly with the national confederation of fishers and support of the marine scientific community, developed the TURF policy and eventually navigated the transition (Meltzoff et al., 2002). Furthermore, the shift within the coastal fishery governance panarchy was significantly drawing on changes and experiences happening outside the coastal fisheries, in other panarchies. This is reflected in the political and value changes as well as inspiration for how to organize collective action from other sectors and countries brought by the scientists into the coastal fisheries (Gelcich et al 2010).

The process of navigating the scaling of the new approach reflects how a revolt locally changes with its emergence across levels of a panarchy. This dance between revolt and remembrance within the panarchy as well as the dynamic interactions with parallel panarchies in Chile and elsewhere together generated the transformation of the coastal fishery system of Chile.

South Africa

During the apartheid era in South Africa, water governance was directed by the Water Act of 1956 and therefore was embedded in a value system of white supremacy and the oppression of non-Whites (Herrfahrdt-Pähle and Pahl-Wostl 2012). The highly centralized and technocratic water governance regime served the economic interest of the White minority. Institutions, including those enforcing land rights, led to unequal access to water and practices that favored White farmers and miners and excluded non-Whites. This governance system also led to water scarcity and pollution. The unequal access to safe drinking water supply and sanitation resulted in poor health in Black communities, particularly in rural areas (Eales 2011; van Koppen and Schreiner 2014).

These environmental and social problems related to freshwater increased in the 1970s and triggered the status quo regime to find solutions to deal with them through the Commission of Enquiry into Water Matters (Herrfahrdt-Pähle and Pahl-Wostl 2012). More specifically, it triggered a response among concerned water managers and scientists in the 1980s. They formed an informal collaboration or shadow network to develop a new vision for water governance, discuss a new law that supported this vision, and start experimenting with alternative approaches based on the ideas of integrated water resource management (IWRM).

The global IWRM discourse (Conca 2005) provided the remembrance mechanism for framing and guiding the experiments. The discourse together with the work of the shadow networks was the start of an emerging parallel panarchy. A readiness was developed and through a revolt mechanism eventually resulted in a new panarchy, representing a new water governance system. This means that ideas of innovative governance arrangements from the shadow network were readily available and introduced into the negotiation forum on water, the Standing Committee on Water Supply and Sanitation (SCOWSAS), which served as a learning platform, influencing and shaping the adoption of the new regime (Muller 2014; Herrfahrdt-Pähle et al. 2020). SCOWSAS represented members from the democratic movement, central, regional, and local government, research organizations, water industry, development funding organizations, and nongovernment service organizations (Muller 2014).

The South Africa case is similar to the Chile case in that increasing environmental and social problems triggered collaboration and experimentation. But it differs in that the actors in South Africa were the same that held authority and power in the status quo regime. This created a fairly tight interaction between the status quo system and the associated panarchy and the emerging parallel panarchy.

The decision to transform from the apartheid system to a democratic system, along with the search for new water governance regimes, provided the opportunity to include the IWRM ideas in the new constitution that was written in the early 1990s and combine it with the focus on human rights that had created the dismantling of the apartheid system (Herrfahrdt-Pähle and Pahl-Wostl 2012). However, the shift of the governance system also meant institutional turbulence when creating new institutional structures and processes. These included local governments as a new administrative level and land reforms to which the new water legislation had to be linked and matched. The water sector also experienced several changes of ministers and a high turnover of staff working at different levels (Herrfahrdt-Pähle and Pahl-Wostl 2012). On the positive side, the change of people also weakened the remembrance mechanism of the old water governance system of the apartheid regime. On the negative side, the release phase and the revolt mechanism created turbulence in the governance system that slowed down the institutionalization of the new approach.

Uzbekistan

In Uzbekistan, a Soviet Socialist Republic during the Soviet Union era, the water governance regime was a highly centralized, technocratic system embedded in Marxist–Leninist ideology and a command economy (Wegerich 2005). This resulted in industrialized and large-scale irrigation agriculture practices that in the 1960s became specialized on cotton production. This had a negative impact on the water resources, which resulted in a number of environmental problems. The most well-known case is the Aral Sea, where cotton production created a shortage of water, an ecological crisis, and the collapse of the fisheries in the 1980s (Micklin 2007). The overuse of water and the shrinking of the Aral Sea resulted in soil salinization and other environmental problems.

The collapse of the Soviet Union provided an opportunity to transform the water governance regime that was producing unsustainable and unjust outcomes, but this did not happen (Schlüter and Herrfahrdt-Pähle 2011).

Unlike what happened in South Africa and Chile, the early signs of a looming crisis did not trigger groups of actors to start experimenting with alternative approaches. This means that the Soviet Union example lacked the revolt mechanism that radical niches can provide.

Emerging in the wake of the Soviet Union collapse was a very similar system to the precollapse one with strong top-down control that obstructed the establishment of new formal institutions (Jones Luong 2003; Pahl-Wostl 2015). More specifically, obstructions included informal institutions that enabled corruption, including nepotism and patronage. From a remembrance point of view the regime was not weakened enough and did not seem to go through a release phase, although the larger Soviet Union did. Similar to South Africa, the IWRM idea was also introduced to the Uzbek context, in this case by donors. But there were no radical niche experiments that it could align with, and instead the dominant system coopted the idea of IWRM into the dominant discourse of top-down governance.

In all three cases it was not the initiatives and experiments around water and fisheries themselves that caused the status quo to change (Herrfahrdt-Pähle et al. 2020). There were other forces in play that destabilized the dominant systems and associated panarchies. However, in Chile and South Africa the radical niches challenged values and socioeconomic paradigms at the macro level, rules and regulations at the meso level, and practices and behaviors at the micro level and provided novelty for the reorganization phase and contributed to the development of new governance systems.

The sociopolitical change in Uzbekistan and the collapse of the Soviet Union provided opportunity for transformative reconfigurations (Herrfahrt-Pähle et al. 2020), but there was no apparent parallel panarchy emerging with new practices and associated values. Thus, the remembrance dynamics of the existing system were strong enough that changes were adaptations or maladaptations. This confirms the fact that a system can go through a release phase at the macro or landscape level without resulting in agency-guided transformations to more desired states (Gunderson and Holling 2002). It seems as if the weakening of the remembrance mechanism is important at this stage for the system to enter a new configuration with new feedbacks through transformative change.

These insights also point to the role of shadow networks that operate at the margin of the dominant systems and give direction to system transformation (Olsson et al. 2006). As shown in the South Africa and Chile cases, these networks operate in the interface between and connect different panarchies at multiple levels. For example, the informal collaborations in South Africa and the fishing network in Chile helped connect the regime and niche levels and with macro trends such as the global discourse on IWRM. The revolt mechanism was supported by agency of the shadow networks in the form of strategies to navigate the disruptive impact of the radical niche. The shadow networks were also important in creating solutions where new and old components of the system were recombined with some elements of novelty, a process called bricolage. If navigated carefully this process can result in transformative change (Chapin et al. 2012; Olsson et al. 2017), but there is always a risk in such processes that new ideas become coopted by the dominant status quo system.

Transformative Capacities for Large-Scale System Change

The three preceding examples of revolt and remembrance for transformations all occur at a national level. But in the context of Anthropocene risks and challenges (Keys et al. 2019) we still need to understand transformative capacities to achieve transformations with global impact, beyond the level of communities and nations. Carl Polanyi's (1944) work on transformation to neoliberalism and a market-based economy is useful for considering the type of scalar challenge that confronts us in the Anthropocene. Evidence shows that neoliberal ideals and approaches have been taken up by so many that they have become a fundamental part of nearly everyone's daily lives, whether people are aware of it or not (McCarthy and Prudham 2004). As Eakin et al. (2017) demonstrated, even a single cup of coffee is tied into all kinds of global connections that are not necessarily visible to the person drinking it, all hinged on trade in a market-based economy. As a consequence, that cup of coffee also holds links to the embedded inequality and unsustainable patterns of human actions that reflect the downsides of the rapid expansion of the global economy over a few decades. At the moment, the alternative systems that could help us move toward equitable and sustainable pathways in tune with the Earth system are operating only at the margins. Concerned people need to make an effort and be creative in order to even support or find access to participate in such alternatives; that is, it is not nearly as easy or embedded in system dynamics as the dominant neoliberal system. Overall, society seems engaged in multifaceted experimentation. Maintaining such experimentation may help inspire novel pathways to desirable futures, but there is a risk of societies becoming trapped in backward-looking narratives that threaten long-term sustainable outcomes due to remembrance dynamics (Carpenter et al. 2019).

In the next section we apply the parallel panarchy model to the case of the emergence of neoliberalism and the market-based economy as a macro trend and its interactions with new practices for derivatives and option pricing (Geobey 2017; Olsson 2017; Westley et al. 2017). It provides an example that can help us understand the cross-scale interactions between practices, rules and regulations, values, and cultures that may be key to large-scale, transformative changes in the Anthropocene (Westley et al 2011).

Although the Keynesian economic paradigm was widely popular after World War I and into the 1950s and 1960s, a growing tension emerged that this particular model put nations on a path to becoming totalitarian states, which infringed on people's freedom. Because of this growing fear, and a growing concern that the Bretton Woods international monetary systems did not provide sufficient economic stability, a revolt began to appear (Geobey 2017). The initial development of option pricing and the derivatives market was tightly linked to a larger movement toward a neoclassical economic paradigm.

The economic instability and discomfort with the dominant economic pathway led early market proponents such as Friedrich Hayek and his predecessor and mentor, Ludwig von Mises, to start publicly challenging the Keynesian paradigm (Olsson 2017). For example, in the 1930s Hayek participated in a series of debates with Keynes about macroeconomics at several locales including the London School of Economics and Cambridge University. A niche began to form around this countercurrent discourse.

Hayek, with prominent thinkers in the field such as Frank Knight, Karl Popper, Ludwig von Mises, George Stigler, and Milton Friedman, later convened the Mont Pelerin Society meetings beginning in 1947. The idea behind the Mont Pelerin meetings was to organize and mobilize a diverse and fragmented set of pro-market thinkers in order to identify common interests and agree on some basic liberal principles (Mirowski and Plehwe 2009). The Mont Pelerin Society, the Chicago School of Economics with Milton Friedman, and other new economic departments around the world all provided platforms for developing a new economic paradigm. This included research agendas and think tanks that contributed to a shift into a new economic paradigm (Geobey 2017). This work in combination with the failure of the Bretton Woods system to provide economic stability opened up opportunities for new ideas and paved the way for economists to develop models and formulas such as the Black–Scholes–Merton model. This mathematical model helped strengthen derivatives, including option pricing, as an economic instrument and supported the emergence of a market-liberal discourse as well (Geobey 2017).

One of the reasons for the early success and rapid spread of this model was that the Chicago Board Options Exchange (where the model was first introduced in 1973) and Chicago's International Monetary Market provided markets that allowed derivative exchanges. The Black–Scholes–Merton model was also helped by technological advances in computing power, which provided the capacity to carry out complex calculations to make the model work (Olsson 2017).

Hence, the relationship between the Black–Scholes–Merton model and the neoliberal paradigm was a symbiotic one. There was a growing demand for risk mitigation options in order to create stability in an uncertain and changing economic environment (Geobey et al. 2017). The eventual transformation that unfolded did not emerge from a linear set of cause–effect factors. Instead, the new platform and paradigm were the result of distributed agency working across different scales and in different parts of the system (Olsson 2017). Interestingly, the academic platform provided by the Chicago School of Economics later became connected with the Pinochet regime in Chile, which was implementing market liberalism in that country (Fischer 2009). This connection happened because Chilean economists, also referred to as the Chicago Boys, were trained by Milton Friedman and others at the University of Chicago School of Economics. Upon their return to Chile some of them found economic advisory positions in Pinochet's government. Initiatives in both Chile and Chicago became leading examples and played a major role in creating a reason for the niche to begin interacting with different scales within the panarchy and began to interact across the parallel panarchies. For example, the lessons from Chile were important to the Reagan and Thatcher administrations in adopting and implementing market-based economies in the United States and the United Kingdom (Geobey 2017). This means that the bricolage process that formed the version of neoliberalism and market-based economies that were implemented in the United States and the United Kingdom involved interactions across different countries. It spread from a niche in the United States to becoming part of a regime in Chile, which in turn became important in the United States for the Reagan administration to accelerate the implementation of the paradigm that Nixon had started to open up.

In the case of option pricing, the Keynesian and Bretton Woods panarchical system was weakened by the fact that it did not deliver economic stability (Geobey 2017). This failure shook the norms and values that were attached to this system and weakened the remembrance mechanisms, and the instability—or fears of it—resulted in a crisis-driven dismantling and replacement of that system. Agency was still important, though, and Milton Friedman's ideas appealed to the Nixon administration, which led to the Nixon shocks in 1971. *Nixon shocks* refers to a number of significant measures implemented in 1971 to address increasing inflation, such as wage and price freezes, and abandoning the gold standard by no longer allowing the conversion of U.S. dollars to gold (Geobey 2017). This step contributed to making the Bretton Woods system inoperative and eventually led to its abandonment. The emerging parallel panarchy with the new practices (micro) such as the Black–Scholes–Merton models, embedded in and supported by the neoliberal paradigm (macro), had developed a level of readiness to scale. This meant that they could seize the opportunity and target the rules and regulations of the Keynesian panarchy when the dominant system started to fail and go through a release phase. That is, the movement that was created around the ideas of neoliberalism both contributed to building a new panarchy and contributed to breaking down the dominant system when certain actors and organizations targeted the decision makers to change the rules, and the discourse shifted the norms and values of both these decision makers, and academics and the public at large.

Again, crises in panarchy open up spaces for novelty, and in these spaces actions may adapt the panarchy to the new circumstances and continue development on the same pathways. For example, the financial crisis of 2008 was seized on by some as a moment to transform the finance sector, but the strongest forces pushed the system back to something resembling the precrash status quo (Loorbach and Huffenreuter 2013). Alternatively, the panarchy may transform into completely new pathways of development based on interactions with the parallel panarchy. The sources of such transformations, we argue, are derived predominantly from other panarchies ripe for influencing the open space, becoming the seeds for new disruptive innovations, or providing the window of opportunity for shadow networks to become established and take over. These may cascade all the way to the global level, as witnessed in the neoliberal case above or in the global adaptive governance of the Southern Ocean fisheries (Österblom and Folke 2013).

Implications for People–Planet Futures

In the context of panarchy theory, we will highlight some critical issues related to large-scale transformations that are needed in order to embark on just and safe trajectories where humans and nature thrive together (Westley et al. 2011). Recent work clarifies that within recent decades the economic machinery of the global economy has emerged into a hyperefficient global system generating goods, services, capital, material well-being, and improved health for many (Steffen et al. 2015a). However, this rapid change has occurred at the expense of resilience of the biosphere and human equality (Leach et al. 2018; Nyström et al. 2019). Examples of this erosion of resilience include simplifications of landscapes and seascapes, pervasive loss of biodiversity (Bar-On et al. 2018; IPBES 2019), inequalities in human development (United Nations Development Program 2019), increased resistance to antibiotics, insecticides, and herbicides (Jørgensen et al. 2020), human dominance of ecosystems and Earth system processes (Vitousek et al. 1997; Steffen et al. 2018), and an aggregate human imprint that challenges planetary boundaries for well-being (Rockström et al. 2009; Steffen et al. 2015b).

The dynamics of people and the planet have become deeply intertwined with a complex, interconnected, and cross-scale interplay recently referred to as Anthropocene risks (Keys et al. 2019). The Great Acceleration of human activity on Earth has self-organized into, for instance, dominance by a few transnational corporations shaping the planet (Folke et al. 2019) and with major discrepancies in equality (Scheffer et al. 2017; Hamann et al. 2018). Many scholars have expressed concerns that nations and institutions have not been able to keep up with, deal with or even recognize this development (Scholte 2005; Walker et al. 2009; Biermann et al. 2012). In many cases, policy and practice have actively supported the shift toward hyperefficiency. At the same time, the significance of the biosphere foundation for well-being has been out of sight and out of mind (Folke et al. 2011).

While parts of contemporary global society were already moving away from the Industrial Revolution era into a new yet undefined space before the recent global pandemic, COVID-19 heightened the attention and emphasis on the need for transformations to sustainable and equitable futures. Examples of experimentation and social innovation have been emerging in recent decades (Bennett et al. 2016; Olsson et al. 2017), sometimes supported through advances in technology (Carpenter et al. 2019). New social movements have been rapidly spreading across the planet in response to the climate crisis and social injustices (Johnson 2019; Stuart et al. 2020), and there are some signs of large, transnational corporations redirecting their businesses towards sustainability (Folke et al. 2019). This precrisis mobilization of alternative systems in parallel panarchies follows the patterns and processes outlined in this chapter. Along with this earlier activity, observations of efforts during COVID-19 point to, at least in principle, a growing commitment to far-reaching economic and social change, such as efforts by the COVID-19 Task Force established by the UN Global Compact Action Platform for Sustainable Ocean Business and the R3 Coalition of the Global Impact Investment Network. These responses to the pandemic can help accelerate and amplify the precrisis efforts, especially now when the resilience of the current status quo systems is weakened and innovative alternatives are sufficiently well developed. For example, the European Green Deal has brought together private, public, and civil society actors to create effective public policies and improved government regulations, contributing to a hopeful shift from excessive, wasteful, and imbalanced consumption founded on a fossil fuel–driven economy. The Green Deal, which was developed before the COVID-19 crisis, has been put at the center of the European Union's response to the pandemic to help navigate a green and fair recovery. Such efforts may prove important for moving into a renewable energy–based economy of low waste and circularity within a broader value foundation beyond profit alone and with the purpose of shepherding and safeguarding the resilience of the biosphere for human well-being (Folke et al. 2019). Perhaps these are indications of a deeper transformation toward reconnecting economic, cultural, and human development with the biosphere in the direction of an equitable and sustainable future.

We posit that panarchy theory can provide guidance by investigating and clarifying the dynamic interplay of revolt and remembrance within a panarchy, but perhaps even more so across panarchies. The case of neoliberalism described above shows how transformations can be connected across the globe and how changes in one part of the world—even changes in discourse or ideas, later followed by policy—can have an impact somewhere else. A telecoupling is a process that refers to socioeconomic and ecological interactions between distant coupled social-ecological systems, but these couplings are often not easily visible. Such cross-continental connections can happen through a range of social-ecological relationships, including economic activities, trade connections, transportation networks, and hydrologic cycles but can also occur through information flows, discourse, policies, and more (Hull and Liu 2018). The panarchy and more specifically the parallel panarchy framework can help us explore the features and mechanisms of telecoupling processes and how transformations in one place is connected to transformations in other places.

Clearly, there is a need for a combined understanding of the interplay of fast change and slow change, of fast and slow variables and processes, and the complex and dynamic timing of those in multiple parallel panarchies. This implies interpreting when the time for change is more transparent, being prepared for that change, and finding and proposing new attractors, new narratives, and visions that can support transforming actions toward sustainable and equitable futures.

Conclusions

A panarchy framework—and specifically, the parallel panarchies idea developed in this chapter—can help us understand the cross-scale dynamics of large-scale change to equitable and sustainable futures. Specifically, and as illustrated by the three case studies in this chapter, applying the panarchy model provides a more nuanced picture of the revolt and remembrance mechanisms that continuously shape, and reshape, how transformations unfold over time and space while also providing insights into collapse and renewal. Together, these dynamics help to provide a clearer picture of what is meant by *nonlinear dynamics* with regard to how radical ideas within niches can have an impact at other scales.

The parallel panarchy idea can also help us understand agency, and transformative capacities help navigate transformations at larger scales in the Anthropocene. Previous scholarship has shown that agency is critical in both the breakdown of an existing system and the invention or buildup of another (Olsson 2017). A panarchy framework allows us to recognize how agency interacts to promote or hamper remembrance and revolt mechanisms in such a way as to influence processes and outcomes at other scales. Agency works by connecting different systems and associated panarchies on temporal and spatial scales, even globally between countries and over decades. Instead of a rapid and total collapse, often there is more of an untangling of systems, sometimes very slowly, which produces unsustainable and unjust outcomes, which requires strategic actions at several levels.

As a response to the COVID-19 crisis, several initiatives and movements (Stuart et al. 2020) are proposing alternatives to the current neoliberal, market-based, capitalistic economic systems. These include new practices that are embedded in new thinking and mental models. However, as the cases in this chapter show, the early development of novelty within a niche is never easy because it also involves the struggles to establish an organizational platform that enables the (re)combination and incubation of ideas and actions to experiment that test ideas, generate practical examples, and mobilize resources. Although the niche ideas may eventually begin to be established as a panarchy in their own right, it is the interactions with the existing dominant system that will reconfigure and shape ultimate transformations. Without transformative capacities, though, it will be difficult to navigate the promising, radical solutions and initiatives that emerge to deal with global challenges and have a disruptive impact on the systems to bring about needed change in the Anthropocene.

Large-scale change is widely understood as needed, and many are suggesting that these transformations need to occur soon. There is hope that the pandemic will create additional momentum for transformations toward more equitable and sustainable societies at the scale and speed that are needed. The parallel panarchy framework can help us understand the mechanisms that contribute to called-for transformations, and the capacities to navigate them, without leading to collapse. This framework also suggests a warning that tensions, conflicts, and overt power struggles, sometimes violent, will also exist between parallel, coexisting panarchies. But these challenges should be confronted, not avoided in our search for sustainable and equitable Anthropocene.

Literature Cited

Allen, C.R., D.G. Angeler, G. Cumming, C. Folke, D. Twidwell, and D.R. Uden. 2016. Quantifying spatial resilience. *Journal of Applied Ecology* 53: 625–635.

Allen, C.R., D.G. Angeler, A.S. Garmestani, L.H. Gunderson, and C.S. Holling. 2014. Panarchy: Theory and application. *Ecosystems* 17: 578–589.

Avelino, F. 2017. Power in sustainability transitions: Analysing power and (dis) empowerment in transformative change towards sustainability. *Environmental Policy and Governance* 27: 505–520.

Avelino, F., and J.M. Wittmayer. 2016. Shifting power relations in sustainability transitions: A multi-actor perspective. *Journal of Environmental Policy & Planning* 18: 628–649.

Bar-On, Y.M., R. Phillips, and R. Milo. 2018. The biomass distribution on Earth. *Proceedings of the National Academy of Sciences* 115: 6506bio.

Barthel, S., C. Folke, and J. Colding. 2010. Social-ecological memory in urban gardens: Retaining the capacity for management of ecosystem services. *Global Environmental Change* 20: 255–265.

Bennett, E. M., M. Solan, R. Biggs, T. McPhearson, A. V. Norström, P. Olsson, L. Pereira, G. D. Peterson, C. Raudsepp-Hearne, F. Biermann, S. R. Carpenter, E. C. Ellis, T. Hichert, V. Galaz, M. Lahsen, M. Milkoreit, B. M. Lopez, K. A. Nicholas, R. Preiser, G. Vince, J. M. Vervoort, and J. Xu. 2016. Bright spots: Seeds of a good Anthropocene. *Frontiers in Ecology and Environment* 14(8): 441–448.

Berkes, F., and C. Folke. 2002. Back to the future: Ecosystem dynamics and local knowledge. In *Panarchy: Understanding Transformations in Human and Natural Systems*, ed. L.H. Gunderson and C.S. Holling, 121–146. Island Press, Washington, DC.

Biermann, F., K. Abbott, S. Andresen, K. Bäckstrand, S. Bernstein, M. M. Betsill, H. Bulkeley, B. Cashore, J. Clapp, C. Folke, A. Gupta, J. Gupta, P. M. Haas, A.

Jordan, N. Kanie, T. Kluvánková-Oravská, L. Lebel, D. Liverman, J. Meadowcroft, R. B. Mitchell, P. Newell, S. Oberthür, L. Olsson, P. Pattberg, R. Sánchez-Rodríguez, H. Schroeder, A. Underdal, S. Camargo Vieira, C. Vogel, O. R. Young., A. Brock, and R. Zondervan. 2012. Navigating the Anthropocene: Improving Earth system governance. *Science* 335: 1306–1307. http://dx.doi.org/10.1126/science.1217255

Blythe, J., J. Silver, L. Evans, D. Armitage, N.J. Bennett, M.L. Moore, T.H. Morrison, and K. Brown. 2018. The dark side of transformation: Latent risks in contemporary sustainability discourse. *Antipode* 50: 1206–1223.

Carpenter, S.R., C. Folke, M. Scheffer, and F. Westley. 2019. Dancing on the volcano: Social exploration in times of discontent. *Ecology and Society* 24(1): 23. doi.org/10.5751/ES-10839-240123

Castilla, J. C. 1994. The Chilean small-scale benthic shellfisheries and the institutionalization of new management practices. *Ecology International Bulletin* 21: 47–63.

Chaffin, B.C., A.S. Garmestani, L.H. Gunderson, M.H. Benson, D.G. Angeler, C.A. Arnold, B. Cosens, R.K. Craig, J.B. Ruhl, and C.R. Allen. 2016. Transformative environmental governance. *Annual Review of Environment and Resources* 41: 399–423.

Chapin, F.S. III, S.R. Carpenter, G.P. Kofinas, C. Folke, N. Abel, W.C. Clark, P. Olsson, D.M. Stafford Smith, B.H. Walker, O.R. Young, F. Berkes, R. Biggs, J.M. Grove, R.L. Naylor, E. Pinkerton, W. Steffen, and F.J. Swanson. 2010. Ecosystem stewardship: Sustainability strategies for a rapidly changing planet. *Trends in Ecology and Evolution* 25: 241–249.

Chapin, F.S. III, A.F. Mark, R.A. Mitchell, and K.J. Dickinson. 2012. Design principles for social-ecological transformation toward sustainability: Lessons from New Zealand sense of place. *Ecosphere* 3: 1–22.

Conca, K. 2005. *Governing Water: Contentious Transnational Politics and Global Institution Building.* MIT Press, Cambridge, MA.

Eakin, H., X. Rueda, and A. Mahanti. 2017. Transforming governance in telecoupled food systems. *Ecology and Society* 22(4): 32. https://doi.org/10.5751/ES-09831-220432

Eales, K., 2011. Water services in South Africa 1994–2009. In *Transforming Water Management in South Africa: Designing and Implementing a New Policy Framework, Global Issues in Water Policy,* ed. B. Schreiner, R. Hassan, J. Albiac, E. D. Mungatana, V. Pochat, and R. M. Saleth, 33–72. Springer, London.

Fischer, K. 2009. The influence of neoliberals in Chile before, during, and after Pinochet. In *The Road from Mont Pelerin. The Making of the Neoliberal Thought Collective,* ed. P. Mirowski and D. Plehwe, 305–346. Harvard University Press, Cambridge, MA.

Folke, C., S.R. Carpenter, B.H. Walker, M. Scheffer, F.S. Chapin III, and J. Rockström. 2010. Resilience thinking: Integrating resilience, adaptability and transformability. *Ecology and Society* 15(4): 20. http://www.ecologyandsociety.org/vol15/iss4/art20/

Folke, C., J. Colding, and F. Berkes. 2003. Synthesis: Building resilience and adaptive capacity in social-ecological systems. In *Navigating Social-Ecological Systems: Building Resilience for Complexity and Change,* ed. F. Berkes, J. Colding, and C. Folke, 352–387. Cambridge University Press, Cambridge, UK.

Folke, C., Å. Jansson, J. Rockström, P. Olsson, S.R. Carpenter, F.S. Chapin, A.-S. Crepín, G. Daily, K. Danell, J. Ebbesson, T. Elmqvist, V. Galaz, F. Moberg, M. Nilsson, H. Österblom, E. Ostrom, Å. Persson, G. Peterson, S. Polasky, W. Steffen, B. Walker, and F. Westley. 2011. Reconnecting to the biosphere. *Ambio* 40: 719–738.

Folke, C., H. Österblom, J.-B. Jouffray, E. Lambin, M. Scheffer, B.I. Crona, M. Nyström, S.A. Levin, S.R. Carpenter, W.N. Adger, J.M. Anderies, F.S. Chapin III, A.-S. Crépin, A. Dauriach, V. Galaz, L.J. Gordon, N. Kautsky, B.H. Walker, J.R. Watson, J. Wilen, and A. de Zeeuw. 2019. Transnational corporations and the challenge of biosphere stewardship. *Nature Ecology & Evolution* 3: 1396–1403.

Frantzeskaki, N., V.C. Broto, L. Coenen, and D. Loorbach. 2017. Urban sustainability transitions: The dynamics and opportunities of sustainability transitions in cities. In *Urban Sustainability Transitions*, 1–20. Routledge, Milton Park, UK.

Garmestani, A.S., and M.H. Benson. 2013. A framework for resilience-based governance of social-ecological systems. *Ecology and Society* 18(1): 9. http://www.ecologyandsociety.org/vol18/iss1/art9/

Garmestani, A., D. Twidwell, D.G. Angeler, S.M. Sundstrom, C. Barichievy, B.C. Chaffin, T. Eason, N.A.J. Graham, D. Granholm, L. Gunderson, M. Knutson, K.L. Nash, M.J. Nelson, M. Nystrom, T.L. Spanbauer, C.A. Stow, and C.R. Allen. 2020. Panarchy: Opportunities and challenges for ecosystem management. *Frontiers in Ecology and the Environment* 18(10): 576–583.

Geels, F.W. 2010. Ontologies, socio-technical transitions (to sustainability), and the multi-level perspective. *Research Policy* 39: 495–510.

Geels, F.W. 2018. Low-carbon transition via system reconfiguration? A socio-technical whole system analysis of passenger mobility in Great Britain (1990–2016). *Energy Research & Social Science* 46: 86–102.

Gelcich, S., T.P. Hughes, P. Olsson, C. Folke, O. Defeo, M. Fernández, S. Foale, L.H. Gunderson, C. Rodríguez-Sieker, M. Scheffer, R. Steneck, and J.C. Castilla. 2010. Navigating transformations in governance of Chilean marine coastal resources. *Proceedings of the National Academy of Sciences* 107: 16794–16799.

Geobey, S. 2017. The global derivatives market as social innovation. In *The Evolution of Social Innovation*, ed. F. Westley, K. McGowan, and O. Tjörnbo, 146–175. Edward Elgar Publishing, Cheltenham, UK. http://dx.doi.org/10.4337/9781786431158.00009

Gunderson, L.H., and C.S. Holling (Eds.). 2002. *Panarchy: Understanding Transformations in Human and Natural Systems*. Island Press, Washington, DC.

Hamann, M., K. Berry, T. Chaigneau, T. Curry, R. Heilmayr, J.G.P. Henriksson, J. Hentati-Sundberg, A. Jina. E. Lindkvist, Y. Lopez-Maldonado, E. Nieminen, M. Piaggio, J. Qiu, J.C. Rocha, C. Schill, A. Shepon, A.R. Tilman, I. van den Bijgaart, and T. Wu. 2018. Inequality and the biosphere. *Annual Review of Environment and Resources* 43(1): 61–83. doi:10.1146/annurev-environ-102017-025949

Herrfahrdt-Pähle, E., and C. Pahl-Wostl. 2012. Continuity and change in social-ecological systems: The role of institutional resilience. *Ecology and Society* 17(2): 8. http://dx.doi.org/10.5751/ES-04565-170208

Herrfahrdt-Pähle, E., M. Schlüter, P. Olsson, C. Folke, S. Gelcich, and C. Pahl-Wostl. 2020. Sustainability transformations: Socio-political shocks as opportunities for governance transitions. *Global Environmental Change* 63: 102097.

Holling, C.S., L.H. Gunderson, and G.D. Peterson. 2002. Sustainability and panarchies. In *Panarchy: Understanding Transformations in Human and Natural Systems*, ed. L.H. Gunderson and C.S. Holling, 63–102. Island Press, Washington, DC.

Hull, V., and J. Liu. 2018. Telecoupling: A new frontier for global sustainability. *Ecology and Society* 23(4): 41. https://doi.org/10.5751/ES-10494-230441

IPBES. 2019. *The Global Assessment Report on Biodiversity and Ecosystem Services of the Intergovernmental Science-Policy Platform on Biodiversity and Ecosystem Services*, ed. S. Díaz, J. Settele, E.S. Brondizio, H. T. Ngo, M. Guèze, J. Agard, A. Arneth, P. Balvanera, K.A. Brauman, S.H.M. Butchart, K.M.A. Chan, L.A. Garibaldi, K. Ichii, J. Liu, S.M. Subramanian, G.F. Midgley, P. Miloslavich, Z. Molnár, D. Obura, A. Pfaff, S. Polasky, A. Purvis, J. Razzaque, B. Reyers, R. Chowdhury, Y.J. Shin, I.J. Visseren-Hamakers, K.J. Willis, and C.N. Zayas. IPBES Secretariat, Bonn, Germany.

Johnson, A. E. 2019. Ocean conservation and climate action. *Journal of International Affairs* 73(1): 243–248.

Jones Luong, P. 2003. *Political Obstacles to Economic Reform in Uzbekistan, Kyrgyzstan, and Tajikistan: Strategies to Move Ahead*. World Bank Working Paper. World Bank, Washington, DC.

Jørgensen, P.S., C. Folke, P.J. Henriksson, K. Malmros, M. Troell, and A. Zorzet. 2020. Coevolutionary governance of antibiotic and pesticide resistance. *Trends in Ecology and Evolution* 35(6): 484–494.

Keys, P., V. Galaz, M. Dyer, N. Matthews, C. Folke, M. Nyström, and S. Cornell. 2019. Anthropocene risk. *Nature Sustainability* 2: 667–673.

Kingdon, J. 1984. *Agendas, Alternatives, and Public Policies*. Little Brown, Boston, MA.

Köhler, J., F.W. Geels, F. Kern, J. Markard, E. Onsongo, A. Wieczorek, F. Alkemade, F. Avelino, A. Bergek, F. Boons, L. Fünfschilling, D. Hess, G. Holtz, S. Hyysalo, K. Jenkins, P. Kivimaa, M. Martiskainen, A. McMeekin, M.S. Mühlemeier, B. Nykvist, B. Pel, R. Raven, H. Rohrachers, B. Sandén, J. Schot, B. Sovacool, B. Turnheim, D. Welch, and P. Wells. 2019. An agenda for sustainability transitions research: State of the art and future directions. *Environmental Innovation and Societal Transitions* 31: 1–32.

Leach, M., B. Reyers, X. Bai, E.S. Brondizio, C. Cook, S. Díaz, G. Espindola, M. Scobie, M. Stafford-Smith, and S.M. Subramanian. 2018. Equity and sustainability in the Anthropocene: A social-ecological systems perspective on their intertwined futures. *Global Sustainability* 1: e13, 1–13. doi:10.1017/sus.2018.12

Leach, M., I. Scoones, and A. Stirling. 2010. *Dynamic Sustainabilities: Technology, Environment, Social Justice*. Routledge, Milton Park, UK.

Leichenko, R., and K. O'Brien. 2019. *Climate and Society: Transforming the Future*. John Wiley & Sons, Hoboken, NJ.

Loorbach, D. 2010. Transition management for sustainable development: A prescriptive, complexity-based governance framework. *Governance* 23: 161–183. http://dx.doi.org/10.1111/j.1468-0491.2009.01471.x

Loorbach, D., N. Frantzeskaki, and F. Avelino. 2017. Sustainability transitions research: Transforming science and practice for societal change. *Annual Review of Environment and Resources* 42: 599–626.

Loorbach, D.A., and R.L. Huffenreuter. 2013. Exploring the economic crisis from a transition management perspective. *Environmental Innovation and Societal Transitions* 6: 35–46.

Loorbach, D., R. van der Brugge, and M. Taanman. 2008. Governance in the energy transition: Practice of transition management in the Netherlands. *International Journal of Environmental Technology and Management* 9(2–3): 294–315.

McCarthy, J., and S. Prudham. 2004. Neoliberal nature and the nature of neoliberalism. *Geoforum* 35: 275–283.

Meltzoff, S.K., W. Stotz, and Y.G. Lichtensztajn. 2002. Competing visions for marine tenure and co-management: Genesis of a marine management area system in Chile. *Coastal Management* 30: 85–99.

Micklin, P.P. 2007. The Aral Sea disaster. *Annual Review of Earth and Planetary Sciences* 35: 47–72.

Mirowski, P., and D. Plehwe (Eds.). 2009. *The Road from Mont Pèlerin: The Making of the Neoliberal Thought Collective.* Harvard University Press, Cambridge, MA.

Moore, M.-L. 2017. Synthesis: Tracking transformative impacts and cross-scale dynamics. In *The Evolution of Social Innovation*, ed. F. Westley, K. McGowan, and O. Tjörnbo, 218–238. Edward Elgar Publishing, Cheltenham, UK. http://dx.doi.org/10.4337/9781786431158.00009

Moore, M.-L., and M. Milkoreit. 2020. Imagination for transformations to sustainable and just futures. *Elementa: Science of the Anthropocene* 8(1): 81.

Moore, M.-L., P. Olsson, W. Nilsson, L. Rose, and F.R. Westley. 2018. Navigating emergence and system reflexivity as key transformative capacities: Experiences from a Global Fellowship program. *Ecology and Society* 23(2): 38. https://doi.org/10.5751/ES-10166-230238

Moore, M.-L., D. Riddell, and D. Vocisano. 2015. Scaling out, scaling up, scaling deep: Strategies of non-profits in advancing systemic social innovation. *Journal of Corporate Citizenship* 58: 67–84.

Moore, M.-L., O. Tjornbo, E. Enfors, C. Knapp, J. Hodbod, J.A. Baggio, A. Norström, P. Olsson, and D. Biggs. 2014. Studying the complexity of change: Toward an analytical framework for understanding deliberate social-ecological transformations. *Ecology and Society* 19(4): 54. http://dx.doi.org/10.5751/ES-06966-190454

Muller, M. 2014. Allocating power and functions in a federal design: The experience of South Africa. In *Federal Rivers: Managing Water in Multi-Layered Political Systems*, ed. D. Garrick. Edward Elgar Publishing, Cheltenham, UK.

Mylan, J., C. Morris, E. Beech, and F.W. Geels. 2019. Rage against the regime: Niche-regime interactions in the societal embedding of plant-based milk. *Environmental Innovation and Societal Transitions* 31: 233–247.

Nyborg, K., J.M. Anderies, A. Dannenberg, T. Lindahl, C. Schill, M. Schlüter, W.N. Adger, K.J. Arrow, S. Barrett, S.R. Carpenter, F.S. Chapin III, A.-S. Crépin, G. Daily, P. Ehrlich, C. Folke, W. Jager, N. Kautsky, S.A. Levin, O.J. Madsen, S. Polasky, M. Scheffer, B.H. Walker, E.U. Weber, J. Wilen, A. Xepapadeas, and A. de Zeeuw. 2016. Social norms as solutions. *Science* 354: 42–43.

Nyström, M., and C. Folke. 2001. Spatial resilience of coral reefs. *Ecosystems* 4: 406–417.

Nyström, M., J.-B. Jouffray, A. Norström, P. Sogaard-Jørgensen, V. Galaz, B.E. Crona, S.R. Carpenter, and C. Folke. 2019. Anatomy and resilience of the global production ecosystem. *Nature* 575: 98–108.

Olsson, P. 2017. Synthesis: Agency and opportunity. In *The Evolution of Social Innovation*, ed. F. Westley, K. McGowan, and O. Tjörnbo, 58–72. Edward Elgar Publishing, Cheltenham, UK. http://dx.doi.org/10.4337/9781786431158.00009

Olsson, P., Ö. Bodin, and C. Folke. 2010. Building transformative capacity in social-ecological systems: Insights and challenges. In *Adaptive Capacity and Environmental Governance*, ed. D. Armitage and R. Plummer, 263–286. Springer Verlag, New York.

Olsson, P., C. Folke, V. Galaz, T. Hahn, and L. Schultz. 2007. Enhancing the fit through adaptive comanagement: Creating and maintaining bridging functions for matching scales in the Kristianstads Vattenrike Biosphere Reserve Sweden. *Ecology and Society* 12(1): 28. http://www.ecologyandsociety.org/vol12/iss1/art28/

Olsson, P., C. Folke, and T.P. Hughes. 2008. Navigating the transition to ecosystem-based management of the Great Barrier Reef, Australia. *Proceedings of the National Academy of Sciences* 105: 9489–9494.

Olsson, P., V. Galaz, and W. Boonstra. 2014. Sustainability transformations: A resilience perspective. *Ecology and Society* 19(4). https://www.ecologyandsociety.org/vol19/iss4/art1/

Olsson, P., L.H. Gunderson, S.R. Carpenter, P. Ryan, L. Lebel, C. Folke, and C.S. Holling. 2006. Shooting the rapids: Navigating transitions to adaptive governance of social-ecological systems. *Ecology and Society* 11(1): 18. http://www.ecologyandsociety.org/vol11/iss1/art18/

Olsson, P., M.-L. Moore, F.R. Westley, and D.D. McCarthy. 2017. The concept of the Anthropocene as a game-changer: A new context for social innovation and transformations to sustainability. *Ecology and Society* 22(2): 31. https://doi.org/10.5751/ES-09310-220231

Österblom, H., and C. Folke. 2013. Emergence of global adaptive governance for stewardship of regional marine resources. *Ecology and Society* 18(2): 4. http://dx.doi.org/10.5751/ES-05373-180204

Pahl-Wostl, C. 2015. *Water Governance in the Face of Global Change: From Understanding to Transformation.* Springer, New York. https://doi.org/10.1007/978-3-319-21855-7

Patterson, J., K. Schulz, J. Vervoort, S. van der Hel, O. Widerberg, C. Adler, M. Hurlbert, K. Anderton, M. Sethi, and A. Barau. 2017. Exploring the governance and politics of transformations towards sustainability. *Environmental Innovation Social Transition* 24: 1–16. https://doi.org/10.1016/j.eist.2016.09.001

Polanyi, K. 2001. [1944]. *The Great Transformation.* Beacon Press, Boston, MA.

Preiser, R., R. Biggs, A. De Vos, and C. Folke. 2018. Social-ecological systems as complex adaptive systems: Organizing principles for advancing research methods and approaches. *Ecology and Society* 23(4): 46. https://doi.org/10.5751/ES-10558-230446

Repetto, R. 2006. *Punctuated Equilibrium and the Dynamics of U.S. Environmental Policy.* Yale University Press, New Haven, CT.

Rockström, J., W. Steffen, K. Noone, Å. Persson, F.S. Chapin III, E.F. Lambin, T.M. Lenton, M. Scheffer, C. Folke, H.J. Schellnhuber, B. Nykvist, C.A. de Wit, T. Hughes, S. van der Leeuw, H. Rodhe, S. Sörlin, P.K. Snyder, R. Costanza, U. Svedin, M. Falkenmark, L. Karlberg, R.W. Corell, V.J. Fabry, J. Hansen, B.H. Walker, D. Liverman, K. Richardson, P. Crutzen, and J.A. Foley. 2009. A safe operating space for humanity. *Nature* 461: 472–475.

Rotmans, J., R. Kemp, and M. Van Asselt. 2001. More evolution than revolution: Transition management in public policy. *Foresight* 3: 15–31.

Scheffer, M., B. van Bavel, I.A. van de Leemput, and E.H. van Nes. 2017. Inequality in nature and society. *Proceedings of the National Academy of Sciences* 114: 13154–13157.

Schlüter, M., and E. Herrfahrdt-Pähle. 2011. Exploring resilience and transformability of a river basin in the face of socioeconomic and ecological crisis: An example from the Amudarya River basin, Central Asia. *Ecology and Society* 16(1): 32. http://www.ecologyandsociety.org/vol16/iss1/art32/

Scholte, J.A. 2005. *Globalization: A Critical Introduction*. Macmillan International Higher Education, London.

Schot, J., and F.W. Geels. 2008. Strategic niche management and sustainable innovation journeys: Theory, findings, research agenda, and policy. *Technology Analysis & Strategic Management* 20: 537–554.

Scoones I., A. Stirling, D. Abrol, J. Atela, L. Charli-Joseph, H. Eakin, A. Ely, P. Olsson, L. Pereira, R. Priya, P. van Zwanenberg, and L. Yang. 2020. Transformations to sustainability: Combining structural, systemic and enabling approaches. *Current Opinion in Environmental Sustainability* 42: 65–75. https://doi.org/10.1016/j.cosust.2019.12.004

Smith, A. 2007. Translating sustainabilities between green niches and socio-technical regimes. *Technology Analysis & Strategic Management* 19: 427–450.

Smith, A., and R. Raven. 2012. What is protective space? Reconsidering niches in transitions to sustainability. *Research Policy* 41: 1025–1036.

Steffen, W., W. Broadgate, L. Deutsch, O. Gaffney. and C. Ludwig. 2015a. The trajectory of the Anthropocene: The Great Acceleration. *Anthropocene Review* 2: 81–98.

Steffen, W., K. Richardson, J. Rockström, S. Cornell, I. Fetzer, E. Bennett, R. Biggs, S.R. Carpenter, W. de Vries, C.A. de Wit, C. Folke, D. Gerten, J. Heinke, G.M. Mace, L.M. Persson, V. Ramanathan, B. Reyers, and S. Sörlin. 2015b. Planetary boundaries: Guiding human development on a changing planet. *Science* 347: 1259855-1-10.

Steffen, W., J. Rockström, K. Richardson, T.M. Lenton, C. Folke, D. Liverman, C.P. Summerhayes, A.D. Barnosky, S.E. Cornell, M. Crucifix, J.F. Donges, I. Fetzer, S.J. Lade, M. Scheffer, R. Winkelmann, and H.J. Schellnhuber. 2018. Trajectories of the Earth system in the Anthropocene. *Proceedings of the National Academy of Sciences* 115: 8252–8259.

Stuart, D., R. Gunderson, and B. Petersen. 2020. The climate crisis as a catalyst for emancipatory transformation: An examination of the possible. *International Sociology* 35(4): 433–456.

United Nations Development Program (UNDP). 2019. *Human Development Report 2019: Beyond Income, beyond Averages, beyond Today: Inequalities in Human Development in the 21st Century*. UNDP, New York.

van Koppen, B., and B. Schreiner. 2014. Moving beyond integrated water resource management: Developmental water management in South Africa. *International Journal of Water Resources Development* 30: 543–558.

Vitousek, P.M., H.A. Mooney, J. Lubchenco, and J.M. Melillo. 1997. Human domination of Earth's ecosystems. *Science* 277: 494–499.

Waddell, S., S. Waddock, S. Cornell, D. Dentoni, M. McLachlan, and G. Meszoely. 2015. Large systems change: An emerging field of transformation and transitions. *Journal of Corporate Citizenship* 58: 5–30.

Walker, B.H., S. Barrett, S. Polasky, V. Galaz, C. Folke, G. Engström, F. Ackerman, K. Arrow, S.R. Carpenter, K. Chopra, G. Daily, P. Ehrlich, T. Hughes, N. Kautsky, S.A. Levin, K.-G. Mäler, J. Shogren, J. Vincent, T. Xepapadeous, and A. de Zeeuw. 2009. Looming global-scale failures and missing institutions. *Science* 325: 1345–1346.

Walker, B.H., C.S. Holling, S.R. Carpenter, and A. Kinzig. 2004. Resilience, adaptability and transformability in social-ecological systems. *Ecology and Society* 9: 5. http://www.ecologyandsociety.org/vol9/iss2/art5

Wegerich, K. 2005. *Institutional Change in Water Management at Local and Provincial Level in Uzbekistan*. European University Studies: Series IV Geography. Peter Lang, Bern, Switzerland.

Westley, F., K. McGowan, and O. Tjörnbo. 2017. *The Evolution of Social Innovation: Building Resilience through Transitions*. Edward Elgar Publishing, Cheltenham, UK.

Westley, F., P. Olsson, C. Folke, T. Homer-Dixon, H. Vredenburg, D. Loorbach, J. Thompson, M. Nilsson, E. Lambin, J. Sendzimir, B. Banarjee, V. Galaz, and S. van der Leeuw. 2011. Tipping towards sustainability: Emerging pathways of transformation. *Ambio* 40: 762–780.

Westley, F., M. Patton, and B. Zimmerman. 2006. *Getting to Maybe: How the World Is Changed*. Random House, Toronto, Canada.

Westley, F., O. Tjörnbo, L. Schultz, P. Olsson, C. Folke, B. Crona, and Ö. Bodin. 2013. A theory of transformative agency in linked social-ecological systems. *Ecology and Society* 18(3): 27. http://dx.doi.org/10.5751/ES-05072-180327

Whyte, K. 2018. Settler colonialism, ecology, and environmental injustice. *Environment and Society* 9: 125–144.

CHAPTER 9

Panarchy and Law in the Anthropocene

Robin Kundis Craig, Barbara Cosens,
Ahjond Garmestani, and J. B. Ruhl

Panarchy is the discontinuous scaled structure of complex systems that models the reality that social-ecological systems continually adapt and change in response to disturbances and cross-scale interactions (both temporal and spatial). In the Anthropocene, stressors such as climate change, pervasive pollution, and biodiversity loss provide constant disturbance and change. However, law has as its ultimate goal social stability. Achieving this goal over time requires a nuanced balance of legal predictability and legal flexibility that can and should vary by context: Law must adapt to social-ecological changes, but it cannot be so continually and pervasively in flux that it becomes an impediment to social and economic investment and security. Thus, some aspects of law, like the basics of criminal law, need to remain dependable—we want our governments to deter, catch, and punish murderers (although the definition of "murder" might evolve) regardless of social-ecological change. Other aspects of law, especially the facets of law that deal most directly with the ecological components of social-ecological systems (e.g., natural resources and environmental law), may need to become more adaptive and more cognizant of cross-scale and temporal interactions to continue to be effective.

We explore the ways in which law can better reflect an increasingly changing world. We begin with a discussion of the structure of law (centered on U.S. law), arguing that law is more panarchical than is generally acknowledged but could still do more to incorporate panarchy's insights. We then summarize some emerging perspectives on how the law can both better reflect and better deal with the realities of panarchy and social-ecological change.

Because the concept of panarchy emerged from the ecological sciences, theorists initially gave little thought to its implications for law. However, those implications have become increasingly important, particularly in those subfields of law that interact directly with the ecological facets of social-ecological systems, such as environmental law, natural resource law, and energy law (Garmestani et al. 2009; Benson and Garmestani 2011; Ruhl 2012; Garmestani and Benson 2013). In the twenty-first century, climate change, ocean acidification, and other aspects of the Anthropocene are accelerating the inherent dynamism of social-ecological systems that panarchy models and promoting threshold crossings and system change. Because, ultimately, all of society will be changing in reaction to these Anthropocene forces, a legal system that understands and incorporates panarchy and its insights into social-ecological change is likely to become critical to all aspects of a well-governed society. Thus it is important to understand both how the basic structure and goals of law relate to the panarchy model and why there is a need for some types of law to more explicitly reflect that model. This chapter begins with those issues, then offers a concrete example of how a current law, the Clean Water Act, could be reframed to better reflect the insights of panarchy and resilience theory. It ends with a survey of three emerging governance approaches that better incorporate the panarchy model but have not yet been fully implemented through law.

Panarchy and Law's Structures and Goals

Gunderson and Holling (2002, 74) coined the term *panarchy* to "capture the adaptive and evolutionary nature of adaptive cycles that are nested one within each other across space and time scales." They expressly distinguished panarchies from hierarchies, which are more rigid, top-down systems in which "the larger, slower levels constrain the behavior of faster levels" (Gunderson and Holling 2002, 74). The question, therefore, is which structure law most resembles.

There is no universal answer to that question, because legal structures are human constructs that can vary from nation to nation; authoritarian regimes do not operate through the same legal structures as pluralistic democracies. Nevertheless, it is generally far easier to make the case that law is hierarchical rather than panarchical, a model where international law constrains national-level law, which in turn constrains lower levels of government down to municipalities. To take the United States as a particularly multilayered system of governance, international treaties to which the United States is a party do in fact dictate the contents of many domestic federal laws; the U.S. Constitution proclaims in its Supremacy Clause (art. VI, cl. 2) that it, federal statutes, and treaties are the supreme law of the land and thus can preempt state and municipal law; and state constitutions and statutes generally dictate what municipalities can and cannot do.

Nevertheless, legal structure can also be panarchical, because law can operate through nested, changing, and interacting scales of government authority from local to national and national to local. Staying with the United States as an example, every level of law includes mechanisms for changing itself, and moments of particular social crisis can spark flurries of legal activity and adoption of new directions and norms in law—the law's version of release and reorganization. At the federal level, the end of the Civil War prompted the adoption of the Thirteenth, Fourteenth, and Fifteenth Amendments into the U.S. Constitution, added between 1864 and 1870. These amendments not only outlawed slavery and enfranchised the newly freed slaves but also significantly altered the federal–state relationship, constraining states' authority to interfere with civil rights. President Franklin D. Roosevelt's New Deal, implemented in response to the Great Depression in the 1930s, not only generated new federal statutes and a variety of federal programs but also almost single-handedly invented the modern administrative state—including much of the law that allows this extraconstitutional expansion of the executive branch to exist. The Civil Rights Movement of the 1960s induced the U.S. Supreme Court not only to reverse its own position on the legality of segregation but also to become an active driver of desegregation.

Moreover, like a panarchy, different scales of law change at different rates. Amending the U.S. Constitution is a long and laborious process that has been completed only twenty-seven times in 230 years—fewer if one considers the multiple-amendment packages that gave us the Bill of Rights and the Civil War amendments as single events. In contrast, sixteen states allow their constitutions to be changed through a popular vote in a single election (National Conference of State Legislatures 2020). Municipal laws, generally taking the form of ordinances, building codes, and land use plans, can change far more rapidly than state or federal law because change generally requires only a quick public process and a majority vote of the relevant municipal council.

Perhaps the least acknowledged panarchical aspect of legal structure, however, is the ability for change at lower levels to effect change at higher ones. The meanings of several aspects of the federal constitution, for example, depend on evolving norms of state practice. Perhaps the most famous—and arguably most important—of these provisions is the Eighth Amendment's prohibition on "cruel and unusual punishment," a standard that has evolved since 1789 to eliminate torture and drastically limit the crimes for which the death penalty is allowable. As the U.S. Supreme Court has explained,

> To determine whether a punishment is cruel and unusual, courts must look beyond historical conceptions to "'the evolving standards

> of decency that mark the progress of a maturing society.'" *Estelle v. Gamble*, 429 U.S. 97, 102 . . . (1976) (quoting *Trop v. Dulles*, 356 U.S. 86, 101 . . . [1958] [plurality opinion]). "This is because '[t]he standard of extreme cruelty is not merely descriptive, but necessarily embodies a moral judgment. The standard itself remains the same, but its applicability must change as the basic mores of society change.'" *Kennedy v. Louisiana*, 554 U.S. 407, 419 . . . (2008) (quoting *Furman v. Georgia*, 408 U.S. 238, 382 . . . [1972] [Burger, C.J., dissenting])
>
> [In interpreting the Eighth Amendment,] [t]he Court first considers "objective indicia of society's standards, *as expressed in legislative enactments and state practice*," to determine whether there is a national consensus against the sentencing practice at issue. (*Graham v. Florida*, 560 U.S. 48, 58, 61 [2010] [emphasis added])

The Court's due process decisions can also turn on whether states allow or prohibit certain practices (e.g., *Burnham v. Superior Court of California*, 495 U.S. 604, 619 [1990]). Thus, legal changes at the state level in certain arenas drive changes in the meaning of the U.S. Constitution.

Perhaps the most extreme example of bottom-up change in the law is through citizen referendums and citizen initiatives to change state law, a process that allows a changed norm within a populace at large to change substantive law at the state level. All but three states and territories allow popular referendums for state statutes, and sixteen states and territories allow for direct citizen initiatives to amend the state constitution (NCSL 2020). A recent example of this popular "revolt" at the smallest scale of governance has been the decriminalization of marijuana. As of August 2020, forty-two states had legalized marijuana at least for medical use (DISA 2020), and seventeen of those legalizations in place by 2019 came through citizen-initiated ballot measures (Ballotpedia 2019). More telling are the statistics for recreational use, where nine of the current ten legalizations came about through citizen-initiated ballot measures (Ballotpedia 2019). This citizen revolt may eventually bring about legal change even at the federal level, which has criminalized all marijuana use since 1970.

Law's Goal of Social Stability

As the previous section illustrates, panarchy can help illustrate complex changes in legal systems as well as in social-ecological systems. Although panarchy and its basis in resilience theory are criticized for failure to account for social manifestations of human consciousness such as power and agency (Olsson et al. 2015), and law is the manifestation of power and agency, the question to consider is whether panarchy is nevertheless a useful model for law (or at least some forms of law) to acknowledge and incorporate. Panarchy is most useful in legal studies as an aid to thinking about points for legal intervention and assistance and to considering different legal approaches that might be needed for different parts of the adaptive cycle and from different levels of government. In other words, an understanding of the panarchy model of social-ecological systems can help legal systems better achieve their societal goals.

Like legal structure, legal systems' goals have varied in different places and over time. However, it is fair to say that the overall goals of a contemporary democracy's legal system are social stability and security, collective productivity, and individual flourishing—that is, a society largely free from violence, with enough predictability in substantive legal rules and standards (e.g., allowed and proscribed activities, consequences) and enough protection of individual rights and liberties that individuals, groups, and businesses are and feel secure enough to invest in their own and the nation's present and future social, cultural, and economic welfare.

Social stability, collective productivity, and individual flourishing do not emerge from an overly rigid legal system. Instead, achieving those goals requires a nuanced approach to balancing legal stability and persistence with legal flexibility and change to reflect the differing realities of different societal contexts. Overall financial security requires quick responsiveness to market fluctuations, such as the Federal Reserve's ability in the United States to change interest rates with minimal procedure (Craig and Ruhl 2014). In contrast, individual liberty requires secure civil rights that both legislatures and courts tend to change only slowly, after extensive consideration of how individuals exercising their rights can affect each other, and almost always in the direction of expansion rather than contraction.

In this context, *stability* refers "to the persistence over time, in the same or similar form, of governance structures (e.g., the branches of government and their implementing bodies at all levels, the rule of law, nongovernmental governance institutions such as industry groups); of substantive rules, standards, and norms; and of procedural requirements and opportunities, including public and interest group participation" (Craig et al. 2017, 2). Legal and government stability, including adherence to the rule of law, provide important benefits to society, including avoidance of societal chaos, which in the extreme forms of instability, military coups and revolutions, benefits neither most individuals nor overall social progress; predictability of rules and protection of property and labor that allow for fruitful investment, financially and socially, at both the individual and societal levels; trust in social norms and government, which encourages cooperation, democratic participation, and compliance with the law; and individual and societal security (Craig et al. 2017).

Nevertheless, too much emphasis on legal stability can become problematic when social and cultural norms are evolving or social-ecological systems are changing, a problem sometimes called a rigidity trap (Gunderson and Holling 2002; Carpenter and Brock 2008; Craig et al. 2017). In democracies, contemporary law generally recognizes this fact and incorporates a certain amount of inherent flexibility to cope with changing realities. For example, the normal legal processes of judge-made common law and statutory reform were generally sufficient to evolve state landlord–tenant law in the United States, which moved over the course of the twentieth century from a focus on the landlord's right to control rented property to a focus on tenants' rights, similar to the emergence of consumer protection law (Quester 2006). Sometimes, however, social change demands more profound legal revision, profitably viewed as revolutionary rather than evolutionary. In the Civil Rights Movement, for example, the U.S. Supreme Court's decision to overturn its outdated views of equal protection was a key recasting of the Fourteenth Amendment that allowed the movement to make social headway, albeit still often with violence.

Despite law's many existing modes of flexibility, however, in the twenty-first century the Anthropocene's impacts on social-ecological systems pose challenges to law's ability to balance legal stability and flexibility appropriately while still promoting the overall goals of a contemporary democracy. For example, climate change and ocean acidification, alone and in synergistic interaction with other ecological stressors, are rapidly and often unpredictably altering social-ecological systems across the world as the planet warms, hydrologic and precipitation patterns change, increasingly hot summers and milder winters become the norm, sea levels rise, agricultural capacity changes, species migrate toward the poles, and food webs shift (Intergovernmental Panel on Climate Change 2014). Nevertheless, even if it were currently technologically possible at a global or even national scale, an abrupt legal decision to decarbonize a nation could be as immediately disruptive to that society as the Anthropocene is likely to be in the longer run. Changing the law to cope with changing social-ecological realities while shielding society from the most disruptive impacts of either kind of change is arguably the greatest governance challenge of this century.

More manageable is the need to update specific kinds of laws to accommodate what the Anthropocene means for a panarchical reality. For example, most environmental and natural resource laws in the United States, written mostly between the 1930s and 1980, embody an outdated balance-of-nature model of ecological systems (Benson and Craig 2017). These laws codify assumptions that social-ecological systems operate within bounded envelopes of change and retain consistent properties (stationarity) and that preservation and restoration of these systems are always possible (Craig 2010; Benson and Craig 2017). Resilience theory and the panarchy model indicate that these assumptions were wrong from the beginning, but they have become even more problematic now, warranting a reframing of these statutes to better reflect current understandings of how social-ecological systems actually function (Craig 2010; Benson and Craig 2017). Administrative law—the law that governs federal and state agencies and hence much resource management—is a manifestation of the preference of economic actors for certainty (Cosens 2013). Thus, law governing agency action emphasizes front-end decision making and finality in decision making, which make it difficult for agencies to engage in innovative and responsive forms of management (Craig and Ruhl 2014). As noted, federal law can preempt state law innovation; a recent example is the Trump administration's attempt to annul California's regulation of greenhouse gas emissions from automobiles (NBC News 2019). At the same time, the plurality of scales of governance in the United States often operate simultaneously on the same landscape but with little integration of purpose, goal, or method, leading to regulatory fragmentation and an inability for law to comprehensively deal with change. One prominent example in the United States is regulation of the ocean, its uses, and its resources. Such regulation is divided between multiple federal departments, hundreds of federal agencies, thirty coastal states and five oceanic territories, and thousands of state and territorial agencies and coastal municipalities (U.S. Commission on Ocean Policy 2004). All of these concrete legal impediments to coping more effectively with the Anthropocene could benefit from the panarchy model's illumination of how social-ecological systems actually function.

Using Panarchy to Improve the Law by Changing How We Frame the Goals of System Management

Current environmental and natural resource laws in the United States embody simplistic understandings of how social-ecological systems function. One intersection of law and panarchy, therefore, is to update how these laws are implemented and how their goals are framed on the ground in order to reflect new understandings of how social-ecological systems actually work.

A particularly striking example comes from the Federal Water Pollution Control Act of 1972 (33 U.S.C. §§ 1251-1388), known more colloquially from its 1977 amendments as the Clean Water Act. This statute provides the federal backbone of water quality law in the United States. Its primary goal is "to restore and maintain the chemical, physical, and biological integrity of the Nation's waters" (33 U.S.C. § 1251[a]).

In keeping with resilience theory and the panarchy model, federal and state agencies using a regulatory approach to implement the Clean Water Act based on managing resilience would seek to maintain the general chemical, physical, and biological integrity of a lake or river while allowing the system to evolve and even transform. In other words, the regulatory and management goals would be framed in terms of maintaining a functional and productive social-ecological system within and around the waterbody despite its adaptations and transformations. Indeed, keeping pollutants, especially toxics, out of these systems—that is, steady pursuit of the act's no discharge goal (33 U.S.C. § 1251[a][1])—would strengthen aquatic systems' resilience to other changes, such as climate change.

However, this is not how the U.S. Environmental Protection Agency (EPA) currently frames regulation under the act. Instead, the EPA's regulations commit the agencies implementing the act to the pursuit of stability and stationarity and to an outdated balance-of-nature model of ecological systems. Under its framing, aquatic ecosystems can always be restored to historical system states, and desirable features and uses can always be maintained. Thus, maintaining the chemical, physical, and biological integrity of a waterbody means freezing it in a particular historical state.

Illuminating that current framing requires a bit of background on the statute. The Clean Water Act operates through cooperative federalism, meaning that it provides significant roles for both the states and the federal government, the latter operating through two agencies: the EPA and the U.S. Army Corps of Engineers (USACE). The act's regulatory program begins with a declaration that, except as in compliance with the statute, "the discharge of any pollutant by any person shall be unlawful" (33 U.S.C. § 1311[a]). A "discharge of a pollutant" is "any addition of any pollutant to navigable waters from any point source" or "any addition of any pollutant to the waters of the contiguous zone or the ocean from any point source other than a vessel or other floating craft" (33 U.S.C. § 1362[12]). People engaged in activities that meet this definition must get one of the act's two types of permits: a section 404 dredge and fill permit if the person is discharging dredged or fill material or a section 402 National Pollutant Discharge Elimination System (NPDES) permit if the person is discharging anything else (33 U.S.C. §§ 1342[a], 1344[a]). The USACE still issues most dredge and fill permits. The EPA received the initial authority to issue NPDES permits, but most states have taken over this permitting authority, as the act encourages (33 U.S.C. § 1342[b]). However, the EPA still establishes many of the effluent limitations—discharge standards—that become part of the NPDES permits.

In contrast, states establish the ultimate water quality goals for waterbodies, the water quality standards. For each waterbody within the state, the appropriate state agency determines what uses the waterbody should eventually support (the designated uses) and what water quality criteria are necessary to support those uses, including temperature, pH, and turbidity (33 U.S.C. § 1313[c]). The act's permit programs are ultimately tied to these water quality goals, allowing effluent limitations to become more stringent if standard limitations aren't enough to allow the waterbody to meet its water quality standards (33 U.S.C. §§ 1312, 1313[d]).

Importantly for the Anthropocene, however, the state's water quality standards also must include an antidegradation policy (40 C.F.R. § 130.12). Under this policy, the state must maintain waterbodies' existing uses—"those uses actually attained in the water body on or after November 28, 1975, whether or not they are included in the water quality standards"—forever. There is no legal way to allow a waterway to degrade below the status it had in November 1975 (40 C.F.R. § 130.10[g]; Craig 2013).

The EPA clearly intended its regulatory existing use requirements to protect water quality, forbidding states from allowing their waterways to degrade further. However, those requirements assume that immediate human activities, particularly pollution, are the only source of water degradation. In the Anthropocene, the existing use requirements are coming increasingly into conflict with climate change and its impacts on waterbodies—sources of system alteration that are not amenable to correction through the Clean Water Act's permitting requirements.

Many effects of climate change can interfere with a waterbody's ability to maintain existing uses, including reductions in precipitation, prolonged drought, or more frequent and more violent flooding. The impact that has become most problematic so far, however, is increasing temperature and its interference with coldwater fish species. For example, Alaska, California, Oregon, Washington, Idaho, and Montana include salmon and trout habitat, spawning, and rearing as Clean Water Act designated uses for many of the rivers in these states. Even if not designated, moreover, many of these rivers supported salmon and trout in 1975, making coldwater fish habitat and reproduction legally enforceable existing uses for those rivers. In the twenty-first century, however, many of these rivers are becoming too warm to support coldwater fish, leading to salmon die-offs and the potential extinction of some species and runs (Craig 2013; Berwyn 2019) and rendering many streams unfit for trout (NPR 2017).

The complicated social-ecological interactions leading to climate change can be described through the framework of panarchy. The local burning of fossil fuels that started in the Industrial Revolution first created smaller-scale legal and public health problems in the form of acute air pollution. However, as the greenhouse gas (especially carbon dioxide) byproducts of such fossil fuel use accumulated globally in the atmosphere, that accumulation perturbed the much larger-scale planetary climate adaptive cycle, shifting it out of the relative stability (conservation phase) that humans have enjoyed during the last roughly 12,000 years of the Holocene epoch. The result for waterbodies in the United States is that the Clean Water Act's attempt to ensure perpetual protection of water quality no longer matches social-ecological reality. The act now constrains state and community flexibility to adapt to changing conditions, such as warmer water that can no longer support cold-adapted species of fish, and to facilitate a transformation of these rivers to new but still productive system states.

New Approaches to Governance in a Panarchical World

The changes to social-ecological systems during the Anthropocene challenge the ability of existing legal systems in most contemporary democracies to achieve ultimate goals of social stability and security, collective productivity, and individual flourishing. Nevertheless, new approaches to governance are emerging that may have more capacity to cope effectively with changing realities without destabilizing and disrupting society in the process. These new approaches, adaptive management and adaptive governance, hold promise in part because they incorporate the insights of panarchy and hence accept complex social-ecological change as a continual reality, as described in the following sections.

Adaptive management

Resilience theory and the concept of ecological resilience began with the work of ecologists who relied on the emerging science of complexity to understand the nonlinear behavior of systems (Holling 1973). From almost the beginning, these researchers recognized that traditional forms of management, implementing laws grounded in outdated models of ecological stationarity, were insufficient to cope with a world operating through ecological resilience and panarchical dynamism.

In 1978, the late C.S. (Buzz) Holling, the father of ecological resilience theory, recognized the need for a new form of ecosystem and natural resource management to handle this complexity and uncertainty, calling this new approach adaptive management (Holling 1978; Walters 1986). Adaptive management is an experimental approach to managing resilience through incremental learning by doing. "Unlike a traditional trial and error approach, adaptive management has explicit structure, including careful elucidation of goals, identification of alternative management objectives and hypotheses of causation, and procedures for the collection of data followed by evaluation and reiteration" (Allen et al. 2011, 1339).

Adaptive management has received considerable attention from both agency scientists (Szaro et al. 2009) and legal scholars (Ruhl and Fischman 2010; Craig and Ruhl 2014). Although promising in controlled situations (Flessa et al. 2013; Craig and Ruhl 2014), application of adaptive management to extensive landscapes has not worked well (Volkman and McConnaha 1973; Lee 1999). In situations with multiple competing interests and multiple jurisdictions, a technocratic implementation of adaptive management simply lacks legitimacy (Cosens 2010, 2013). The uncertainty in the trajectory of iterative changes in management and the need to make value judgments as conditions challenge the political will of competing interests to hand over decision making to agency scientists (Lee 1999).

As a result, scholars and managers began to search for a governance framework within which managers could implement true adaptive management (Dietz et al. 2003; Folke et al. 2005). One focus in this effort has been reform of administrative law, a leading legal impediment to government adaptive management (Craig and Ruhl 2014). Another has been on adaptive governance, an important emerging innovation in its own right.

Adaptive governance

Law's tendency to favor stability is not its only limitation in fully engaging with social-ecological panarchy. Law also tends to favor top-down solutions. Lawmakers dictate what actors can and cannot do, as in the Clean Water Act. In the wake of the environmental movement that brought about major new regulatory legislation in the 1970s, however, it became clear that the complexity of environmental problems and their lack of uniformity exceeded the capacity of a legal system based on top-down uniform processes, requirements, and solutions (Dorf and Sabel 1998; Cosens et al. 2020). In response to the real and perceived failures of top-down bureaucracies, legal reform emerged that applied in a broader arena than just environmental regulation and, in fact, often was aimed primarily at regulation of economic activities (Dorf and Sabel 1998). Beginning in the 1980s, this neoliberal reform reduced the role of government through deregulation and market-based approaches to achieving regulatory goals, such as mitigation banking for wetlands and species habitat (Salzman and Ruhl 2000) or tradeable allowances to deal with acid rain (U.S. Environmental Protection Agency 2020).

Neoliberal reforms arguably swung the legal pendulum too far away from government, however. Beginning in the 1990s, people began to recognize that many such reforms created gaps in governance that in turn led to negative unintended consequences. As a result, citizens began to self-organize in search of greater citizen involvement in governance and regulation (Karkkainen 2004; Burris et al. 2008; Bevir 2009) and to call for greater government oversight. New problems such as terrorism and climate change helped to prompt this reappreciation of government's role in regulation and management. Perhaps more importantly, however, the increasing private involvement in governance created issues of corruption, democratic accountability, legitimacy, and equity, and these major concerns also became reasons to prefer government oversight (Bevir 2009).

The increased role for private actors in the delivery of public services and policymaking, accompanied by a new role for government in steering and overseeing, gave rise to what some scholars call new governance (Bevir 2009), which includes but goes well beyond the environmental sector (Karkkainen 2004; Lockwood et al. 2010). A focus on governance broadens the inquiry of how to deal with changing social-ecological systems beyond the law itself, because "[i]n its most general form governance refers to the construction of social orders, social coordination, or social practices through governmental rule, private or market activity, and includes emergent networks between government and society" (Bevir 2009, 1). This broadened focus therefore encompasses not just what government can do but how it may interact with private actors and exercise discretion and flexibility to allow new forms of governance to emerge.

Coincident with the work of political scientists and legal scholars tracking the rise of new governance, scientists and social scientists who focused on resilience scholarship began to describe emergent forms of governance arising as an alternative to (or complement of) environmental regulation (Cosens et al. 2020). Such adaptive governance seemed particularly suited to landscape-scale environmental management in the face of change and uncertainty (Dietz et al. 2003; Folke et al. 2005; Chaffin et al. 2014).

The narrowest definition of adaptive governance is the governance needed to implement adaptive management. Nevertheless, empirical work to identify governance with high adaptive capacity has moved beyond this definition to more broadly consider adaptive governance as the governance necessary to manage resilience (Folke et al. 2005; Lebel et al. 2006; Huitema et al. 2009; Cosens et al. 2017; Cosens and Gunderson 2018). Although adaptive management remains appropriate when the ability to control experiments exists (Craig and Ruhl 2014), large landscapes are generally "characterized by competing interests, jurisdictional complexity, and multiple drivers of change, and thus the ability to identify single management goals and to control experimentation is limited" (Cosens et al. 2018, 4).

The work of Elinor Ostrom and her lab at Indiana University significantly advanced the empirical work documenting the value of adaptive governance in complex settings. Ostrom and her lab studied resource-dependent communities from fishing villages to irrigation districts, finding that even in the absence of regulation, communities are capable of self-organizing to ensure the resource's sustainability (Ostrom 1990). Importantly, Ostrom's Nobel Prize–winning work refuted Garrett Hardin's position in his famous essay, "The Tragedy of the Commons" (Hardin 1968), that only private ownership or government regulation could prevent the overuse of common-pool resources (Locher 2018). Ostrom participated with Dietz et al. (2003) in coining the term *adaptive governance* and identifying the conditions under which locally based adaptive governance arises.

The Resilience Alliance (RA) formed in 1999 as "an international, multidisciplinary research organization that explores the dynamics of social-ecological systems. RA members collaborate across disciplines to advance the understanding and practical application of resilience, adaptive capacity, and transformation of societies and ecosystems in order to cope with change and support human well-being" (Resilience Alliance 2020). This collaboration brought social scientists together with the architects of ecological resilience theory, and they began to work on bridging the empirical observation of adaptive governance with resilience theory (Folke et al. 2005).

Folke et al. (2005, 441) "explore[d] the social dimension that enables adaptive ecosystem-based management," relying on empirically based research. They recognized that social systems exhibit behavior of complex adaptive systems and that social and ecological systems are intertwined in ways that result in emergent properties. The authors relied on empirical evidence to identify the conditions that appear to lead to the emergence of adaptive governance, including:

- the need to devolve management to the resource scale, while nevertheless setting policy and maintaining legitimacy within existing governmental structures;
- the importance of social capital, including trust, leadership, networks connecting polycentric governmental and non-governmental institutions, and broad public participation;
- the existence of social memory in the form of experience with adaptation, and the capacity to combine science and local knowledge;
- the need for bridging organizations across sectors, levels, and public/private entities;
- the importance of a process for organizational learning so that learning occurs not only within management implementation, but at the level of policy setting; and
- the capacity to deal with change and surprise, including system redundancy.

Like the governance scholarship by political scientists, resilience scholars recognized the need for attributes of good governance—accountability, equity, and justice—to ensure legitimacy in democratic societies (Lebel et al. 2006). Nevertheless, the importance of legal reform to allow the emergence of adaptive governance while also authorizing government participation and ensuring good governance in implementation did not receive attention until legal scholars entered the field (Garmestani et al. 2009; Cosens 2010, 2013; Craig and Benson 2013; Garmestani et al. 2013; Benson and Craig

2014, 2017; Garmestani and Allen 2014; Green et al. 2015; Camacho and Glicksman 2016; DeCaro et al. 2017; Cosens and Gunderson 2018; Craig 2019; Garmestani et al. 2019).

The Legal Move to a Circular Economy

A circular economy is one in which current modes of economic production based on using resources to make one-use products, resulting in significant waste, are replaced by cyclical production models in which industry designs products for long life and reuse and feeds biologically based materials back into ecological processes through processes such as composting and anaerobic digestion (Ellen MacArthur Foundation n.d.). A circular economy "can be seen as an embodiment of *ecological modernization*—the idea that conflicts between environment and economy can be overcome through innovation both technical, but also social (e.g., new business models)" (McDowall et al. 2017, 653). Driven initially by European desires to reduce waste (McDowall et al. 2017), the concept of a circular economy intersects with panarchy by promoting a system approach to the social–ecological–economic system that seeks to increase the resilience of these systems.

Webster (2013) drew the most direct connections between the circular economy and resilience theory, panarchy, adaptive governance, and the law. For instance, he noted,

> Like all living systems, a circular economy must be ***dynamic but adaptive***, and if enduring, it must be ***effective***, neither courting disaster by over-emphasizing efficiency (***brittleness***) or too resistant to change (***stagnation***). It ***celebrates diversity***—of scale, culture, place, connection and time because a dynamic system is full of change, by definition, and thriving in such an environment requires diversity—a fount of ***creative adaption***, a means of ***resilience***, a source of redundancy or back up. It is led by ***business for a profit*** within the "rules of the game" decided by an active citizenship in a flourishing democracy. (Webster 2013, 543)

Furthermore, "there will be ***redundancy*** and resilience and *attention to scale*" (Webster 2013, 545). The transition to the circular economy will even operate through something like panarchy's adaptive cycle: At the end of the growth and exploitation phase (r-phase), the global economy matured (K-phase) into its current state, from which it will reorganize and renew into the circular economy (Webster 2013).

As Webster (2013) noted, the rules of the game are critical to businesses operating in a circular economy, and hence law is an important driver of the social, economic, and industrial transformations necessary to bring about that economy. For example, both China and the European Union have implemented such legal reforms. China's Circular Economy Promotion Law came into force in 2009, generating a variety of action plans, and the European Union initiated a series of policies known as the Circular Economy Package in 2011, replacing them with An Action Plan for a Circular Economy in 2015 (McDowall et al. 2017). These are contrasting approaches, with China taking a more top-down approach where the national government is dictating implementation, whereas the European Union emphasizes a more bottom-up approach in which it supplies overarching principles but allows individual member nations to experiment with implementation. Scholars are already contrasting the two as interesting case studies of legal function and agency (e.g., McDowall et al. 2017).

Like the science of complex adaptive systems and panarchy itself, the circular economy requires a reframing of how the world works—in this case, from linear modes of production to circular design for reuse and recycling ab initio (Raworth 2017). Thus, the circular economy is, by anyone's estimation, a challenge to the law's ability to promote or force profound changes in social, economic, and environmental norms—that is, to actually bring about the cultural adoption of a new narrative about how the economy should work. Considerable doubt is warranted about whether the law can accomplish so much equitably and legitimately by top-down fiat. Instead, new legal frameworks that allow the circular economy to emerge as a form of adaptive governance may stand a much better chance of success with far less transitional social disruption—thus, somewhat ironically, better promoting law's core interest in social stability.

Conclusions

Rethinking how legal systems can continue to support societal security, productivity, and individual flourishing in a dramatically changing world is, without doubt, a challenge. However, in a world where climate change and other anthropogenic stressors are actively pushing adaptive cycles into release and reorganization phases, and where transformation of social-ecological systems is increasingly likely in many parts of the world, the continuation of societal stability increasingly if somewhat ironically depends on law's ability both to adapt itself and to allow for societal adaptation and transformation (Craig et al. 2017; Garmestani et al. 2019). The maladaptiveness of much law is currently most obvious in the laws that directly mediate humans' interactions with their environments: environmental laws and natural resources laws that, premised on a balance-of-nature model of stationarity, continue to assume that restoration and preservation of historical (or even contemporary) ecological and social-ecological states are both desirable and possible.

However, more comprehensive visions of how society should operate in a panarchical world, like the circular economy, extend the potential maladaptiveness of law into many other arenas. Clearly, the circular economy challenges current legal conceptions of industry and would transform manufacturing to meet legal conceptions of product longevity and recyclability, not just legal constraints on resource extraction, pollution control, and waste management. Circular economy scholars themselves emphasize that their vision also depends on a rapid transition to renewable energy (Ellen MacArthur Foundation n.d.; Raworth 2017). Corporate legal duties may need to devolve away from shareholder profit and toward social responsibility, including long-term responsibility for products and biological reintegration. Antitrust law may need to evolve to allow greater homogeneity and interchangeability of products and their components. Even basic legal conceptions of property may need to change, eliminating private property rights in natural resources in favor of stronger private property rights in valuable resources already incorporated into products.

At the same time, however, the law must remain the guardian of other values that democratic society holds dear, values that together add up to good governance, equity, and social stability but that can easily be sacrificed in times of change and transition. Legal procedure remains one of the most important means of protecting these values even as substantive requirements and norms evolve (Craig et al. 2017). For example, as long as any change to the U.S. Constitution receives a two-thirds majority vote in both the House and the Senate, or is proposed through a constitutional convention supported by two-thirds of the states, and then is ratified by three fourths of the states (U.S. Constitution, article V), we can be reasonably certain that the resulting change is necessary and fair and that anyone or any group with a stake in the change has had plenty of opportunity to make their views known. Thus, even as we amend laws' substantive flexibility and discretion to cope with the

Anthropocene and a panarchical world, the legal system must still insist that even the government follow the rules—the very essence of the rule of law.

Literature Cited

Allen, C.R., J.J. Fontaine, K.L. Pope, and A.S. Garmestani. 2011. Adaptive management for a turbulent future. *Journal of Environmental Management* 92: 1339–1345.

Ballotpedia. 2019. History of marijuana on the ballot. https://ballotpedia.org/History_of_marijuana_on_the_ballot

Benson, M.H., and R.K. Craig. 2014. The end of sustainability. *Society and Natural Resources* 27: 777–782.

Benson, M.H., and R.K. Craig. 2017. *The End of Sustainability: Resilience and the Future of Environmental Governance in the Anthropocene*. University of Kansas Press, Lawrence, KS.

Benson, M.H., and A.S. Garmestani. 2011. Embracing panarchy, building resilience and integrating adaptive management through a rebirth of the National Environmental Policy Act. *Journal of Environmental Management* 92: 1420–1427.

Berwyn, B. 2019. Global warming is pushing Pacific Salmon to the brink, federal scientists warn. *InsideClimateNews*. https://insideclimatenews.org/news/29072019/pacific-salmon-climate-change-threat-endangered-columbia-river-california-idaho-oregon-study

Bevir, M. 2009. *Key Concepts in Governance*. Sage Publications, London.

Burris, S., M. Kempa, and C. Shearing. 2008. Changes in governance: A cross-disciplinary review of current scholarship. *Akron Law Review* 41: 1–66.

Camacho, A.E., and R.L. Glicksman. 2016. Legal adaptive capacity: How program goals and processes shape federal land adaptation to climate change. *University of Colorado Law Review* 87: 711–826.

Carpenter, S.R., and W.A. Brock. 2008. Adaptive capacity and traps. *Ecology and Society* 13(2): 40. http://www.ecologyandsociety.org/vol13/iss2/art40/

Chaffin, B.C., H. Gosnell, and B.A. Cosens. 2014. A decade of adaptive governance scholarship: synthesis and future directions. *Ecology and Society* 19(3): 56. https://doi.org/10.5751/ES-06824-190356

Cosens, B.A. 2010. Transboundary river governance in the face of uncertainty: Resilience theory and the Columbia River Treaty. *University of Utah Journal of Land, Resources, and Environmental Law* 30: 229–265.

Cosens, B.A. 2013. Legitimacy, adaptation, and resilience in ecosystem management. *Ecology and Society* 18(1): 3. http://dx.doi.org/10.5751/ES-05093-180103

Cosens, B.A., R.K. Craig, S.L. Hirsch, C.A. Arnold, M.H. Benson, D.A. DeCaro, A.S. Garmestani, H. Gosnell, J.B. Ruhl, and Edella Schlager. 2017. The role of law in adaptive governance. *Ecology and Society* 22(1): 30. http://www.ecologyandsociety.org/vol22/iss1/art30/

Cosens, B.A., and L. Gunderson. 2018. An introduction to practical panarchy: Linking law, resilience, and adaptive water governance of regional scale social-ecological systems. In *Practical Panarchy for Adaptive Water Governance: Linking Law to Social-Ecological Resilience*, ed. B. Cosens and L. Gunderson, 1–16. Springer International Publishing, Gewerbestrausse, Switzerland.

Cosens, B.A., L. Gunderson, and B.C. Chaffin. 2018. Introduction to the special feature practicing panarchy: Assessing legal flexibility, ecological resilience, and adaptive governance in regional water systems experiencing rapid environmental change. *Ecology and Society* 23(1): 4. https://doi.org/10.5751/ES-09524-230104

Cosens, B.A., J.B. Ruhl, N. Soininen, and L. Gunderson. 2020. Designing law to enable adaptive governance of modern wicked problems. *Vanderbilt Law Review* 73: 1687–1732.

Craig, R.K. 2010. "Stationarity is dead"—Long live transformation: Five principles for climate change adaptation law. *Harvard Environmental Law Review* 34: 9–73.

Craig, R.K. 2013. The Clean Water Act, climate change, and energy production: A call for principled flexibility regarding "Existing Uses." *George Washington Journal of Energy and Environmental Law* 4: 26–45.

Craig, R.K. 2019. Trickster law: Promoting resilience and adaptive governance by allowing other perspectives on natural resource management. *Arizona Journal of Environmental Law and Policy* 9: 140–157.

Craig, R.K., and M.H. Benson. 2013. Replacing sustainability. *Akron Law Review* 46: 841–880. http://ideaexchange.uakron.edu/akronlawreview/vol46/iss4/2

Craig, R.K., A.S. Garmestani, C.R. Allen, C.A. Arnold, H. Birgé, D.A. DeCaro, A.K. Fremier, H. Gosnell, and E. Schlager. 2017. Balancing stability and flexibility in adaptive governance: An analysis of tools available in U.S. environmental law. *Ecology and Society* 22(2): 3. https://doi.org/10.5751/ES-08983-220203

Craig, R.K., and J.B. Ruhl. 2014. Designing administrative law for adaptive management. *Vanderbilt Law Review* 67: 1–87.

DeCaro, D.A., B.C. Chaffin, E. Schlager, A.S. Garmestani, and J.B. Ruhl. 2017. Legal and institutional foundations of adaptive environmental governance. *Ecology and Society* 22(1): 32. https://doi.org/10.5751/ES-09036-220132

Dietz, T., E. Ostrom, and P.C. Stern. 2003. The struggle to govern the commons. *Science* 302: 1907–1912.

DISA. 2020. *Map of Marijuana Legality by State*. https://disa.com/map-of-marijuana-legality-by-state

Dorf, M.C., and C.F. Sabel. 1998. A constitution of democratic experimentalism. *Columbia Law Review* 98: 267–473.

Ellen MacArthur Foundation. n.d. Circular economy: Concept. https://www.ellenmacarthurfoundation.org/circular-economy/concept

Flessa, K.W., E.P. Glenn, O. Hinojosa-Huerta, C.A. de la Parra-Rentería, J. Ramírez-Hernández, J.C. Schmidt, and F.A. Zamora-Arroyo. 2013. Flooding the Colorado River delta: A landscape-scale experiment. *Transactions of the American Geophysical Union* 94: 485–496.

Folke, C., T. Hahn, P. Olsson, and J. Norberg. 2005. Adaptive governance of social-ecological systems. *Annual Review of Environment and Resources* 30: 441–473.

Garmestani, A.S., and C.R. Allen. 2014. *Social-Ecological Resilience and Law*. Columbia University Press, New York.

Garmestani, A.S., C.R. Allen, and M.H. Benson. 2013. Can law foster social-ecological resilience? *Ecology and Society* 18 (2): 37. http://www.ecologyandsociety.org/vol18/iss2/art37/

Garmestani, A.S., C.R. Allen, and H. Cabezas. 2009. Panarchy, adaptive management and governance: Policy options for building resilience. *Nebraska Law Review* 87: 1036–1054.

Garmestani, A.S., and M.H. Benson. 2013. A framework for resilience-based governance of social-ecological systems. *Ecology and Society* 18 (1): 9. http://www.ecologyandsociety.org/vol18/iss1/art9/

Garmestani, A., J.B. Ruhl, B.C. Chaffin, R.K. Craig, H.F.M.W van Rijswick, D.G. Angeler, C. Folke, L. Gunderson, D. Twidwell, and C.R. Allen. 2019. Untapped capacity for resilience in environmental law. *Proceedings of the National Academy of Sciences* 116: 19899–19904.

Green, O.O., A.S. Garmestani, C.R. Allen, L.H. Gunderson, J.B. Ruhl, C.A. Arnold, N.A.J. Graham, B. Cosens, D.G. Angeler, B.C. Chaffin, and C.S. Holling. 2015. Barriers and bridges to the integration of social-ecological resilience and law. *Frontiers in Ecology and the Environment* 13: 332–337.

Gunderson, L.H., and C.S. Holling. 2002. *Panarchy: Understanding Transformations in Human and Natural Systems*. Island Press, Washington, DC.

Hardin, G. 1968. The tragedy of the commons. *Science* 162: 1243–1248.

Holling, C.S. 1973. Resilience and stability of ecological systems. *Annual Review of Ecology and Systematics* 4: 1–23.

Holling, C.S. (Ed.). 1978. *Adaptive Environmental Assessment and Management*. Wiley, New York.

Huitema, D., E. Mostert, W. Egas, S. Moellenkamp, C. Pahl-Wostl, and R. Yalcin. 2009. Adaptive water governance: Assessing the institutional prescriptions of adaptive (co) management from a governance perspective and defining a research agenda. *Ecology and Society* 14(1): 26. http://www.ecologyandsociety.org/vol14/iss1/art26/

Intergovernmental Panel on Climate Change (IPCC). 2014. *Climate Change 2014: Synthesis Report*. IPCC, Geneva, Switzerland.

Karkkainen, B.C. 2004. New governance in legal thought and in the world: Some splitting as antidote to overzealous lumping. *Minnesota Law Review* 89: 471–497.

Lebel, L., J.M. Anderies, B. Campbell, C. Folke, S. Hatfield-Dodds, T.P. Hughes, and J. Wilson. 2006. Governance and the capacity to manage resilience in regional social-ecological systems. *Ecology and Society* 11(1): 19. http://dx.doi.org/10.5751/ES-01606-110119

Lee, K.N. 1999. Appraising adaptive management. *Conservation Ecology* 3(2): 3. http://www.consecol.org/vol3/iss2/art3/

Locher, F. 2018. Historicizing Elinor Ostrom: Urban politics, international development and expertise in the U.S. context (1970–1990). *Theoretical Inquiries in Law* 19: 533–558.

Lockwood, M., J. Davidson, A. Curtis, E. Stratford, and R. Griffith. 2010. Governance principles for natural resources management. *Society and Natural Resources* 23(10): 986–1001. https://doi.org/10.1080/08941920802178214

McDowall, W., Y. Geng, B. Huang, E. Barteková, R. Bleischwitz, S. Türkeli, R. Kemp, and T. Doménech. 2017. Circular economy policies in China and Europe. *Journal of Industrial Ecology* 21: 651–660.

National Conference of State Legislatures (NCSL). 2020. Initiative and referendum states. http://www.ncsl.org/research/elections-and-campaigns/chart-of-the-initiative-states.aspx

National Public Radio. 2017. In the Rockies, climate change spells trouble for cutthroat trout. *Morning Edition*. https://www.npr.org/sections/thetwo-way/2017/04/18/523406934/in-the-rockies-climate-change-spells-trouble-for-cutthroat-trout

NBC News. 2019. Trump administration set to end California's ability to regulate auto emissions. https://www.nbcnews.com/politics/donald-trump/trump-administration-set-end-california-s-ability-regulate-auto-emissions-n1055616

Olsson, L., A. Jerneck, H. Thoren, J. Persson, and O'Byrne, D. 2015. Why resilience is unappealing to social science: Theoretical and empirical investigations of the scientific use of resilience. *Science Advances* 1(4): e1400217.

Ostrom, E. 1990. *Governing the Commons: The Evolution of Institutions for Collective Action*. Cambridge University Press, Cambridge, UK.

Quester, A. 2006. Evolution before revolution: Dynamism in Connecticut landlord–tenant law prior to the late 1960s. *American Journal of Legal History* 48: 408–452.

Raworth, K. 2017. *Doughnut Economics: Seven Ways to Think Like a 21st-Century Economist*. Chelsea Green Publishing, White River Junction, VT.

Resilience Alliance. 2020. https://www.resalliance.org/about

Ruhl, J.B. 2012. Panarchy and the law. *Ecology and Society* 17(3): 31. http://dx.doi.org/10.5751/ES-05109-170331

Ruhl, J.B., and R.L. Fischman. 2010. Adaptive management in the courts. *Minnesota Law Review* 95: 424–484.

Salzman, J., and J.B. Ruhl. 2000. Currencies and the commodification of environmental law. *Stanford Law Review* 53: 607–694.

Szaro, R.C., B.K. Williams, and C.D. Shapiro. 2009. *Adaptive Management: The U.S. Department of the Interior Technical Guide*. Science and Decisions Center, US Department of the Interior, Washington, DC. https://pubs.er.usgs.gov/publication/70194537

U.S. Commission on Ocean Policy (USCOP). 2004. *An Ocean Blueprint for the 21st Century*. USCOP, Washington, DC.

U.S. Environmental Protection Agency (EPA). 2020. Acid rain program. https://www.epa.gov/acidrain/acid-rain-program

Volkman, J.M., and W.E. McConnaha. 1993. Through a glass, darkly: Columbia River salmon, the Endangered Species Act, and adaptive management. *Environmental Law* 23: 1249–1272.

Walters, C.J. 1986. *Adaptive Management of Renewable Resources*. Macmillan Publishers Ltd., Basingstoke, UK.

Webster, K. 2013. What might we say about a circular economy? Some temptations to avoid if possible. *World Futures* 69: 542–554.

Chapter 10

Panarchy and the Economy

Joshua Farley and Megan Egler

The modern economic system emerged around the turn of the eighteenth century, simultaneous with the transition to fossil fuels and industrial production, resulting in exponential increases in economic output and human population. Economic theory evolved over the same period, with the core elements of mainstream microeconomic theory coming together from the 1870s onward (Colander 2000; Arnsperger and Varoufakis 2006). However, microeconomic theory fell short in explaining the Great Depression (Keynes 1936), resulting in the emergence of macroeconomic theory. Macroeconomic policies facilitated more rapid growth, driving accelerations in resource depletion, waste emissions, and increases in human population (Steffen et al. 2011). The economic growth of the past 250 years has contributed to improvements in some standards of living, measured by general health, longevity, education, and material consumption. However, impacts of this growth also threaten to flip global social-ecological systems into an alternative state with potentially catastrophic impacts on human welfare (Daly 2014). Neither microeconomics nor macroeconomic theory in its current form can explain or address such state changes. Thus there is again a great need for something new. Panarchy concepts may help inform economic models, as illustrated by a classic example in social–ecological–economic systems involving forest management.

Forest ecosystems are typically composed of a matrix of forest patches in different stages of growth and maturation. Such spatial and temporal patterns of biodiversity, nutrient cycling, and carbon storage are related to recurring natural disturbances such as fire or forest pests (Steel et al. 2015). Holling and Meffe (1996) found that attempts to control or dampen natural disturbances, often in the promotion of economic activity, can lead to more homogeneous and therefore less resilient systems. These systems become more vulnerable to disturbances and their potential to cause systemic collapse.

Most forest ecosystems in pre-European North America experienced recurring fires. Under such regimes, individual fires tend to be of low intensity, limited extent, and with less biomass loss because of limited fuel availability. Such a pattern was captured in Holling's (1986) four-phase adaptive cycle of development, conservation, creative destruction, and renewal. Beginning in the early twentieth century, when timber products became an economic commodity, suppression of wildfires (creative destruction phase) on public lands became the dominant management policy. This resulted in a reduced frequency of fires and a long-term accumulation of fuels, which increased fire severity (Steel et al. 2015). Gunderson (1994) presented similar findings in the Florida Everglades, where delayed fire created abnormally high biomass accumulation and an increased probability that the eventual fire would be of higher intensity and with greater loss of vegetation and soil substrate, affecting not only the particular landscape unit but the entire regional ecosystem. In both cases, management for the conservation phase and suppression of renewal cycles resulted in a collapse at the next hierarchical scale. The suppression of adaptive cycles within economic systems at various scales can be likened to fire suppression in forest ecosystems. However, before presenting our new economic model and the role of panarchy theory therein, we first assess how mainstream economic theory currently addresses hierarchically nested adaptive cycles.

Microeconomics

Mainstream microeconomic theory (hereafter simply *microeconomics*) is based on assumptions that systems tend to be stable and operate around equilibrium conditions. The capacity to deal with shocks or disturbances is described by Holling's (1996, 33) engineering resilience, which he defined as "stability near an equilibrium steady state, where resistance to disturbance and speed of return to the equilibrium are used to measure the property." The foundation of microeconomics is a mathematical framework known as competitive equilibrium. External shocks to supply and demand can disturb this equilibrium while the price mechanism stabilizes it. The basic idea is that price is a measure of scarcity; when a resource becomes scarce, the price increases. Consumers respond to this price increase by demanding less, and producers respond by providing more or by developing substitutes.

After World War II, there was general concern that the economy might face a shortage of raw materials. This led the U.S. government to commission an economic study on resource shortages, which found that technological innovation could replace natural resources as scarcity increased their price (President's Materials Policy Commission 1952), so there was no reason to worry. *Scarcity and Growth*, a highly influential book published in 1963, came to the same conclusion (Barnett and Morse 1963). However, in the 1960s and 1970s, some economists began to worry about environmental amenities, also known as ecosystem services, which do not have prices (Carson 1962; Boulding 1966; Ayres and Kneese 1969). These resources are outside the price mechanism because they are primarily nonexcludable, meaning they have no owners, or nonrival, meaning that use by one person does not leave less for others, and therefore they are not scarce in the economic sense. When resources are nonexcludable, they cannot be sold in markets. When resources are nonrival, prices impose artificial scarcity and are inefficient (Daly and Farley 2011). In response to this concern, several studies concluded that with appropriate policies forcing market actors to pay attention to these nonmarket variables, for example by internalizing external costs into taxes and hence into prices, the market would still allocate resources efficiently (Simpson et al. 2005). Again, there was no need for concern.

Demand-side shocks to the economy are quite different and occur when consumers no longer want to purchase an older product or suddenly shift their preferences toward a new one. Early microeconomic theories were primarily static, focusing only on minor disturbances to equilibrium, but Joseph Schumpeter, building off Marx, recognized that real economies were highly dynamic and that technological advances could make industries obsolete through a process he called creative destruction (Schumpeter 1927). Although they predate panarchy, dynamic models of the microeconomy implicitly incorporate several of its core elements. For example, at the lowest hierarchical scale in the economy, new technologies can make large groups of previously skilled workers obsolete. Firing these workers is analogous to release, and hiring new ones is analogous to reorganization and growth of the firm. Suppressing the adaptive cycle of worker turnover, for example through laws that limit the ability of companies to fire workers, helps increase individual worker resilience to economic shocks but at the expense of firm resilience, especially in an economy predicated on growth. If the firm goes bankrupt, all the workers lose their jobs.

Another adaptive cycle, as described in almost every microeconomic textbook, functions at the level of the firm. Firms using outmoded technologies or producing less desirable products fail to earn a profit. When this happens, the famed "invisible hand of the market" leads owners of the unprofitable firm to reallocate the resources at their disposal toward a more profitable technology or industry. As some firms exit, supply decreases and prices increase, restoring equilibrium and increasing the resilience of the industry, a form of release and reorganization. Creative destruction can also make entire industries obsolete. As old industries become unprofitable, the invisible hand of the market naturally reallocates the factors of production toward more profitable industries—release and reorganization at a larger scale. As more and more new firms enter a profitable industry it increases output—the growth phase of the adaptive cycle—and drives prices down, hence reducing profits. But as output increases in an industry, prices and hence profits are driven down, leading to a conservation phase. The net result is that factors of production will be allocated to the sector where they produce most economic value at the margin. Consumers will be able to purchase products at the lowest price that ensures a fair return on factors of production. Welfare is maximized. Any effort to interfere with the market, to suppress its "natural cycles," threatens the health of the economy as a whole. In short, microeconomics focuses on the price mechanism as a negative feedback loop that maintains economic equilibrium in an ever-growing economy. This ideology rejects any government interventions in markets, any limits to growth, or any possibility of the market economy flipping into an alternative state.

One problem with this theory is that it is extremely simplistic. We know of no complex systems that can be driven to equilibrium by a single feedback loop. Another problem with microeconomics is that individual firms, in general, focus on efficiency over resilience. The goal of a firm is to maximize profits, and most production is financed by interest-bearing credit. Firms therefore can reduce costs through just-in-time production, which means they hold no stocks or factors of production for very long and avoid interest payments incurred by holding them. Firms also have a major incentive to pay as little as possible for factors of production, including raw materials and labor. This often means sourcing inputs from distant countries with weak environmental and labor standards (Daly 2001). The focus on efficiency undermines economic resilience.

Food systems illustrate another major problem with microeconomic theory and practice. According to the literature on planetary boundaries, some of the greatest threats to global ecosystems are biodiversity loss, climate change, nitrogen and phosphorus emissions, land use change, freshwater use, and the emission of toxic chemicals (Steffen et al. 2015). Although agriculture is a major driver of all of these, mainstream economics considers them to be externalities, which is to say external to the economic decision-making process. Environmental economists (a subfield of mainstream economics) claim that we can internalize these externalities by incorporating their estimated monetary values into market prices. However, even if it were possible to accurately monetize such values and convince policymakers to adjust prices accordingly, which we sincerely doubt, the values would change not only with the supply of ecosystem services but also with incomes and with the supply and demand of all market goods and services, demanding continual reestimation. We see little difference between this approach and a centrally planned economy.

Within the market economy, microeconomics accepts adaptive cycles of growth, conservation, release, and reorganization at different scales, with the potential for disruption of these cycles at one scale to threaten resilience at higher scales. However, the highest hierarchical scale within microeconomic theory is the economy as a whole, which is believed to automatically maintain a dynamic equilibrium. There is no recognition that the economy itself is sustained and contained by our finite planetary ecosystem. Although in nonhuman natural systems a finite quantity of resources means that reorganization can only follow release, microeconomic theory assumes that technological innovation can create new resources, such as metals, fossil fuels, or nuclear power. Resources must be released from older firms and industries, and a continual influx of brand new resources allows the whole economy to remain in competitive (engineering) equilibrium even as it grows exponentially and infinitely.

Furthermore, we must stress that despite the nested adaptive cycles of the economy characteristic of complex adaptive systems (e.g., social-ecological systems), the modern market economy did not emerge spontaneously or naturally. Polanyi (1944) argued that so-called laissez faire economies were planned, whereas social uprisings to protest the harm they caused were spontaneous. Premodern economies tended to be far less productive but far more diverse and resilient. Social norms and obligations drove economic activities and relationships, not market forces. In this, the economy was embedded in social relations at a higher level. A confluence of events—large-scale use of fossil fuels, the emergence of the modern state, and nascent capitalism—led to a release and reorganization within the economic system that disrupted society (Hagens 2020). In panarchy theory, this is often called a revolt (Allen et al. 2014). Rather than providing the memory needed for the economy to return to its previous regime, the state and nascent capitalists helped guide the reorganization of the economy at a higher scale, in which social norms became subservient to the market economy. Spontaneous social uprisings fought this revolt, but state efforts to suppress or mitigate them further strengthened the market (Polanyi 1944). The state acquired the memory necessary to sustain capitalism after release and reorganization. Such state efforts are the domain of macroeconomics, to which we now turn.

Macroeconomics

It can be difficult to make sweeping generalizations about macroeconomics because there are so many theories and so little consensus about which are correct. The competitive equilibrium model failed to explain the Great Depression, and before this, macroeconomics in its modem form did not even exist. Keynes (1936), perhaps the most famous economist of the twentieth century, developed a highly influential approach to macroeconomics that acknowledged the possibility of shocks to aggregate demand. A drop in aggregate demand leads firms to reduce production, buy fewer inputs, and lay off workers, which further reduces demand. This can create a positive feedback loop that flips an economy into an entirely different regime of behavior, another stability domain from which it would be difficult to return to the original state without significant government intervention. Macroeconomic theory recognizes that economies can exhibit multiple equilibria, such as one with full production and employment and one with excess capacity and high unemployment.

After World War II, leading economists integrated neoclassical microeconomics with Keynes's theories to develop the neoclassical–Keynesian synthesis, or neo-Keynesianism, as the most widely accepted theory, in which neoclassical economics was used to explain the microeconomy and Keynesian economics to explain the macroeconomy (Blanchard 2008). But this theory proved incapable of dealing with the stagflation of the 1970s, leading to a proliferation of macroeconomic theories. For example, monetarism argued that only monetary policy should be used in response to economic crises (Friedman 2008); new neoclassical macroeconomics tried to reduce macroeconomics to the general equilibrium roots of microeconomics, coupled with rational expectations (Lucas and Sargent 1981). Post-Keynesian economics builds from Keynesianism but rejects neo-Keynesianism (Lavoie 2009). There are many other theories as well. Nobel Laureate Paul Romer summarized our views on most of these theories when he stated that "For more than three decades, macroeconomics has gone backwards" because "objective fact [has been displaced] from its position as the ultimate determinant of scientific truth" (Romer 2016, 1).

All these theories have had some impact on macroeconomic policy, but none were fully able to explain the Great Recession starting in 2007. As further evidence of the problems with macroeconomics, Ben Bernanke, former chair of the Federal Reserve who had conducted research on the Great Depression, stated in 2004 that we had achieved a great moderation—a reduction in macroeconomic volatility driven by structural changes and better understanding of macroeconomic policy (Bernanke 2004). Ironically, Irving Fisher, a famous economist from the 1920s, just days before the market crash of 1929 stated that "stock prices have reached what looks like a permanently high plateau. I do not feel there will be soon, if ever, a 50 or 60 point break from present levels, such as (bears) have predicted. I expect to see the stock market a good deal higher within a few months" (cited in Rapp 2009, 34). Bernanke was presumably aware of this quote yet made the same mistake. Astonishingly few mainstream economists predicted the great recession (Colander et al. 2009), although several heterodox economists did so (e.g., Baker 2006; Keen 2011).

Despite the deep disagreement on how the macroeconomy works, macroeconomics acknowledges the potential for positive feedback loops that can drive the economy out of equilibrium and flip it into another stable regime from which it is difficult to return. Resilience under these conditions is known as ecological resilience, the amount of change a system can absorb before shifting to a new regime (Holling 1973; Gunderson 2000). Macroeconomic policy attempts to manage or suppress such positive feedback loops. Governments have created numerous stabilizing mechanisms, ranging from safety nets that stabilize aggregate demand, such as Social Security and unemployment insurance, to implicit guarantees of bailouts for the financial sector, as exemplified most recently by Fed chairman Jerome Powell's March 23, 2020, commitment to buy unlimited financial assets (Smialek 2020). Not coincidentally, Powell's commitment triggered a record surge in the stock market in the midst of the worst economic crisis since the Great Depression (Stevens 2020). Central banks also use interest rates as a policy. When interest rates are high, firms are unlikely to invest in increased capacity and consumers are unlikely to borrow. In the event of an economic shock, central banks typically lower interest rates to stimulate job-creating investments and consumer borrowing. However, firms have no incentive to invest in new production when aggregate demand is low, so lowering interest rates has little impact (Keynes 1936).

The explicit goals of macroeconomic policy are generally to maintain economic growth, to ensure full employment, and to manage inflation. There are two basic categories of tools at the disposal of macroeconomists. The first of these is fiscal policy, which includes taxation and other forms of capturing revenue for the government, as well as government expenditures. The second is monetary policy, which primarily uses interest rates to influence investment, savings, and consumption decisions. The basic idea is that lowering taxes or increasing government expenditures increases aggregate demand, hence employment. Lowering interest rates encourages firms to spend more, speculators to invest more, and consumers to purchase more, also increasing aggregate demand. If aggregate demand grows faster than supply, it can trigger inflation (Daly and Farley 2011).

The state that macroeconomics tries to maintain is that of continuous growth. It therefore focuses on suppressing the release and reorganization cycles of the economic subsystem as a whole. Before the 1970s, the government intentionally engaged in Keynesian policies, increasing government expenditures and reducing taxes when the economy was sputtering in order to increase aggregate demand, and doing the opposite to suppress economic bubbles. Keynesian economic theories were unable to explain the recession plus inflation, known as stagflation, triggered by the oil price shocks of the 1970s, leading to a resurgence of an extreme form of free market economics known as neoliberalism, with the basic tenet that free markets are inherently efficient and government intervention should be avoided (Wasserman 2019). This led to major deregulation of the financial sector (Evanoff 1985; Sherman 2009).

The financial sector plays a critical role in the modern economy. Most money is loaned into existence as interest-bearing debt and destroyed when that debt is repaid. Governments can also spend money into existence and destroy it through taxes, but outstanding debt to private banks is by far the primary source of our money supply (Jackson and Ryan-Collins 2011; Farley et al. 2013; Wray 2015). However, finance is particularly prone to positive feedback loops driven by debt and default. As we will explain, excessive accumulation of debt can lead to financial conflagrations that can flip the system into an alternative state, in the same way that the excessive accumulation of dead wood in forests can cause conflagrations that irreparably damage forest ecosystems.

Before deregulation in the 1980s, the financial sector typically loaned money to businesses, which invested in increasing real physical output (Hudson 2012). When labor and other factors of production are not fully employed, investments create new jobs and increase aggregate demand, potentially creating demand for yet more investments in a positive feedback loop (a growth phase) that is naturally dampened as full employment returns (a conservation phase). Within any given industry, however, increased investments increase supply and lower prices, decreasing demand for new investments, in a negative feedback loop that again facilitates a transition from growth to conservation. If the value of increased production exceeds payments of principle plus interest, all works well for firms and financiers, but this is unlikely for two reasons. First, our money supply is unpaid debt, which grows by the inexorable laws of compound interest, whereas real physical output is subject to biophysical constraints that make continuous exponential growth impossible (Soddy 1935; Daly 1980). Debt tends to grow faster than real output (figure 10.1), increasing the likelihood of default. Second, negative feedback loops are less effective when there are long time lags between investments and output (Meadows 2008). The problem is compounded for essential resources (such as food or energy) that exhibit price-inelastic demand, meaning that demand is highly insensitive to price, and price is highly sensitive to supply (Farley et al. 2015). If too many firms respond to high prices by increasing investment, the resulting increase in supply a year (e.g., agriculture) or two (e.g., oil) later can cause prices to crash, in which case investors may be unable to recoup sufficient revenue to repay loans. This mechanism was driving record bankruptcies in the hydraulic fracturing industry even before its collapse because of the demand side shock of COVID-19. Since 2008, loans to the fracking sector have exceeded revenue by $250 billion (McLean 2020; Mikulka and Kelly 2020; Phillips and Krauss 2020). When debt cannot be repaid, banks stop lending new money. The resulting decrease in money supply can cause recession or depression—the release phase of the adaptive cycle. The government provides the accumulated memory needed for subsequent recovery—the reorganization and growth phases of the cycle. However, when banks lend more to industries than they recover, they are subsidizing production, with all its ecological harm, threatening the larger and slower cycles of the global ecosystem.

Hyman Minsky (1986), a post-Keynesian economist, proposed a theory of a macroeconomic cycle of growth, release, and reorganization. Investors look to the recent past as a predictor of future events and are cautious after a crisis. Cautious behavior makes economic crisis unlikely but also forgoes potential money-making opportunities. As time passes without crisis, investors borrow money for increasingly riskier investments. Greater investment stimulates the economy and inflates asset prices, stimulating more demands for loans in a positive feedback loop. Eventually, investors take on loans whose repayment depends on rising asset prices and a growing economy, at which point even a small disruption to the economy can halt the rise in asset prices. Investors must now sell their assets to meet payment obligations, driving asset prices down, leading to even more sales and lower prices. Many investors

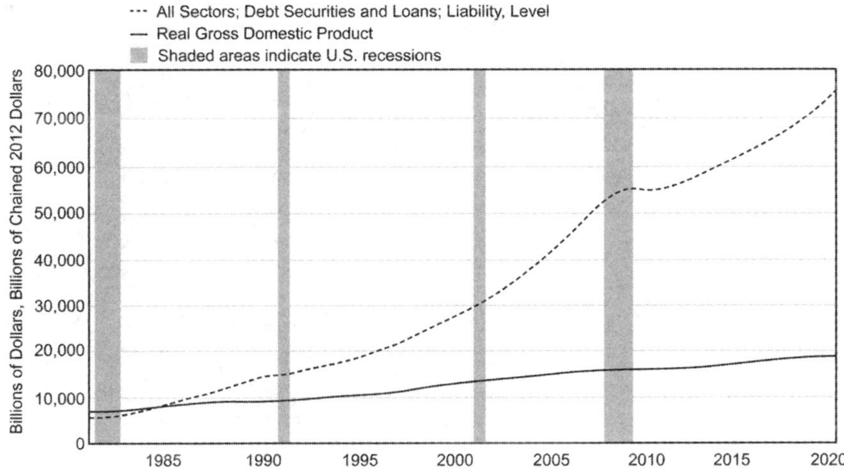

Figure 10.1. The growth of debt and real gross domestic product from 1980 to 2020 in the United States. Shaded bars indicate periods of U.S. economic recessions. (Data from U.S. Federal Reserve Bank of St. Louis, https://fred.stlouisfed.org/series/BOGZ1FL894104005A).

default, leading banks to restrict credit, driving asset prices down even further. As credit dries up, the money supply contracts, and even risk averse investors fail to make payments, leading to a full-blown financial and economic crisis with low investment and low employment. If the crisis is severe, government intervention is necessary to guide reorganization and initiate a subsequent growth cycle. The 2007 financial crisis is a recent example of this narrative (Cassidy 2008).

Financial deregulation increases the likelihood and severity of such cycles. After deregulation, most credit has been extended for the purchase of existing assets (Hudson 2012). The supply of land is also fixed, as are stocks, aside from initial public offerings (IPOs). For existing assets, rising prices increase demand and provide more collateral to back new loans, further increasing prices in a positive feedback loop. Whereas loans for productive investments affect supply more than demand and result in lower prices (especially in essential commodities such as oil and food), loans for purchasing existing assets affect demand, not supply, and result in higher prices, as depicted in figure 10.2.

Before 1982, firms were prohibited from buying back their own stocks, considered a form of market manipulation (Saez and Zucman 2019). This changed with financial deregulation. Firms can now either invest in real production, driving down the price of what they produce, or buy back their own

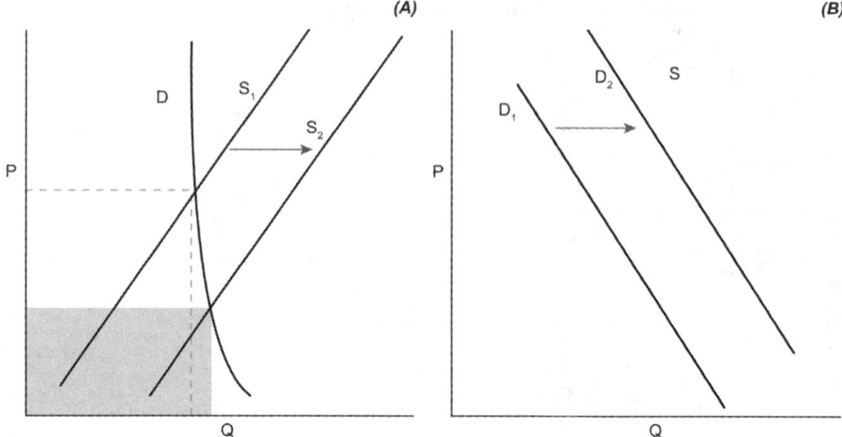

Figure 10.2. Supply and demand curves for two different resources. (A) An essential resource with inelastic demand, such as oil. Financing investments in greater production shifts supply from S_1 to S_2, leading to only a small increase in demand. Revenue declines from the square marked by the dashed line to the gray square, making repayment difficult. (B) Supply of an existing asset such as land or stocks. An increase in credit shifts the demand curve from D_1 to D_2, increasing the price, which further increases the demand for credit—a positive feedback loop. Any subsequent decrease in demand will reduce prices, making repayment difficult or impossible.

stocks, driving up the price. Aggravating matters, taxes on capital (also known as unearned income) were cut in 1997 and 2001 relative to taxes on earned income (Saez and Zucman 2019). Interest payments are also exempt from taxes. Clearly, under these rules, purchasing existing assets for later resale at a higher price is likely to earn higher profits than investing in new production. The increased demand for assets drives up the price of those assets, further increasing demand in a positive feedback loop with no obvious biophysical limit (Saez and Zucman 2019). When investors buy IPOs, it provides money for firms to invest in real production. The value of IPOs from 2010–2019 reached almost $600 billion (Brewer 2020). But corporate stock buybacks in 2019 alone, funded primarily by debt (Light 2019), were in the vicinity of $800 billion (Lazonick et al. 2020), exceeding IPOs by an order of magnitude. Money spent on buybacks cannot be invested elsewhere, and the stock market has become a net drain on capital that could be invested in real production.

Investors have been borrowing at record levels to buy stock, with margin debt now exceeding $500 billion (Mislinski 2020). The value of stocks and land is now determined largely by the availability of credit and the interest rate on that credit (Hudson 2012). When the stock market falls, many investors are unable to repay their debt, forcing them to sell their stocks, which in turn drives stock values even lower in a process known as debt deflation. With old loans defaulting and the stock market falling, the financial sector will dramatically reduce loans for stock market purchases, and demand for stocks will fall further (Fisher 2012). If defaults continue, banks reduce lending to producers as well, which can trigger recession. Enough defaults threaten the entire financial system and economy. Central banks typically respond by lowering interest rates and increasing credit in order to prolong growth and avoid release and reorganization. However, just as suppressing forest fires only worsens future conflagrations, suppressing debt defaults leads to an increasingly vulnerable financial sector capable of doing far more harm when it inevitably collapses (Hagens, 2020).

Recent events support this argument. In response to the COVID-19 crisis, the stock market dropped by a third from February 20 to March 23, 2020. On March 15 Fed chairman Jerome Powell slashed real interest rates to below zero, and on March 23 he promised to purchase assets in whatever quantity necessary to stabilize the economy. This was widely interpreted as including purchases of stocks (Cox 2020), and in May the Fed began purchasing exchange traded funds (ETFs). Although the Fed so far has focused primarily on buying bonds and mortgage backed securities, this has freed up money for investing in stocks. As a result of these policies, the stock market has soared (Sizemore 2020) even as economic news grows increasingly dismal. Whereas classical economists argued that speculators did not contribute to economic production, neoliberals asserted that markets are inherently efficient, and the generation of profits is equivalent to wealth creation. From this perspective there is no difference between profits from money invested in physical production and money chasing existing assets (Mazzucato 2018). These macroeconomic policies follow the same playbook as in 2008. We believe the resulting increase in debt only creates more fuel for a future financial meltdown (Mazzucato 2018).

So far, we have focused on monetary debt, which can trigger economic collapse, but abundant credit at low interest also stimulates industrial production, which in turn consumes raw materials and energy, unavoidably generating waste in the process. When we consume raw materials faster than they regenerate and spew waste (e.g., greenhouse gases) faster than it can be absorbed, we also incur ecological debt. Earth Overshoot Day is the calendar day by which the global economy uses an entire year's worth of planetary production. In 2019, this took place on July 29. For the remainder of the year, we "borrowed" from the planet's capital stocks, equivalent to deficit spending of 40 percent of our ecological income. If we sum up this ecological deficit since we first experienced overshoot in the 1970s, we now have an ecological debt of 1,700 percent of our ecological income. This implies that if we halted all economic activity, it would take only seventeen years for the Earth's ecosystems to recover, no doubt a conservative estimate (Wackernagel et al. 2019). Even so, it dwarfs the scale of credit market debt shown in figure 10.1, which was some 360 percent of U.S. GDP before the massive economic contraction and debt expansion triggered by COVID-19. We believe our ecological debt is far more alarming than monetary debt because it directly reduces the resilience of the global ecosystems on which we depend but is largely ignored by mainstream macroeconomists and economic policy.

To summarize, the goal of modern macroeconomics is to suppress adaptive cycles in the economic system at various scales to maintain economic growth. Just as the suppression of forest fires in individual stands can lead to a buildup of fuel, creating the conditions for extreme fires that lead to collapse of the larger forest system, the effort to sustain rising asset prices, particularly stocks, can lead to a buildup of debt that results in the collapse of the financial system. Over recent decades, the size of the financial system relative to the economy as a whole has nearly quadrupled (Philippon 2019). As we saw in 2008, the collapse of the financial system threatens the economy as whole. Much more dire, however, has been the role of macroeconomics in sustaining continuous economic growth since 1950. The rate at which we increase our ecological debt threatens to flip the entire global ecosystem into an alternative state that may not be suitable for humankind. The longer we remain in this growth phase, the greater the threat of a revolt that disrupts planetary ecosystems. We need new economic theories and new economic systems capable of restoring adaptive cycles at multiple scales. We need new economics: ecological economics.

Ecological Economics

Ecological economics starts from the assumption that the economy is a subsystem of society, similar to macroeconomics, but also recognizes the economy is a subsystem of our finite planetary ecosystem (Daly 1973). Economy, society, and ecosystem are all coevolving complex adaptive systems (Norgaard 1994). In short, ecological economics recognizes that the economy is a rapidly evolving complex adaptive system nested within and dependent on larger, slower-moving complex adaptive systems—in short, a panarchy. Continued economic growth—the goal of mainstream economics and policymakers—threatens to flip the global ecosystem into a new state with potentially catastrophic impacts for humans and other species.

The goals of ecological economics derive from this biophysical paradigm. One of the primary goals is to ensure that the material scale of the human economy does not exceed the capacity of global ecosystems to sustain it while also supporting well-being for the broader community of life (Daly and Farley 2011). Given a finite planet, solar energy flows and terrestrial stocks of raw materials and energy must be justly distributed between human societies and the rest of nature. The raw materials transformed into economic products alternatively serve as the structural building blocks of ecosystems. Therefore, efficient allocation, the third goal of ecological economics, has two components. First, we must efficiently allocate ecosystem structure between economic production and the ecosystem functions that sustain all life. Second, we must allocate raw materials toward the economic products that generate the greatest well-being for this and future human generations. The goal of ecological economics is not to promote endless growth but rather to balance "what is biophysically possible with what is socially, ethically and morally desirable" (Farley and Voinov 2016, 1).

It is therefore important to emphasize both sociopolitical and ecological processes (Spash 2011) rather than focusing only on markets as the central framework to organize society. Ecological economists expand the definition of the economy beyond markets to include all forms of social provisioning. This is in distinct contrast to mainstream environmental economics, which theorizes that negative impacts on society or the environment due to economic activity can be corrected by internalizing them within the economy through the expansion and creation of markets. The market economy is conceptualized as the whole, with nature and society as nested subsystems. Ecological economics' framing of economies as being nested within societies nested within ecosystems is better equipped to understand how interdependencies between the three play into the systems' resilience. Although the core ideas of ecological economics were developed before panarchy, both were heavily influenced by system ecology, and we see the two as closely related. Below, we offer suggestions for how panarchy can inform ecological economics as we strive to create a just and sustainable future.

Institutional Plurality as a Path to Resilience

Forest ecosystems are composed of different stands in different stages of regrowth with different species compositions. As some stands reach the release stage of the adaptive cycle, others help provide the "memory" for them to reorganize into their previous state. Diversity is also a prerequisite for evolutionary adaptation as conditions change at larger hierarchical scales. In an analogous fashion, human societies and economies should strive toward institutional plurality, meaning a variety of economic, social, and political institutions. Global society has been converging toward growth-dependent market economies, which both increases the likelihood of abrupt changes to global ecosystems and reduces the variation needed for evolutionary adaption to these inevitable changes, some of which may already be imminent and unavoidable (Hughes et al. 2013; Steffen et al. 2015). We believe that numerous independent socioeconomic experiments, varying between contexts and redundant in function, will prove better able to adapt to changing conditions. We can never be sure how our environment will evolve. Different socioeconomic structures might collapse under different scenarios, while others will thrive. Plurality prepares us for unpredictable events, as when manufacturing firms are more diverse within and across scales in a regional social-ecological system (Garmestani et al. 2006). When some structures collapse, remaining structures provide the adaptive capacity of a system, allowing a system to rebuild and new structures to take seed (Angeler et al. 2019).

Schumpeter (1927) uses the term *creative destruction* to describe the importance of collapse and renewal of firms and industries in order for the larger economic system to sustain itself. In other words, resilience requires adaptation, which happens through a process of creative destruction and renewal. This is in contrast to the notion of sustainability, which would seek to sustain, preserve, and maintain systems at their status quo (Voinov and Farley 2007). Sustainability in the socioeconomic sense, then, occurs when higher hierarchical levels seek to slow or stall adaptive cycles in lower levels in order to inhibit regime change, or revolt. Applied to economics, panarchy theory helps explain why economies sustained in this manner, as in modern macroeconomics, become more spatially homogeneous and less resilient over time. Since the Industrial Revolution, market economies have displaced a variety of alternative institutions, many of which exhibited greater dependence on institutions of reciprocity, redistribution, and householding (production for one's own use). Financial systems have become ubiquitous, and when they collapse, they threaten market-dependent communities' ability to meet their essential needs. This is because there no longer exists a strong redundancy in systems of social provisioning. The system is more likely to degrade under disturbances that previously could have been absorbed, and these disturbances become increasingly hard to predict. It becomes reasonable to anticipate that an eventual release will come at an unacceptable cost to humankind: a collapse in the functioning of the complex and interdependent institutions that maintain the structure of modernity (Gunderson and Holling 2002).

Similar to the twentieth-century management of forest ecosystems, we have made the wrong decisions about what to sustain in the economy. For example, during the COVID-19 crisis the U.S. Federal Reserve has been pumping trillions of dollars in new debt into financial markets to avoid complete collapse of asset prices (Stein and Whoriskey 2020), even when those prices have little relationship to the underlying value of the assets themselves. The federal government's $500 billion in emergency grants and loans to large American corporations was nominally dedicated to the more important goal of maintaining employment, but the absence of restrictions on how these funds might be spent and absence of oversight meant that the lion's share was transferred to executives and shareholders, essentially passing the support on to the already wealthy (Stein and Whoriskey 2020). This was remarkably similar to the government response to the 2007 financial crisis (Barofsky 2012). Some of the $350 billion relief fund intended for small businesses across America in 2020 ended up in the hands of publicly traded companies with millions of dollars in annual sales, including chain restaurants and the energy sector (Sullivan et al. 2020). This represents a largely debt-based stimulus package for companies already dependent on economic growth rather than locally rooted and supported businesses that also maintain interdependent relationships with community and local social-ecological systems. This makes economic sense only in terms of a theory that not only positions the economy as separate from ecosystems and capable of infinite growth but also prioritizes growth as an end in itself.

Within panarchy theory, higher levels set the conditions within which lower levels reorganize through the connection of "memory." This means when lower levels (e.g., financial markets) do eventually collapse, higher levels (e.g., central banks and national governments) can prevent the system from reorganizing into new system configurations. The power to adapt to changing conditions is lost. Attempts to limit and correct recurrent crises have retained the same system structure and thus laid the foundation for new, and often worse subsequent crises. In essence, we have been putting out the financial forest fires but adding more fuel (debt) for future, larger collapse. At the higher hierarchical scale of the global economy, we have been rapidly increasing ecological debt to maintain growth, threatening resilience at the highest hierarchical level relevant to humans: the global scale. The growth-oriented market economy now requires release and reorganization into new system configurations. Accumulated financial debt will eventually drive a collapse of the financial sector severe enough to drive release at the scale of the global economy. Even worse, accumulated ecological debt may drive release at the planetary scale. Our unprecedented challenge is to achieve release at the scale of the global economy that minimizes negative impacts without causing undue suffering at the scale of the household and community. Economic stimulus aimed instead toward adaptive cycles at lower hierarchical levels of the economic panarchy, such as context-specific supports that focus on community or individual well-being, would do more to build resilient societies in the face of crisis while release and reorganization of the system at higher levels is permitted.

Planned Release to Achieve Sustainable Scale

The prevention of forest fires in forest ecosystems can result in fuel accumulation, and because not every fire can be stopped, an intensification, both horizontally and vertically, of the eventual fire event can generate disturbances to higher-level systems. Similarly, ecological economics recognizes that preventing collapse of the financial sector by increasing debt threatens disturbance to the global economy, and maintaining economic growth by increasing ecological debt threatens profound disturbances to global ecosystems.

Analogous to controlled forest fires designed to reduce the excessive accumulation of fuel, societies can reduce their economic and ecological debt through carefully planned release, without triggering conflagration. Within ecosystems, we care about the overall health of the system or of the species within, rarely about specific individuals, but in human systems justice and individual rights are fundamental. Financial debt redistributes wealth from debtor individuals and countries to creditors (owners of capital), from the poor to the rich (Hudson 2012; Piketty 2014). Ecological debt is disproportionately incurred by the wealthy, whereas the poor, who have benefited least from the resulting economic growth, suffer disproportionately from its costs (Oxfam 2017; https://ejatlas.org/). Controlled release must strive to address this injustice.

Ecological economists call for a steady-state economy, in which the resource extraction, waste emissions, human-made artifacts, and the human population are all nongrowing. Since Herman Daly first proposed these ideas in the 1970s (Daly 1973, 1974), it has become obvious that a period of degrowth must precede the steady state. Degrowth scholars, allies of ecological economists, call for the voluntary, democratic, and equitable downscaling of societies' material throughput, supported by fundamental social transformation (Latouche 2010; Schneider et al. 2010). This is profoundly different from unplanned economic recession; it means reconfiguring systems to be independent of economic growth and finding new ways of achieving well-being for all beings. Feasible proposals for addressing financial debt will probably include some combination of time-honored debt jubilees (Graeber 2011; Hudson 2018, 2020), a public banking system that transforms finance into a public utility (Brown 2013), alternatives to debt-based currencies (Douthwaite 2012), and a deeper understanding of the connections between money creation, human well-being, and ecological debt (Lietaer et al. 2012; Farley et al. 2013; Jackson et al. 2014; Ament 2019, 2020; Svartzman et al. 2019). Addressing ecological debt demands that we use material flows from ecosystems at a rate slower than they regenerate, until the debt has been repaid. Wackernagel et al. (2019) described several specific options for this transition, but this transition is also a central goal of ecological economics and degrowth.

In the context of panarchy theory, higher-level hierarchies would maintain themselves by drawing on the adaptive capacities of lower-level hierarchies but by remaining within the material, spatial, and temporal scale of lower levels' capacities to renew. Within current economic systems' organization, the adaptive cycles of ecological systems used for economic production are hampered by overuse. For example, an agricultural system compatible with the soil ecosystem's capacity to renew itself can be maintained from cycle to cycle. When the level of production exceeds this capacity, the farmer must resort to additives—mostly mined from finite stocks of nonrenewable resources—or bring new land under production in order to maintain the same output. On the demand side, the resilience of the current industrial agricultural system is contingent on the persistent growth of consumer markets and human populations, and on the supply side, a steady stream of primarily nonrenewable inputs to replace degrading soils and changing weather patterns and to adjust to pest evolution. Just as a buildup of fuel in forests threatens a broad-scale fire, the buildup of financial and ecological debt threatens revolt that catastrophically disrupts higher levels of both economic and ecosystem panarchies.

Degrowth, as both a scholarship and a social movement, promotes social practices that replace overconsumptive behaviors, such as the sharing, repairing, and reusing of material products, the building of reciprocal relationships that span beyond human communities, and a reconceptualization of well-being away from the ownership of things and more toward conviviality, sufficiency, and connection. Degrowth also challenges institutions that privilege the logics of productivism, position economic growth as an imperative, and homogenize conceptualizations of possible futures. Release and reorganization in the human system, or degrowth, must be tailored to local societies, ecosystems, and circumstances, restoring the high pluralism that once characterized human societies. Variation is a prerequisite for the evolutionary changes needed to adapt to our continuously changing global ecosystems. What is necessary to recalibrate human societies that have outgrown the material flows of ecosystems is nothing short of social-ecological transformation (see Walker et al. 2004; Chaffin et al. 2016).

Cultural Evolution toward Antifragile Strategies: Final Reflections

Rebuilding the resilience of the human economy and the higher-level hierarchical systems on which it depends is a serious challenge, but the history of cultural evolution gives us reasons to be hopeful. Most species expand their populations to use available resources with subpopulations periodically passing through the adaptive cycles of release, reorganization, growth, and conservations. However, humans accumulate knowledge, technologies, norms, and behaviors beyond the capacity of any single person to create or understand (Wilson 2012; Henrich 2016; Moffett 2018). This represents a major evolutionary transition, in which individual humans have come to cooperate to such a degree that many cannot survive (Szathmáry 2015; West et al. 2015) without the accumulated cultural knowledge dispersed among many individuals (Henrich 2016). Natural selection in humans acts on the group, and cultures evolve (E.O. Wilson 2012; D.S. Wilson 2019). Humans populated the planet by developing myriad new technologies and cultures far more rapidly than we could evolve genetically. In the past, societies have repeatedly grown too large for the ecosystems that sustained them. With other species this might have triggered release and reorganization, and this certainly happened for individual human societies. However, culture, including technology, allowed us to develop new resources, from agricultural production to fossil fuels, that allowed new cycles of reorganization and growth without first experiencing release. Most of these new technologies were developed with a prioritization toward economic growth and tapped finite stocks, not continuous flows. They were, and are, therefore inherently unsustainable. By tapping these stocks, we have increased our ecological debt enormously. We can longer afford to increase our ecological debt and must begin to pay it down. We believe that convivial technologies (Vetter 2018) and cultural evolution can again help us adapt without undergoing a painful release analogous to a forest conflagration, and allow humankind to continue existing on Earth, but only if we learn from our past mistakes.

Literature Cited

Allen, C.R., D.G. Angeler, A.S. Garmestani, L.H. Gunderson, and C.S. Holling. 2014. Panarchy: Theory and application. *Ecosystems* 17: 578–589.

Ament, J. 2019. Toward an ecological monetary theory. *Sustainability* 11(3): 923. https://doi.org/10.3390/su11030923

Ament, J. 2020. An ecological monetary theory. *Ecological Economics* 171: 106421.

Angeler, D.G., H. Peterson, C.R. Allen, A. Garmestani, D. Twidwell, W. Chuang, V.M. Donovan, T. Eason, C.P. Roberts, S.M. Sundstrom, and C.L. Wonkka. 2019. Adaptive capacity in ecosystems. *Advances in Ecological Research* 60: 1–24.

Arnsperger, C., and Y. Varoufakis. 2006. What is neoclassical economics? The three axioms responsible for its theoretical oeuvre, practical irrelevance and, thus, discursive power. *Panoeconomicus* 53: 5–18.

Ayres, R.U., and A.V. Kneese. 1969. Production, consumption, and externalities. *American Economic Review* 59: 282–297.

Baker, D. 2006. Recession looms for the U.S. economy in 2007. Center for Economic and Policy Research. https://www.cepr.net/documents/publications/forecast_2006_11.pdf

Barnett, H., and C. Morse. 1963. *Scarcity and Growth: The Economics of Natural Resource Availability*. John Hopkins University Press, Baltimore, MD.

Barofsky, N. 2012. *Bailout: An Inside Account of How Washington Abandoned Main Street While Rescuing Wall Street*. Free Press, New York.

Bernanke, B. 2004. The great moderation. https://www.federalreserve.gov/boarddocs/speeches/2004/20040220/

Blanchard, O.J. 2008. Neoclassical synthesis. In *The World of Economics*, 2nd ed., ed. J. Eatwell, M. Milgate, and P. Newman, 504–510. Springer, Dordrecht, The Netherlands.

Boulding, K.E. 1966. The economics of the coming Spaceship Earth. In Environmental Quality in a Growing Economy, ed. H. Jarrett, 3–14. Resources for the Future/Johns Hopkins University Press, Baltimore, MD.

Brewer, P. 2020. Analysis: Three decades of IPO deals (1990–2019). *Bloomberg Law*. https://news.bloomberglaw.com/bloomberg-law-analysis/analysis-three-decades-of-ipo-deals-1990-2019

Brown, E. 2013. *The Public Bank Solution: From Austerity to Prosperity*. Third Millennium Press, Baton Rouge, LA.

Carson, R. 1962. *Silent Spring*. Houghton Mifflin, Boston, MA.

Cassidy, J. 2008. The Minsky moment: Subprime mortgage crisis and possible recession. *The New Yorker*. https://www.newyorker.com/magazine/2008/2002/2004/the-minsky-moment

Chaffin, B.C., A.S. Garmestani, L.H. Gunderson, M.H. Benson, D.G. Angeler, C.A. Arnold, B. Cosens, R.K. Craig, J.B. Ruhl, and C.R. Allen. 2016. Transformative environmental governance. *Annual Review of Environment and Resources* 41: 399–423.

Colander, D. 2000. The death of neoclassical economics. *Journal of the History of Economic Thought* 22: 127–143.

Colander, D., M. Goldberg, A. Haas, K. Juselius, A. Kirman, T. Lux, and B. Sloth. 2009. The financial crisis and the systemic failure of the economics profession. *Critical Review* 21: 249–267.

Cox, J. 2020. "Nothing is out of the question": What it would take for the Fed to start buying stocks. CNBC. https://www.cnbc.com/2020/03/29/what-it-would-take-for-the-fed-to-start-buying-stocks-during-coronavirus-crisis.html

Daly, H.E. 1973. *Toward a Steady-State Economy*. W.H. Freeman and Co., San Francisco, CA.

Daly, H.E. 1974. The economics of the steady state. *American Economic Review* 64: 15–21.

Daly, H.E. 1980. The economic thought of Frederick Soddy. *History of Political Economy* 12: 469–488.

Daly, H.E. 2001. Globalization and its discontents. *Philosophy and Public Policy Quarterly* 21: 17–21.

Daly, H.E. 2014. *From Uneconomic Growth to a Steady-State Economy*. Edward Elgar, New York, NY.

Daly, H.E., and J. Farley. 2011. *Ecological Economics: Principles and Applications*, 2nd ed. Island Press, Washington, DC.

Douthwaite, R. 2012. Degrowth and the supply of money in an energy-scarce world. *Ecological Economics* 84: 187–193.

Evanoff, D.D. 1985. Financial industry deregulation in the 1980s. *Economic Perspectives, Federal Reserve Bank of Chicago* 9: 3–5.

Farley, J., M. Burke, G. Flomenhoft, B. Kelly, D.F. Murray, S. Posner, M. Putnam, A. Scanlan, and A. Witham. 2013. Monetary and fiscal policies for a finite planet. *Sustainability* 5: 2802–2826.

Farley, J., A. Schmitt Filho, M. Burke, and M. Farr. 2015. Extending market allocation to ecosystem services: Moral and practical implications on a full and unequal planet. *Ecological Economics* 117: 244–252.

Farley, J., and A. Voinov. 2016. Economics, socio-ecological resilience and ecosystem services. *Journal of Environmental Management* 183: 389–398.

Fisher, J. 2012. No pay, no care? A case study exploring motivations for participation in payments for ecosystem services in Uganda. *Oryx* 46: 45–54.

Friedman, M. 2008. *Monetary History of the United States, 1867–1960*. Princeton University Press, Princeton, NJ.

Garmestani, A.S., C.R. Allen, J.D. Mittelstaedt, C.A. Stow, and W.A. Ward. 2006. Firm size diversity, functional richness and resilience. *Environment and Development Economics* 11: 533–551.

Graeber, D. 2011. *Debt: The First 5000 Years*. Melville House, Hoboken, NJ.

Gunderson, L. 1994. Fire patterns in the southern Everglades. In *Everglades. The Ecosystem and Its Restoration*, ed. S.M. Davis and J.C. Ogden, 291–305. St. Lucie Press, Delray Beach, FL.

Gunderson, L.H. 2000. Ecological resilience—In theory and application. *Annual Review of Ecology and Systematics* 31: 425–439.

Gunderson, L.H., and C.S. Holling, Eds. 2002. *Panarchy: Understanding Transformations in Human and Natural Systems*. Island Press, Washington, DC.

Hagens, N.J. 2020. Economics for the future: Beyond the superorganism. *Ecological Economics* 169: 106520.

Henrich, J. 2016. *The Secret of Our Success: How Culture Is Driving Human Evolution, Domesticating Our Species, and Making Us Smarter*. Princeton University Press, Princeton, NJ.

Holling, C.S. 1973. Resilience and stability of ecological systems. *Annual Review of Ecology and Systematics* 4: 1–23.

Holling, C. S. 1986. The resilience of terrestrial ecosystems: Local surprise and global change. In *Sustainable Development of the Biosphere*, ed. W. C. Clark and R. E. Munn, 292–317. Cambridge University Press, Cambridge, UK.

Holling, C.S. 1996. Engineering resilience versus ecological resilience. In *Engineering within Ecological Constraints*, ed. P.C. Schulze, 31–44. National Academy Press, Washington, DC.

Holling, C.S., and G.K. Meffe. 1996. Command and control and the pathology of natural resource management. *Conservation Biology* 10: 328–337.

Hudson, M. 2012. *The Bubble and Beyond*. Islet-Verlag, Dresden, Germany.

Hudson, M. 2018. . . . *And Forgive Them Their Debts: Lending, Foreclosure and Redemption From Bronze Age Finance to the Jubilee Year*. Islet-Verlag, Dresden, Germany.

Hudson, M. 2020. A debt jubilee is the only way to avoid a depression. *Washington Post*, March 24.

Hughes, T.P., S. Carpenter, J. Rockström, M. Scheffer, and B. Walker. 2013. Multiscale regime shifts and planetary boundaries. *Trends in Ecology and Evolution* 28: 389–395.

Jackson, A., B. Dyson, and H.E. Daly. 2014. *Modernising Money: Why Our Monetary System Is Broken and How It Can Be Fixed*. Positive Money, London, UK.

Jackson, A., and J. Ryan-Collins. 2011. *Where Does Money Come From?* New Economics Foundation, London, UK.

Keen, S. 2011. *Debunking Economics—Revised and Expanded Edition: The Naked Emperor Dethroned?* Zed Books, New York, NY.

Keynes, J.M. 1936. *The General Theory of Employment, Interest and Money*. Palgrave Macmillan, London.

Latouche, S. 2010. Degrowth. *Journal of Cleaner Production* 18: 519–522.

Lavoie, M. 2009. *Introduction to Post-Keynesian Economics*. Palgrave Macmillan, Basingstoke, UK.

Lazonick, W., M.E. Sakinç, and M. Hopkins. 2020. Why stock buybacks are dangerous for the economy. *Harvard Business Review*. https://hbr.org/2020/01/why-stock-buybacks-are-dangerous-for-the-economy.

Lietaer, B., C. Arnsperger, S. Goerner, and S. Brunnhuber. 2012. *Money and Sustainability: The Missing Link*. Triarchy Press, Axminster, UK.

Light, L. 2019. More than half of all stock buybacks are now financed by debt. Here's why that's a problem. *Fortune*. https://fortune.com/2019/2008/2020/stock-buybacks-debt-financed/

Lucas, R.E., and T.J. Sargent. 1981. *Rational Expectations and Econometric Practice*, Volume 1. University of Minnesota Press, Minneapolis, MN.

Mazzucato, M. 2018. *The Value of Everything: Making and Taking in the Global Economy.* Hatchette Book Group, New York.

McLean, B. 2020. Coronavirus may kill our fracking fever dream. *New York Times.* https://www.nytimes.com/2020/2004/2010/opinion/sunday/coronavirus-texas-fracking-layoffs.html

Meadows, D. 2008. *Thinking in Systems: A Primer.* Chelsea Green, White River Junction, VT.

Mikulka, J., and S. Kelly. 2020. Finances of fracking: Shale industry drills more debt than profit. *Desmog.* https://www.desmogblog.com/finances-fracking-shale-industry-drills-more-debt-profit.

Minsky, H.P. 1986. *Stabilizing an Unstable Economy.* Yale University Press, New Haven, CT.

Mislinski, J. 2020. Margin debt and the market: Up 5.3% in May. *Advisor Perspectives.* https://www.advisorperspectives.com/dshort/updates/2020/2006/2026/margin-debt-and-the-market-up-2025-2023-in-may

Moffett, M. 2018. *The Human Swarm: How our Societies Arise, Thrive and Fall.* Hatchette Book Group, New York, NY.

Norgaard, R. 1994. The coevolution of environmental and economic systems and the emergence of unsustainability. In *Evolutionary Concepts in Contemporary Economics*, ed. R.W. England, 213–225. University of Michigan Press, Ann Arbor, MI.

Oxfam. 2017. *An Economy for the 99%.* Briefing papers, OXFAM.

Philippon, T. 2019. *The Great Reversal: How America Gave Up on Free Markets.* The Belknap Press of Harvard University, Cambridge, MA.

Phillips, M., and C. Krauss. 2020. American oil drillers were hanging on by a thread. Then came the virus. *New York Times.* https://www.nytimes.com/2020/03/20/business/energy-environment/coronavirus-oil-companies-debt.html

Piketty, T. 2014. *Capital in the 21st Century.* Harvard University Press, Cambridge, MA.

Polanyi, K. 1944. *The Great Transformation.* Beacon Press, Boston, MA.

President's Materials Policy Commission. 1952. *Resources for Freedom: A Report to the President. Volume I: Foundations for Growth and Security.* U.S. Government Printing Office, Washington, DC.

Rapp, D. 2009. *Bubbles, Booms, and Busts: The Rise and Fall of Financial Assets.* Copernicus Books, New York, NY.

Romer, P. 2016. *The Trouble with Macroeconomics.* Working paper, Stern School of Business, NYU. https://paulromer.net/trouble-with-macroeconomics-update/WP-Trouble.pdf

Saez, E., and G. Zucman. 2019. *The Triumph of Injustice: How the Rich Dodge Taxes and How to Make Them Pay.* W.W. Norton, New York, NY.

Schneider, F., G. Kallis, and J. Martinez-Alier. 2010. Crisis or opportunity? Economic degrowth for social equity and ecological sustainability. *Journal of Cleaner Production* 18: 511–518.

Schumpeter, J. 1927. The explanation of the business cycle. *Economica* 21: 286–311.

Sherman, M. 2009. *A Short History of Financial Deregulation in the United States.* Center for Economic and Policy Research, Washington, DC.

Simpson, R.D., M.A. Toman, and R.U. Ayres. 2005. Introduction: The "new scarcity." In *Scarcity and Growth Revisited: Natural Resources and the Environment in the New*

Millenium, ed. R.D. Simpson, M.A. Toman, and R.U. Ayres, 1–32. Resources for the Future, Washington, DC.

Sizemore, C. 2020. The Fed is buying ETFs. Now what? *Kiplinger*. https://www.kiplinger.com/article/investing/T022-C009-S001-the-fed-is-buying-etfs-now-what.html

Smialek, J. 2020. The Fed goes all in with unlimited bond-buying plan. *New York Times*. https://www.nytimes.com/2020/03/23/business/economy/coronavirus-fed-bond-buying.html

Soddy, F. 1935. *The Role of Money: What It Should Be, Contrasted with What It Has Become*. Harcourt, Brace and Co., New York, NY.

Spash, C.L. 2011. Social ecological economics: Understanding the past to see the future. *American Journal of Economics and Sociology* 70: 340–375.

Steel, Z.L., H.D. Safford, and J.H. Viers. 2015. The fire frequency–severity relationship and the legacy of fire suppression in California forests. *Ecosphere* 6(1): art8.

Steffen, W., J. Grinevald, P. Crutzen, and J. McNeill. 2011. The Anthropocene: Conceptual and historical perspectives. *Philosophical Transactions of the Royal Society A: Mathematical, Physical and Engineering Sciences* 369: 842–867.

Steffen, W., K. Richardson, J. Rockström, S.E. Cornell, I. Fetzer, E.M. Bennett, R. Biggs, S.R. Carpenter, W. de Vries, C.A. de Wit, C. Folke, D. Gerten, J. Heinke, G.M. Mace, L.M. Persson, V. Ramanathan, B. Reyers, and S. Sörlin. 2015. Planetary boundaries: Guiding human development on a changing planet. *Science* 347: 6223.

Stein, J., and P. Whoriskey. 2020. The U.S. plans to lend $500 billion to large companies. It won't require them to preserve jobs or limit executive pay. *The Washington Post*. https://www.washingtonpost.com/business/2020/04/28/federal-reserve-bond-corporations/

Stevens, P. 2020. This is the greatest 50-day rally in the history of the S&P 500. *CNBC: Markets*. https://www.cnbc.com/2020/2006/2003/this-is-the-greatest-2050-day-rally-in-the-history-of-the-sp-2500.html

Sullivan, A., H. Schneider, and A. Saphir. 2020. Main Street bailout rewards U.S. restaurant chains, firms in rural states. *Reuters*. https://www.reuters.com/article/us-health-coronavirus-usa-lending-analys/main-street-bailout-rewards-u-s-restaurant-chains-firms-in-rural-states-idUSKBN21Z3FL

Svartzman, R., D. Dron, and E. Espagne. 2019. From ecological macroeconomics to a theory of endogenous money for a finite planet. *Ecological Economics* 162: 108–120.

Szathmáry, E. 2015. Toward major evolutionary transitions theory 2.0. *Proceedings of the National Academy of Sciences* 112: 10104–10111.

Vetter, A. 2018. The matrix of convivial technology: Assessing technologies for degrowth. *Journal of Cleaner Production* 197: 1778–1786. https://doi.org/10.1016/j.jclepro.2017.02.195

Voinov, A., and J. Farley. 2007. Reconciling sustainability, systems theory and discounting. *Ecological Economics* 63(1): 104–113.

Wackernagel, M., B. Beyers, and K. Rout. 2019. *Ecological Footprint: Managing Our Biocapacity Budget*. New Society Publishers, Gabriola Island, Canada.

Walker, B., C.S. Holling, S.R. Carpenter, and A. Kinzig. 2004. Resilience, adaptability and transformability in social–ecological systems. *Ecology and Society* 9(2): 5. http://www.ecologyandsociety.org/vol9/iss2/art5/

Wasserman, J. 2019. *Marginal Revolutionaries : How Austrian Economists Fought the War of Ideas*. Yale University Press, New Haven, CT.

West, S.A., R.M. Fisher, A. Gardner, and E. T. Kiers. 2015. Major evolutionary transitions in individuality. *Proceedings of the National Academy of Sciences* 112: 10112.

Wilson, D.S. 2019. *This View of Life: Completing the Darwinian Revolution*. Pantheon Books, New York, NY.

Wilson, E.O. 2012. *The Social Conquest of Earth*. Liveright Publishing Corporation, New York, NY.

Wray, L.R. 2015. *Modern Money Theory : A Primer on Macroeconomics for Sovereign Monetary Systems*. Palgrave Macmillan, London, UK.

CHAPTER 11

Assessing Panarchy in Food Systems: Cross-Scale Interactions in Flint, Michigan

Jennifer Hodbod and Chelsea Wentworth

The globalized and interconnected nature of food systems provides many examples of panarchy within a social-ecological system (SES) framing. We present an example of food system dynamics within a panarchy framing from our research in the city of Flint, Michigan. Flint is facing significant food security and malnutrition issues because of the city's multidecadal economic downturn. As a result, the current food system in Flint is a patchwork of local, state, federal, and philanthropic programs all attempting to address the many food and health needs of residents from multiple entry points, including education, production, access, and consumption. We present the findings of a community-engaged research project that uses a resilience assessment (Resilience Alliance 2010) to identify panarchical dynamics, cross-scale interactions, and leverage points for systemic change toward a more sustainable and secure food system.

Food Systems

Food systems include the production, distribution, consumption, and waste of food and are developed to achieve a fundamental human objective: individual biological sustenance (Hodbod and Eakin 2015). However, beyond mere survival, humans also depend on food systems to meet social, emotional, community, and economic needs. Food is imbued with highly diverse cultural, spiritual, and familial meaning. Yet choices about food are often constrained by economic limitations facing producers, distributors, and consumers. Our food system has direct and indirect influences on a wide variety of biophysical and ecological processes and has in many ways been a primary driver of ecological change on the planet (DeFries et al. 2006). Food is also political: Governments have collapsed over failures in food provisioning and food system management (Davis 2001).

Food systems can therefore clearly be portrayed as SESs because they "incorporate multiple and complex environmental, social, political and economic determinants encompassing availability, access and utilization" and involve varying spatial, temporal, and institutional scales (Ericksen 2008, 234). Framing food systems in this way means characterizing them differently from the static and linear flow commonly used to describe, for example, a food supply chain, and requires systematic review of inputs, activities, and outcomes associated with food production, processing, distribution, consumption, and waste disposal and the feedback loops between these elements (Ericksen 2008).

We consider the focal scale of a food system to be the spatial and temporal scales at which production, distribution, and consumption activities occur, but we recognize that these are simultaneously influenced by broader factors beyond the bounded system. Figure 11.1 illustrates that there are slower dynamics occurring at larger scales (i.e., global environmental change and socioeconomic drivers) and that the focal scale influences a smaller scale (i.e., the household or individual demonstrating food security outcomes). Therefore, the cross-scale linkages highlighted in the heuristic of panarchy are especially useful for analyzing the dynamic nature of food systems on local, regional, and global scales. Change in food systems can be both episodic and gradual, triggered by fast, external perturbations (such as a price spike or disease outbreak) or slower internal drivers (such as soil nutrient depletion or shifts in consumer values), which also mediate the impact of fast perturbations (Hodbod and Eakin 2015).

Examining food systems through a social-ecological resilience lens requires a different perspective from traditional thinking to acknowledge the normative nature of food systems. In food systems there are clear, desirable states, and regimes that do not achieve this goal disrupt human well-being. Contributions from the social sciences have led to "a much greater engagement and reflection on social dimensions" (Brown 2014, 114) in the resilience literature. These discussions provide greater awareness of the values that are a necessary part of applying resilience theory to address equity in the context of food systems (Fraser et al. 2005; Sundkvist et al. 2005; Ericksen 2008; Thompson and Scoones 2009; Food and Agriculture Organization of the United Nations 2014; Thorén and Olsson 2017). Here, we use the adaptive cycle and panarchy as tools with explanatory power (Folke 2006) to explore these resilience dynamics, including equity and desirability, within food systems.

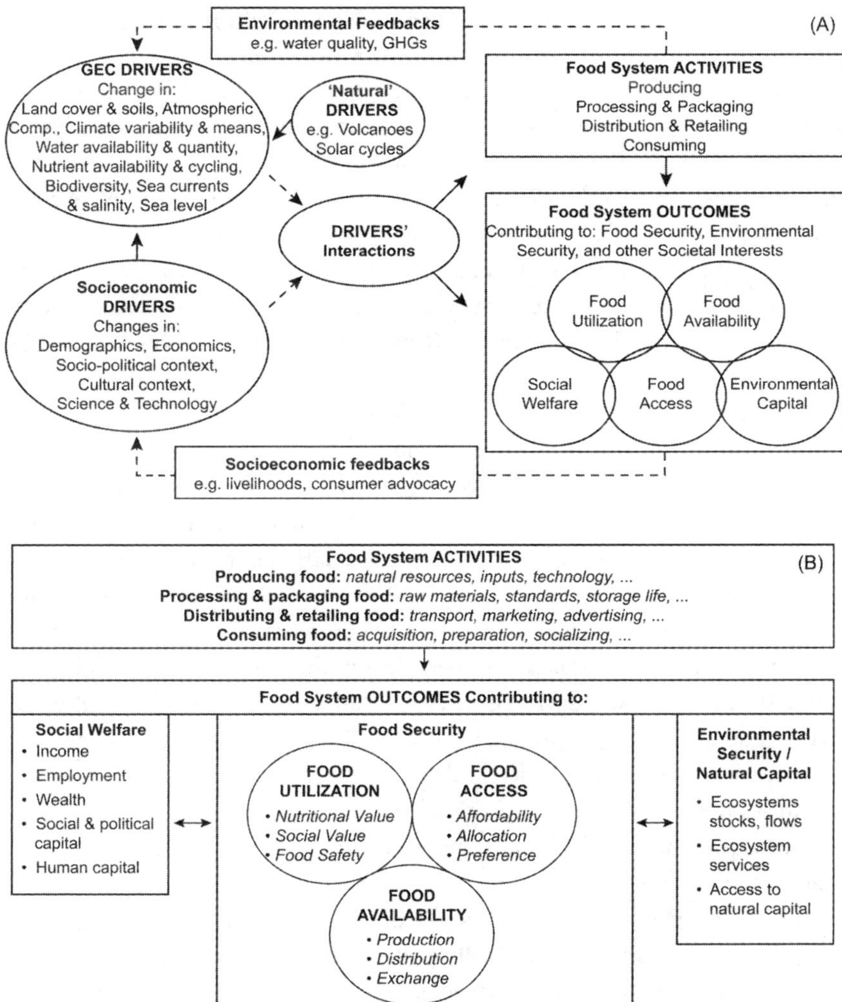

Figure 11.1. (A) Food systems structures include activities (upper right box) that provide socioeconomic outcomes (lower right box). Food system activities generate environmental feedbacks through global environmental drivers that in turn influence activities and outcomes. Food system outcomes generate socioeconomic feedbacks, which also influence environmental changes. (B) Food system activities that result in stabilization of food availability, utilization, and access contribute to food security (lower center box). Environmental and social factors also influence food security. (Reprinted with permission from Ericksen 2008.)

The Adaptive Cycle and Panarchy in Food System Research

Empirical studies of the adaptive cycle are often limited in the scope of social dimensions of environmental change. For example, many studies analyze a regime shift within the ecological system, integrating the social component via actors who have a role in resource management (Beier et al. 2009; Walker et al. 2009; Soane et al. 2012). The majority of studies directly using the adaptive cycle with a focus on social dimensions use historical narratives to identify phase changes, particularly to examine the overexploitation of natural resources, but rarely in food systems (Fraser 2003, 2007; Allison and Hobbs 2004; McAllister et al. 2006; Dugmore et al. 2009; Moen and Keskitalo 2010; Rosen and Rivera-Collazo 2012; Stroink and Nelson 2013; Salvia and Quaranta 2015). There has been a recent emergence of adaptive cycle studies examining the dynamics of institutional arrangements or the influence of potential and connectedness on resilience, but commonly in urban (Bures and Kanapaux 2011; Pelling and Manuel-Navarrete 2011) or conservation systems (Baral et al. 2010)—again, rarely in a food system context. The few examples that do have a social focus in applying the adaptive cycle to food systems examine how rural livelihoods change over time within complex SESs (Robinson 2009; Rasmussen and Reenberg 2012; Goulden et al. 2013; Winkel et al. 2016). The communities studied were small, in both population size and diversity of stakeholders, with clearly defined boundaries. There have also been several studies in urban systems explicitly linked to panarchy, but these studies did not explicitly consider food systems (Garmestani et al. 2009; Herrmann et al. 2016).

In this chapter, we demonstrate for the first time that the adaptive cycle and panarchy (Gunderson and Holling 2002) also lend explanatory power to complex, urban food systems, particularly by focusing on equity dynamics within the analyses. The key focus of political ecology has always been the power dynamics of social-ecological change, given the asymmetrical distribution of resources and power in social systems (e.g., Davidson 2010). Peterson (2000), Robinson (2009), Pelling and Manuel-Navarette (2011), and Beymer-Farris et al. (2012), among others, demonstrated that asymmetries within social groups determine both the outcome and the trajectory of adaptive cycles, for their own perceived self-interest. In this chapter, we build on this work in an urban food system context, exploring how these SESs exhibit cyclical growth, collapse, and reorganization, at what scales different components of the food system are affected, and what factors trigger these changes.

Phases and traps of the adaptive cycle

Here, we use the three axes or dimensions of the adaptive cycle (potential, capital, and resilience) to identify and plot different system states or status (Holling and Gunderson 2002). A three-dimensional depiction of the adaptive cycle (figure 11.2) allows us to identify relative levels of potential (labeled as capacity in this figure), connectedness, and resilience for each phase of the adaptive cycle, as shown in table 11.1.

Building on the original work by Holling and Gunderson (2002), Allison and Hobbs (2004) stated that if each of the three dimensions (potential, connectedness, and resilience) in the adaptive cycle is given two nominal levels,

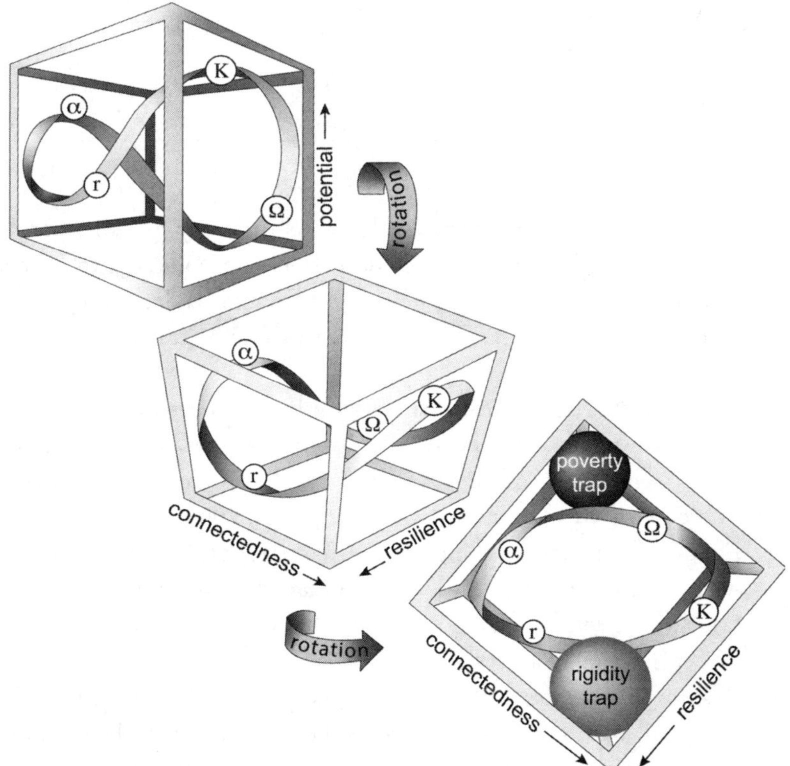

Figure 11.2. A three-dimensional depiction of an adaptive cycle demonstrates relative levels of capacity, connectedness, and resilience for different phases of the cycle. Rotating the view shows where the rigidity trap and poverty trap are located, when systems depart from an adaptive cycle trajectory. (Reprinted with permission from Holling and Gunderson 2002.)

TABLE 11-1. *Characteristics of systems positioned along dimensions of potential, connectedness, and resilience. Each phase (α, r, K, and Ω) of an adaptive cycle has different levels of potential, connectedness, and resilience. Systems not in an adaptive cycle or stuck in one of the phases are considered trapped. Four such traps have been identified. The rigidity trap and poverty trap were described by Holling and Gunderson (2002). The lock-in trap was described by Allison and Hobbs (2004), and the isolation trap is described by Gunderson et al. in chapter 1 of this volume.*

System Position	Potential or Capital Dimension	Connectedness Dimension	Resilience Dimension
α: Reorganization	High	Low	High
r: Exploitation	Low	Low	High
K: Conservation	High	High	Low
Ω: Release	Low	High	Low
Poverty trap	Low	Low	Low
Rigidity trap	High	High	High
Lock-in trap	Low	High	High
Isolation trap	High	Low	Low

either low or high, then the adaptive cycle model uses only four of a possible eight combinations. The remaining combinations are suggested to be maladaptive states, or traps. Traps are departures from the adaptive cycle that maintain the status quo and have thus far been defined to include

- poverty traps: impoverished states with low connectedness, potential, and resilience, where the remember process (i.e., effects cascading down from larger and slower adaptive cycles) is blocked, limiting use of accumulated social memory (Holling et al. 2002);
- rigidity traps: states with high connectedness, potential, and resilience, where the revolt process (i.e., effects cascading up from smaller and faster adaptive cycles) is blocked, limiting innovation and experimentation (Holling et al. 2002);
- lock-in traps: states with high connectedness and resilience but low potential, where the system ceases to be adaptive given the erosion of potential and diversity, and the resulting ecological impoverishment causes resilience to increase because the state is so depauperate it stabilizes (Allison and Hobbs 2004);
- isolation trap: state of has low connectedness and resilience but high potential (Gunderson et al., chapter 1).

In this chapter we test table 11.1 as a model emerging from the heuristic of the adaptive cycle. Such an analysis allows us to identify the four phases of the adaptive cycle and any traps by exploring relative levels of indicators for potential, connectedness, and resilience over time, in this case for the Flint food system and associated systems influenced or influencing this focal scale.

The food system in Flint, Michigan

The city of Flint is located in southeastern Michigan and prospered in the twentieth century while playing a significant role in the production of automobiles in the United States. With the closure of major auto plants from the 1960s to the 1980s and the collapse of many adjacent automotive businesses, Flint's population has been declining. At its peak, in the 1960 census, the population was 196,940, and in 2018 the estimated population was 95,943 (table 11.2). With a median income of $26,330 in 2017, Flint is experiencing higher rates of poverty than midwestern industrial cities of similar size (table 11.2), and 41 percent of Flint residents are living in poverty (U.S. Census Bureau 2019). Both poverty and financial decline have to be framed in the context of the racial dynamics of Flint. Deindustrialization was followed by White population decreases, and currently Flint is a majority Black city (54 percent in 2018) (U.S. Census Bureau 2019).

Today, Flint is most noted to the broader U.S. population for the Flint Water Crisis. After the state-appointed emergency manager switched the municipal water supply from Detroit City Water to a system fed by the Flint River on April 25, 2014 (Pauli 2019), residents noticed a change in color and smell, and many suffered from rashes and hair loss. The emergency manager and state government officials continued to argue that the water was safe to drink, despite the reality that the lack of corrosion control measures in the aging water system resulted in extremely unsafe levels of lead leaching into the water supply (Clark 2018). Even residents who did not see a visible change in their water were consuming water that had elevated levels of lead and had become infected with legionella bacteria (Hanna-Attisha 2018). This contamination was so pervasive that the percentages of elevated blood lead levels in children under age five for the city of Flint increased from 2.4 percent in 2013 to 4.9 percent in 2015 after the change in water source ($P < .05$, $\chi 2$) (Hanna Attisha et al. 2016). Combining these data with geospatial analysis showed that children in the poorest neighborhoods experienced a 6.6 percent increase in their blood lead levels in 2015, illustrating that the most vulnerable children

TABLE 11-2. *Population and median household income of Midwestern U.S. cities with strong industrial workforces (U.S. Census Bureau 2019). Note that the median income of Flint, Michigan, was the lowest among these similar-sized cities.*

City	Population (2018)	Median Income (2017 US$)
Flint, Michigan	95,943	26,330
South Bend, Indiana	101,860	37,441
Lansing, Michigan	118,425	38,642
Green Bay, Wisconsin	104,879	45,473
Peoria, Illinois	111,388	47,697

faced more significant increases in their blood lead levels after the water source switch (Hanna-Attisha et al. 2016). A control group outside the city who were not affected by the change in water source revealed no statistically significant change in blood lead levels. The Michigan Civil Rights Commission (2017) reported that historical, structural, and systemic racism combined with implicit bias led to actions in Flint related to emergency management that would not have been allowed to happen in primarily White communities such as Ann Arbor or Grand Rapids. Thus, the Flint Water Crisis resulted in dramatic negative effects on public health and residents' emotional well-being and a loss of trust in government and outside organizations. When combined with the broader negative images of industrial decline in the Midwest, from the outside Flint is often represented as a city in turmoil. Yet social activism remains strong, and there are a number of community-based programs aimed at promoting public health, improving food security, and promoting social and environmental justice.

It is in this context of historical change and the past social-ecological regimes of the city of Flint that we begin to contextualize the food system in Flint today. In this urban food system, the activities of production and processing are limited, leading to distribution, retailing, and consumption having a greater impact on food system outcomes (figure 11.1). Thus, social welfare factors including high poverty rates and low median household income are essential aspects of our analysis of the food system. Poor nutritional outcomes and low serum nutrient levels in young children and adults can be predicted by poverty rates, "confirming that the traditional poverty measure, whatever its limitations, is still predictive of material hardship in this country" (Bhattacharya et al. 2004, 857). The decreasing population and tax base in a city built with a large infrastructure act as social and economic drivers within the food system, significantly reducing the food security of Flint residents. The remainder of this chapter describes our evaluation of how cross-scale interactions of the food system led to an undesirable regime in Flint.

Methods

We assessed the resilience of the Flint food system by following the approach presented in the Resilience Alliance (2010) Workbook. The first step of our analysis was to define the focal scales of the system before considering adaptive cycles and panarchy. In collaboration with a Community Consultative Panel assembled for our broader project (Flint Leverage Points Project 2020), we determined that the spatiotemporal boundaries of the Flint food system encompass the geographic boundaries of the cities of Flint, Burton, and the unincorporated community of Beecher (all within Genesee County, Michigan), between 1950 and the present day. The long temporal period of focus is necessary to include Flint's industrial peak production and associated social benefits, particularly for communities of color in Flint.

To explore relative levels of potential and connectedness, we created a long list of potential indicators of various types of capital within the Flint SES. Capital is a stock or resource that yields a flow of valuable goods or services into the future (Costanza and Daly 1992). Human well-being arises from the use of various human-derived capitals (financial, social, human, physical [also referred to as manufactured]), all of which are based on stocks and flows of natural capital. Though applicable to all SESs, this framing is suitable for food systems because food security is a result of multiple forms of capital (primarily natural capital and financial capital) and a driver of well-being. Thus, exploring food security dynamics and drivers over time allows us to link capital to food security to well-being in the context of an adaptive cycle.

Our long list identified indicators of natural, financial, physical, and human capital that could be used to assess system potential. Natural capital indicators included food security (categorized here to reflect food availability), pounds of food distributed through food assistance programs, number and type of retail stores, and land under food production. Financial capital indicators included median income, poverty rates, reliance on school feeding programs, and participation in federal food assistance programs (e.g., Electronic Benefits Transfer and Double Up Food Bucks). Physical capital indicators included population, housing and vacant lots, bus routes, community meeting spaces, and food pantries. Human capital indicators included extension programs in production and nutrition. Social capital indicators such as social networks, support networks, functions of institutions, and governance were included in the long list to assess degrees of connectedness within the focal system.

We collected primary and secondary data for each indicator in the long list to establish the longitudinal data available for each indicator. Primary data were elicited through interviews with stakeholders involved in the governance of the food system, both formal and informal; focus groups with participants representing different activities in the food system (consumers, processors, distributors [retail and food assistance], and those involved in production of food within the city); and stakeholder mapping with the same individuals. Secondary data were elicited from peer-reviewed and gray literature and key organizations such as the Food Bank of Eastern Michigan, Feeding America, the U.S. Department of Agriculture Food Access Atlas, the U.S. Census Bureau, and Genesee County. Not surprisingly, we were not able to create a full dataset from 1950 to 2020 for all indicators. Rather than adjust our spatiotemporal focal scale, the scale influences the data available for the analysis. A short list of indicators with longitudinal data going back to 1950 was assembled and analyzed for relative levels over time (table 11.3). Our analysis reflects that to explore cross-scale dynamics, we had to go beyond the initial long list of indicators of potential and connectedness in the food system to also find indicators of potential and connectedness in the systems operating at scales larger and smaller than the focal system of Flint city.

The short list for the food system consists of four main indicators, two for potential and two for connectedness. Potential is based on food insecurity (as an indicator of both natural and financial capital in the system) and donor funds. Although recent secondary data are available from Feeding America for food security in Genesee County, we supplemented it with interview and focus group data about relative levels of both food insecurity and donor funds in Flint. A key indicator for connectedness was the number of food pantries, as informed by quantitative (for present number) and qualitative interview data (for change over time). Our focus group data showed that the number of food pantries was a key indicator of food insecurity in the food system and that while providing a resource (i.e., natural capital, food), the pantries themselves were a demonstration of social capital within the system. We also used the interviews and focus groups to explore collaboration between food system actors over time as an indicator of connectedness. Other data collected for the long list were used to supplement these indicators and provide further context. At the city scale, the shortlisting process identified population, median income,

TABLE 11-3. *Indicators from three scales in a panarchy (city, food system, household) used in the analysis, arranged by adaptive cycle dimensions of potential (capital) and connectedness.*

Adaptive Cycle Axis	Scale	Indicator	Data Source
Potential	City	Population	U.S. Census (1960–2010)
		Median income	U.S. Census (1960–2010)
		Tax base	Interview and focus group data
		Autonomy	Interview and focus group data, participant observation
		Trust	Interview and focus group data, participant observation, Morckel and Terzano (2019), Flint Trust Steering Committee (2020).
	Food system	Food insecurity levels	Feeding America (2011–2015), Gundersen et al. (2016–2019), interview and focus group data
		Donor funds	Interview and focus group data, Pauli (2018)
	Household	Food security	Feeding America (2011–2015), Gundersen et al. (2016–2019), interview and focus group data
		Median income	U.S. Census (1960–2010)
Connectedness	City	Links with actors outside Flint	Interview and focus group data
	Food system	Collaboration	Interview and focus group data
		Number of food pantries	Food Bank of Eastern Michigan (2009–2019), interview data, Pauli (2019)
	Household	Use of social networks for food access	Focus group data

the city's tax base, autonomy, and trust to be key indicators of potential, and links with actors outside Flint were the most relevant indicator of connectedness. At the household scale, we again used median income and food security as indicators of potential.

The analysis used the relative degree (high or low, qualitatively assessed according to the variation over time) of potential and connectedness to identify the current phase of the adaptive cycle of the focal system. Because we were analyzing both qualitative and quantitative data, we did not rely on statistical analyses to identify differences in the time series; our approach was more qualitative. As a validity check, two researchers analyzed the trends separately, identifying relative highs and lows to define the current degree of potential and connectedness. The researchers then collectively completed the analysis to identify the current phase of the adaptive cycle. To triangulate this interpretation, we then used primary data to explore whether our perception of system resilience and the capacity to respond to change aligned with the anticipated high and low for resilience as per table 11.1.

Next, we used the same individual–collective approach to explore the evolution of how the focal system has changed over time (i.e., movement through an adaptive cycle) again using the relative levels of potential and connectedness.

Analysis

In this section we interpret the historical dynamics of food and water systems in Flint by using the adaptive cycle and panarchy as a framework. We first describe patterns of reorganization in the food system that followed the Flint Water Crisis in the summer of 2014. We then examine cross-scale dynamics associated with the Flint food system that explain the transition through the phases of its adaptive cycle and conclude by analyzing the dynamics of the larger and smaller scales in the study. For all subsystems in our analysis, where appropriate, we identify system traps when systems failed to go through the phases of the adaptive cycle.

Water Crisis and reorganization of the food system

The first emergency manager took office on December 1, 2011, sparking a series of cost-saving measures that led to the city of Flint switching its water source on April 25, 2014, the combination of which we identify as a release phase. During this crisis, the system became increasingly connected with outside systems, resulting in changes both in water source and in decision makers who managed water. After the switch in water supply, residents began to complain about body rashes, as well as changes in the color and odor of their drinking water. General Motors announced on October 13, 2014, that it would stop using water sourced from the Flint River because of increased corrosion of machinery in their automotive plants. In September 2015, Flint residents learned about the impacts on water quality and health from both citizen science water testing and the research of public health leaders, such as Dr. Hanna-Attisha's public dissemination of the study of elevated blood lead levels in children (Hanna-Attisha et al. 2016). On October 1, 2015, the Genesee County Health Department issued a public health emergency in Flint, and just over a month into her term, newly elected mayor Karen Weaver declared a state of emergency in Flint on December 14, 2015.

The culmination of these events shifted the system into a reorganization phase. The crisis brought about a rapid increase in national attention, particularly about the greater need for healthful food. Before the Water Crisis in 2014–2015, food insecurity levels were high in Genesee County: 18.0 percent of adults and 23.2 percent of children were food insecure in 2013, exceeding the Michigan average of 16.4 percent, as shown in figure 11.3 (Feeding America 2015). Our interview data demonstrate that there were few financial resources for food assistance programs and thus more competition between organizations running such programs.

After the Water Crisis, the food system transitioned into a growth phase, driven by external funds being transferred to government and nongovernment organizations within the Flint food system. There was also a parallel increase in social capital through collaboration, partly because of the need to collaborate to receive funds but also as priorities aligned between organizations and a common goal was identified. Our interview data demonstrated that since the Water Crisis, community organizations worked together whereas previously they competed for limited resources. There has been a recent increase in food pantries: 141 in 2015, 159 in 2016, and 170 in 2017. The increase is partly due to the Water Crisis: 44 pantries in 2016 and 33 pantries in 2017 were distributing only water. However, it is also an indicator of a food system failing to achieve its primary goal: food security. The Food Bank of Eastern Michigan estimated that 4,163,210 pounds of food (excluding beverages) were distributed in 2012, which grew to 11,457,787 pounds of food (excluding beverages) in 2017 (figure 11.4). It is important to note that weights of food distributed including beverages dramatically increased after the Water Crisis, when water began to be distributed widely.

A key indicator of potential was financial capital in the Flint food system from donors; although it is hard to quantify across all organizations in the focal scale, interview data indicated that it increased significantly after the Water Crisis. This was partially due to the focus on lead-mitigating foods after the Water Crisis (i.e., foods high in iron, calcium, and vitamin C that decrease lead absorption in the body) and partially due to the general increased awareness of Flint and its social challenges, including food insecurity. National news of the

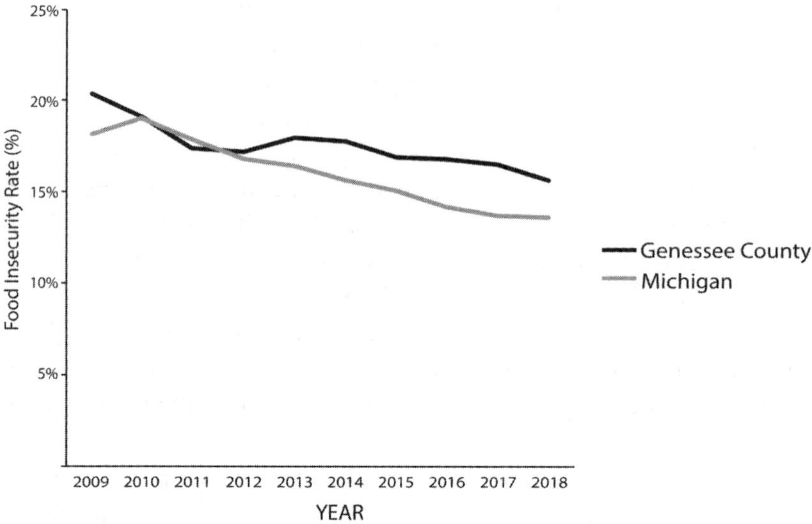

Figure 11.3. Time series of food insecurity rates in Genesee County and Michigan. Food insecurity between 2009 and 2018 declined in Genesee County, although not to the same extent as in Michigan (Feeding America, 2011, 2012, 2013, 2014, 2015; Gundersen et al., 2016, 2017, 2018, 2019, 2020).

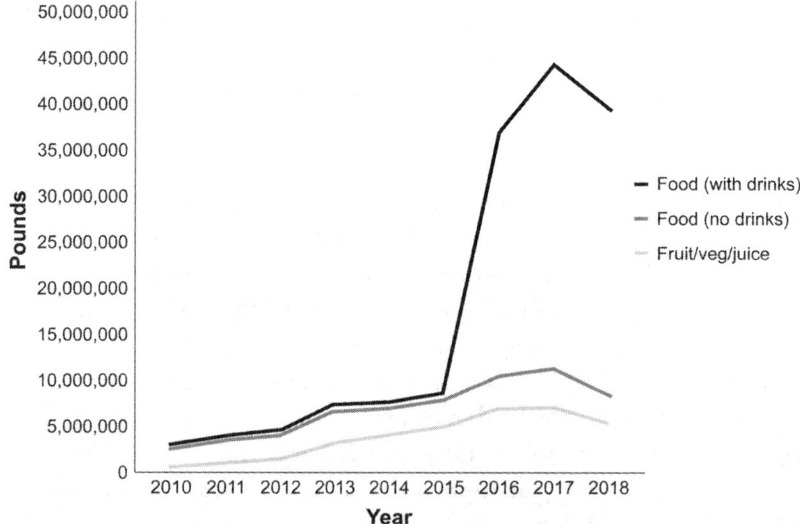

Figure 11.4. Time series of food and drink distribution (pounds) to residents in Flint, Michigan, through the Food Bank of Eastern Michigan from 2010 through 2018. There was a marked increase in the amount of water imported and distributed alongside food after the Water Crisis.

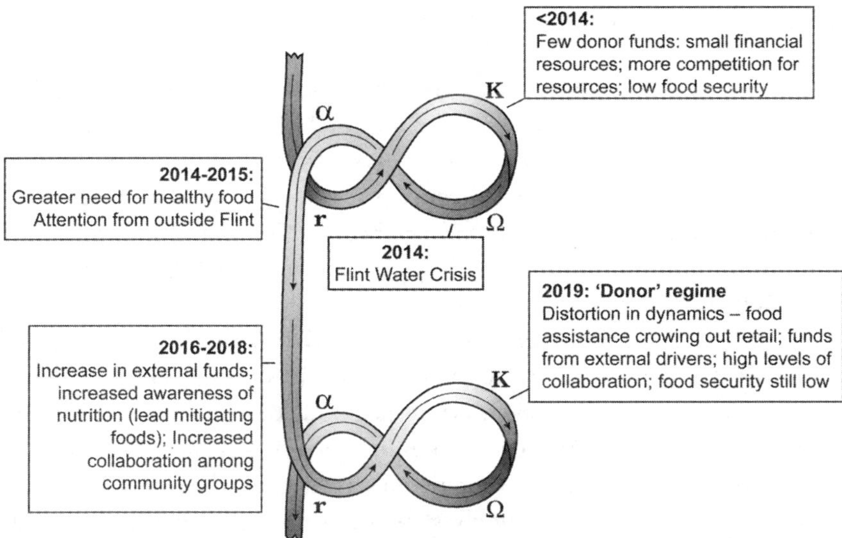

Figure 11.5. Adaptive cycle dynamics of the Flint food system. After the water crisis in 2014 (top cycle), the system reorganized to a donor regime (bottom cycle). In 2019 the food system was in a conservation phase. (Graphic adapted from Goulden et al. 2013.)

Water Crisis generated substantial journalistic interest in Flint (Clark 2018). This generated a significant increase in outside donors, altering the food assistance landscape that existed previously, which was reported in our primary data collection and documented by other scholars (Pauli 2019).

The high and static levels of financial and social capital in 2019 identify a conservation phase (figure 11.5), with high levels of potential and connectedness, and thus low resilience, as resources become vulnerable to outside shocks given the high connectedness. We characterize the 2019 regime as the donor regime and note that there is a distortion in dynamics here, as our interviews report that food assistance appears to be crowding out retail, thus maintaining food availability for the population but in a way that is vulnerable to future shocks from changes in funding and still does not ensure food security, as demonstrated by Genesee County's high levels of food security compared with state levels. The donor regime could therefore reflect a coerced regime (Angeler et al. 2020), whereby external subsidy is maintaining the regime artificially and preventing it from collapse and reorganization.

Cross-scale dynamics: The Flint city system

The above analysis reveals that cross-scale dynamics are essential to understanding the food system. Linking the Water Crisis with political economy of Flint illustrates that the food system is nested within the Flint city system, and events within the city have direct impacts on the food system. Resource indicators reveal financial decline and a higher rate of poverty that corresponds to increased rates of food insecurity. As General Motors began its decades-long divestment in their Flint manufacturing facilities, population declined, and median income decreased (table 11.4). In the 1970s Flint was a city of prosperity and growth, but by 2010, the median income had fallen by nearly two thirds, and more than one third of the population was living in poverty (U.S. Census Bureau 2019). However, infrastructure remained the same, and with a dwindling tax base and steep population decline, the city began to fall into disrepair and the water supply system was not maintained or updated. Additionally, the water supply system was designed for substantially larger usage, and operating the system at lower flow rates increased risks of the spread of waterborne diseases (Morckel 2017).

Our analysis of potential and connectedness indicators at the city scale reveals that Flint has been through multiple iterations of the adaptive cycle, as shown in figure 11.6. From 1945 through the 1970s, Flint moved from through an exploitation (r) into a conservation (K) phase, with a desirable regime characterized by wage and population growth and lower rates of poverty (top cycle, figure 11.6). A release was triggered by the first major General Motors closure in Flint in the 1970s, which led to reorganization into a new, less desirable regime. This adaptive cycle regime (middle cycle, figure 11.6) persisted through the early 2000s and was characterized by declining populations (due to job losses from General Motors plant closures and related industries leaving the area) and a depressed tax base. These changes yielded increased poverty and food insecurity through the early 2000s, resulting in marginalization by race and socioeconomic status.

Another release was triggered in 2011 by the appointment of the first emergency manager. Usurping the democratically elected local government, the emergency manager represented a significant loss of autonomy for Flint residents (representing potential), who were acutely aware of disproportionate political disenfranchisement through the emergency manager system in a state where more than 50 percent of Black people have lived under emergency management as compared with only about 2 percent of Whites (Johnson and Key

Table 11-4. *Time course of population, median income, and poverty rates in Flint, Michigan, from 1940 to 2010 (U.S. Bureau of Labor Statistics 2019; University of Minnesota, 2019; U.S. Census Bureau 2019). Adjusted median income in 2019 dollars from U.S. Bureau of Labor Statistics (2020).*

Year	Population	% African American	Median Family Income (US$)	Adjusted Median Income (2020 US$)	Poverty Rate
1940	151,543	4.35	n/a[a]	n/a[a]	n/a[a]
1950	162,800	8.52	4,002	43,349	n/a[a]
1960	196,940	17.53	6,340	55,217	n/a[a]
1970	193,317	28.06	10,161	67,512	9.5%[b]
1980	159,611	41.43	20,080	62,594	15%[c]
1990	140,761	47.94	25,083	49,779	27.6%[c]
2000	124,943	53.27	31,424	46,990	22.9%[c]
2010	102,434	56.56	23,424	27,705	35.2%[d]

[a] Census data not collected.
[b] Census data measured as percentage of families below a low-income level.
[c] Census data measured as percentage of families below poverty level.
[d] Census data measured as percentage of families below poverty level in the past 12 months.

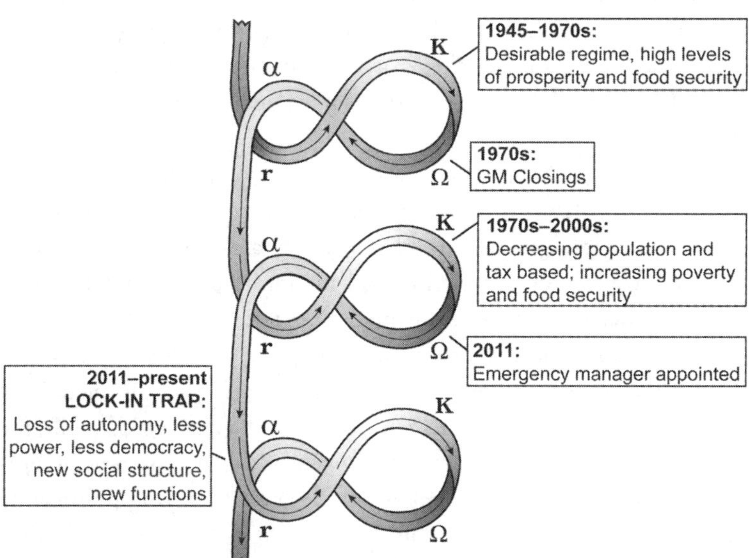

Figure 11.6. Adaptive cycle dynamics of the city of Flint showing that multiple reorganizations led to three different regimes (top, middle, and bottom cycles) between 1945 and 2019. Since 2011, the system has been in a lock-in trap, characterized by less power, less democracy, and a loss of autonomy.

2018). The loss in autonomy was coupled with an increased connectedness given the new relationships with actors from outside Flint, triggering reorganization into a new regime defined by a lack of trust in government leadership (bottom cycle, figure 11.6).

Although some might consider April 25, 2014, the beginning of the Water Crisis because that is when the city switched water sources, for this analysis we mark the beginning of the Water Crisis, and the corresponding release into a new regime, as 2011 with the appointment of the first emergency manager. Primary data from our focus groups, interviews, and participant observation illustrates that Flint residents overwhelmingly associate the Water Crisis with the loss of autonomy and the undermining of locally elected government officials. As per Gray et al. (2017) and the Michigan Civil Rights Commission (2017), participants reported that the history of marginalization by race and socioeconomic status led to this loss of autonomy; many believe that without the emergency manager, the switch in water sources from Detroit city water to the Flint River never would have come to fruition. During this time, local and state government officials denied any problems with the quality and safety of the municipal water system despite repeated resident complaints, but they permitted General Motors to switch back when the water began corroding their machinery. Denials, fraudulent claims, and efforts to conceal reports of water safety by government officials are now being investigated by the court system, and it is no surprise that residents have lost faith in the local and state government.

Subsequent lack of trust in political and external leadership is understandably pervasive and has led to the current state, a lock-in trap. The lock-in trap in Flint is characterized by residents' loss of autonomy, decreased power, decreased trust, and low population and income (i.e., low potential), and a range of new social structures implemented externally (i.e., high connectedness). Nonprofit organizations rushed to help, with celebrity activists promoting #JusticeForFlint and holding concerts, starting relief funds, pledging large donations ($10 million from the Detroit Pistons owner and $3 million from General Motors and the United Auto Workers). Flint hosted a Democratic primary debate, garnering national attention and prompting political promises from both parties. However, this attention from outside Flint has not resulted in significant changes to the key indicators we have studied: population, median income, city tax base, autonomy, or trust. We interpret this to reflect the high resilience of the system: In its depauperate state it is stable.

Foundationally, protecting public health through policy and honest responses to citizen complaints is a basic function of government, and the failure of local and state institutions to fulfill this role damaged trust so severely that a recent study found, "Only 2.2% strongly agreed and 9.1% agreed with the statement, I trust my local government" (Morckel and Terzano 2019, 591). The second new mayor since the water switch in 2014 was elected in November 2019. When we asked participants whether they trusted this new mayor, most residents shook their head no; several said, "We'll see," and one man responded definitively, "I can only trust God now."

Cross-scale dynamics: Food insecure households are in poverty traps

To relate the dynamics at the food system and city scale to the outcomes of the food system and explore any revolt mechanisms, we studied the dynamics of food insecure households, a smaller functional unit nested within the Flint food system. As demonstrated above, the median income of Flint households has decreased over time, is currently lower than comparable cities in the Midwest, and has resulted in above-average poverty and food insecurity rates. The lack of improvement in median income and food security (indicators of potential), particularly in food security compared to Michigan, and even with the increase in funding for food assistance programs, leads us to conclude that currently there are low levels of potential in these households. Additionally, although the food system as a whole is demonstrating high levels of connectedness and there are resources available such as food pantries and nutrition education efforts, above-average food insecurity levels translate to low levels of social capital and connectedness at a household level. Individually, food insecure households reported that they are already fully using their social networks, which are unable to offer more support. We also interpret above-average food insecurity levels as translating into low levels of resilience: Households reported already implementing a range of coping strategies to maintain this (insufficient) level of food security, and any additional shocks to the household would worsen their food security (i.e., they have exhausted their adaptive capacity). Other studies have referred to households in Flint as "trapped" (Sim 2016), where some residents expressed a desire to leave but lacked the financial ability to do so, which is linked to low housing values, a result of historic marginalization in urban planning concentrating the region's Black population in the economically depressed central city (Sadler and Highsmith 2016). Low levels

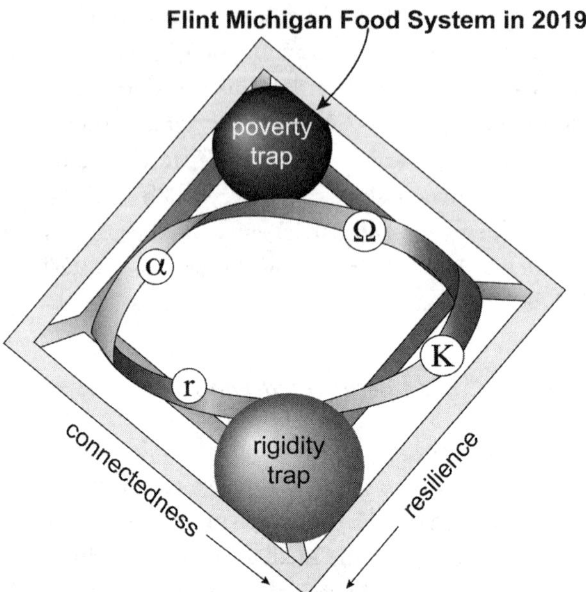

Figure 11.7. Food insecure households are stuck in a poverty trap, with low levels of potential, connectedness, and resilience.

of potential, connectedness, and resilience indicate a poverty trap, as shown in figure 11.7. Poverty traps block the movement of the system through the back loop and maintain the status quo, in this case a low level of well-being.

Conclusions

Using a panarchy framing to synthesize the above findings from this study, figure 11.8 demonstrates the cross-scale dynamics at play in the Flint food system. Most importantly, the traps at the scales above and below the focal scale are blocking revolt and remember connections that have been observed in the past.

The lock-in trap at the city level makes it hard for innovation in the food system to feed up through revolt mechanisms to the city. This is not to say there is not innovation; there are many programs and organizations in Flint that aim to increase access to affordable fresh fruits and vegetables to residents, such as Edible Flint, Flint FARMacy, Flint Fresh, and the North Flint Healthy Food Initiative. However, our data show lack of participation from local communities in engaging in programs designed to help them, preventing the upward

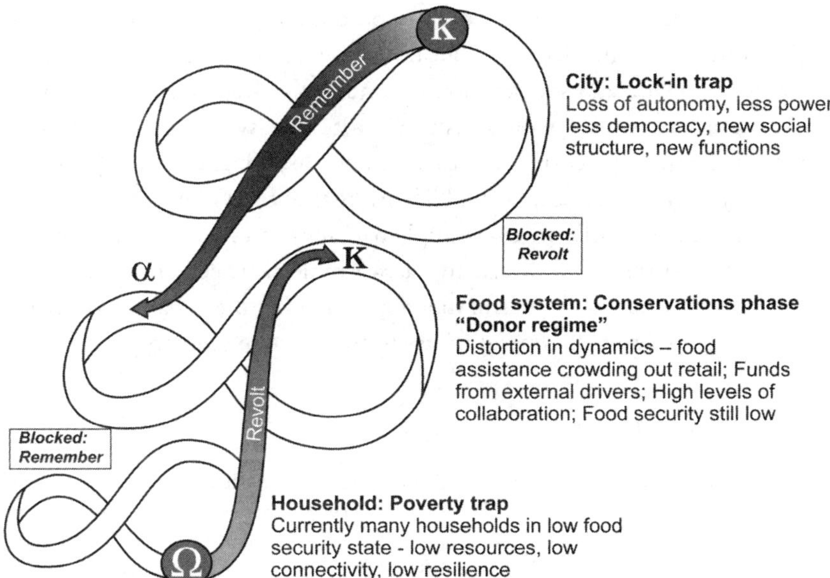

Figure 11.8. Panarchy consisting of the city of Flint, the food system, and individual food insecure household scales. Cross-scale linkages contribute to maladaptive system states at the City and Household levels. Large external inputs from the City maintain the residence of the Donor regime, which are in turn, blocked from reaching Households.

cascade out of individual neighborhoods or organizations to a system-wide uptake. Instead, we suggest it is most likely that unlocking will come from a crisis at a larger scale, although we hope desirable changes occur through urban planning, policy change, and economic development as opposed to the Water Crisis, which can be interpreted as the most recent remember connection. Similarly, the poverty trap for food insecure households makes it difficult for social memory to be transferred from larger and slower scales through remember mechanisms. Translating these barriers to the adaptive cycle into leverage points allows us to identify how to move the systems from these traps, an approach that can be mimicked in studies in any kind of SES.

Overall, our analysis demonstrates that a focus on cross-scale drivers through a panarchy framing is not only feasible in food systems but critical. Such an approach is well suited for food systems given their complexities, but it requires mixed methods and data from multiple disciplines and benefits from a more engaged approach (i.e., integrating the community into the research in a transdisciplinary manner). The result is a rich, longitudinal analysis that allows us to link socioeconomic drivers to both food system activities and food security outcomes, and vice versa, outlining how poor food security outcomes affect social welfare and socioeconomic conditions. Integrating an equity focus into our analysis has allowed us to identify the asymmetries within food systems and how differential power dynamics can influence the trajectory of the food system's adaptive cycle, and in turn that of food insecure households. Although challenges arise when we apply panarchy in urban food system contexts given the diversity of social actors across multiple scales and their notions of desirability, panarchy is a useful tool for elucidating the dynamics of change across scales, and we recommend its use to food system scholars.

Literature Cited

Allison, H.E., and R.J. Hobbs. 2004. Resilience, adaptive capacity, and the "lock-in trap" of the Western Australian agricultural region. *Ecology and Society* 9(1): 3. http://www.ecologyandsociety.org/vol9/iss1/art3/

Angeler, D. G., B. C. Chaffin, S. M. Sundstrom, A. Garmestani, K. L. Pope, D. Uden, D. Twidwell, and C. R. Allen. 2020. Coerced regimes: Management challenges in the Anthropocene. *Ecology and Society* 25(1):4. https://doi.org/10.5751/ES-11286-250104

Baral, N., M.J. Stern, and J.T. Heinen. 2010. Growth, collapse, and reorganization of the Annapurna Conservation Area, Nepal: An analysis of institutional resilience. *Ecology and Society* 15(3): 10. http://www.ecologyandsociety.org/vol15/iss3/art10/

Beier, C., A.L. Lovecraft, and T. Chapin. 2009. Growth and collapse of a resource system: An adaptive cycle of change in public lands governance and forest management in Alaska. *Ecology and Society* 14(2): 5. http://www.ecologyandsociety.org/vol14/iss2/art5/

Beymer-Farris, B.A., T.J. Bassett, and I. Bryceson. 2012. Promises and pitfalls of adaptive management in resilience thinking: The lens of political ecology. In *Resilience and the Cultural Landscape*, ed. T. Plieninger and C. Bieling, 283–302. Cambridge University Press, Cambridge, UK.

Bhattacharya, J., J. Currie, and S. Haider. 2004. Poverty, food insecurity, and nutritional outcomes in children and adults. *Journal of Health Economics* 23: 839–862.

Brown, K. 2014. Global environmental change I: A social turn for resilience? *Progress in Human Geography* 38: 107–117.

Bures, R., and W. Kanapaux. 2011. Historical regimes and social indicators of resilience in an urban system: the case of Charleston, South Carolina. *Ecology and Society* 16(4): 16. http://dx.doi.org/10.5751/ES-04293-160416

Clark, A. 2018. "Nothing to worry about. The water is fine": How Flint poisoned its people. *The Guardian*, July 3. https://www.theguardian.com/news/2018/jul/03/nothing-to-worry-about-the-water-is-fine-how-flint-michigan-poisoned-its-people

Costanza, R., and H.E. Daly. 1992. Natural capital and sustainable development. *Conservation Biology* 6: 37–46.

Davidson, D.J. 2010. The applicability of the concept of resilience to social systems: Some sources of optimism and nagging doubts. *Society & Natural Resources* 23: 1135–1149.

Davis, M. 2001. *Late Victorian Holocausts: El Niño Famines and the Making of the Third World.* Verso, New York, NY.

DeFries, R., G.P. Asner, and J Foley. 2006. A glimpse out the window: Landscapes, livelihoods, and the environment. *Environment: Science and Policy for Sustainable Development* 48: 22–36.

Dugmore, A., C. Keller, T. McGovern, A. Casely, and K. Smiarowski. 2009. Norse Greenland settlement and limits to adaptation. In Adapting to Climate Change: Thresholds, Values, Governance, ed. W. Adger, I. Lorenzoni, and K. O'Brien, 96–113. Cambridge University Press, Cambridge, UK. doi:10.1017/CBO9780511596667.008

Ericksen, P.J. 2008. Conceptualizing food systems for global environmental change research. *Global Environmental Change* 18: 234–245.

Feeding America. 2011. *Map the Meal Gap 2011: A Report on County and Congressional District Food Insecurity and County Food Cost in the United States in 2009.* Feeding America, Chicago, IL.

Feeding America. 2012. *Map the Meal Gap 2012: A Report on County and Congressional District Food Insecurity and County Food Cost in the United States in 2010.* Feeding America, Chicago, IL.

Feeding America. 2013. *Map the Meal Gap 2013: A Report on County and Congressional District Food Insecurity and County Food Cost in the United States in 2011.* Feeding America, Chicago, IL.

Feeding America. 2014. *Map the Meal Gap 2014: A Report on County and Congressional District Food Insecurity and County Food Cost in the United States in 2012.* Feeding America, Chicago, IL.

Feeding America. 2015. *Map the Meal Gap 2015: A Report on County and Congressional District Food Insecurity and County Food Cost in the United States in 2013.* Feeding America, Chicago, IL.

Flint Leverage Points Project. 2020. *Urban Food Systems in Flint, Michigan: Identifying Leverage Points.* College of Agriculture and Natural Resources, Michigan State University. https://www.canr.msu.edu/flintfood

Flint Trust Steering Committee. 2020. *Community Voice on the Flint Water Crisis. A Trust Study, Needs Assessment and Plan of Action.* https://michr.umich.edu/resources/2020/2/18/community-voice-on-the-flint-water-crisis-a-trust-study

Folke, C. 2006. Resilience: The emergence of a perspective for social-ecological systems analyses. *Global Environmental Change* 16: 253–267.

Food and Agriculture Organization of the United Nations. 2014. *FAO's Resilience Index Measurement and Analysis (RIMA) Model, Improved Global Governance for Hunger Reduction.* http://www.foodsec.org/web/resilience/measuring-resilience/en/

Fraser, E.D.G. 2003. Social vulnerability and ecological fragility: Building bridges between social and natural sciences using the Irish Potato Famine as a case study. *Conservation Ecology* 7(2): 9. http://www.consecol.org/vol7/iss2/art9/

Fraser, E.D.G. 2007. Travelling in antique lands: Using past famines to develop an adaptability/resilience framework to identify food systems vulnerable to climate change. *Climatic Change* 83: 495–514.

Fraser, E.D.G., W. Mabee, and F. Figge. 2005. A framework for assessing the vulnerability of food systems to future shocks. *Futures* 37: 465–479.

Garmestani, A.S., C.R. Allen, and L. Gunderson 2009. Panarchy: Discontinuities reveal similarities in the dynamic system structure of ecological and social systems. *Ecology and Society* 14 (1): 15. http://www.ecologyandsociety.org/vol14/iss1/art15/

Goulden, M.C., W.N. Adger, E.H. Allison, and D. Conway. 2013. Limits to resilience from livelihood diversification and social capital in lake social–ecological systems. *Annals of the Association of American Geographers* 103: 906–924.

Gray, S., A. Singer, L. Schmitt-Olabisi, J. Introne, and J. Henderson. 2017. Identifying the causes, consequences, and solutions to the Flint Water Crisis through collaborative modelling. *Environmental Justice* 10(5) 154–161.

Gundersen, C., A. Dewey, A.S. Crumbaugh, M. Kato, and E. Engelhard. 2016. *Map the Meal Gap 2016: A Report on County and Congressional District Food Insecurity and County Food Cost in the United States in 2014*. Feeding America, Chicago, IL.

Gundersen, C., A. Dewey, A. Crumbaugh, M. Kato, and E. Engelhard. 2017. *Map the Meal Gap 2017: A Report on County and Congressional District Food Insecurity and County Food Cost in the United States in 2015*. Feeding America, Chicago, IL.

Gundersen, C., A. Dewey, A. Crumbaugh, M. Kato, and E. Engelhard. 2018. *Map the Meal Gap 2018: A Report on County and Congressional District Food Insecurity and County Food Cost in the United States in 2016*. Feeding America, Chicago, IL.

Gundersen, C., A. Dewey, E. Engelhard, M. Strayer, and L. Lapinski. 2020. *Map the Meal Gap 2020: A Report on County and Congressional District Food Insecurity and County Food Cost in the United States in 2018*. Feeding America, Chicago, IL.

Gundersen, C., A. Dewey, M. Kato, A. Crumbaugh, and M. Strayer. 2019. *Map the Meal Gap 2019: A Report on County and Congressional District Food Insecurity and County Food Cost in the United States in 2017*. Feeding America, Chicago, IL.

Gunderson, L., and C.S. Holling. 2002. *Panarchy: Understanding Transformations in Human and Natural Systems*. Island Press, Washington, DC.

Hanna-Attisha, M. 2018. *What the Eyes Don't See: A Story of Crisis, Resistance, and Hope in an American City*. Penguin Random House, New York, NY.

Hanna-Attisha, M., J. LaChance, R.C. Sadler, and A.C. Schnepp. 2016. Elevated blood lead levels in children associated with the Flint drinking water crisis: A spatial analysis of risk and public health response. *American Journal of Public Health* 106: 283–290.

Herrmann, D.L., W.D. Shuster, A.L. Mayer, and A.S. Garmestani. 2016. Sustainability for shrinking cities. *Sustainability* 8: 911. doi:10.3390/su8090911

Hodbod, J., and H. Eakin. 2015. Adapting a social-ecological resilience framework for food systems. *Journal of Environmental Studies and Sciences* 5: 474–484.

Holling, C.S., and L.H. Gunderson. 2002. Resilience and adaptive cycles. In *Panarchy: Understanding Transformations in Human and Natural Systems*, ed. L.H. Gunderson and C.S. Holling, 25–62. Island Press, Washington, DC.

Holling, C.S., L.H. Gunderson, and G.D. Peterson. 2002. Sustainability and panarchies. In *Panarchy: Understanding Transformations in Human and Natural Systems*, ed. L.H. Gunderson and C.S. Holling, 63–102. Island Press, Washington, DC.

Johnson, J.E., and K. Key. 2018. The Flint Water Community Narrative Group. *Progress in Community Health Partnerships: Research, Education, and Action* 12: 215–221.

McAllister, R.R.J., N. Abel, C.J. Stokes, and I.J. Gordon 2006. Australian pastoralists in time and space: The evolution of a complex adaptive system. *Ecology and Society* 11(2): 41. http://www.ecologyandsociety.org/vol11/iss2/art41/

Michigan Civil Rights Commission. 2017. *The Flint Water Crisis: Systemic Racism through the Lens of Flint.* Michigan Civil Rights Commission, Lansing.

Moen, J., and E.C.H. Keskitalo. 2010. Interlocking panarchies in multi-use boreal forests in Sweden. *Ecology and Society* 15(3): 17. http://www.ecologyandsociety.org/vol15/iss3/art17/

Morckel, V. 2017. Why the Flint, Michigan, USA water crisis is an urban planning failure. *Cities* 62: 23–27.

Morckel, V., and K. Terzano. 2019. Legacy city residents' lack of trust in their governments: An examination of Flint, Michigan residents' trust at the height of the water crisis. *Journal of Urban Affairs* 4: 585–601.

Pauli, B.J. 2019. *Flint Fights Back: Environmental Justice and Democracy in the Flint Water Crisis.* MIT Press, Cambridge, MA.

Pelling, M., and D. Manuel-Navarrete. 2011. From resilience to transformation: The adaptive cycle in two Mexican urban centers. *Ecology and Society* 16(2): 11. http://www.ecologyandsociety.org/vol16/iss2/art11/

Peterson, G.D. 2000. Political ecology and ecological resilience: An integration of human and ecological dynamics. *Ecological Economics* 35: 323–336.

Rasmussen, L.V., and A. Reenberg. 2012. Collapse and recovery in Sahelian agro-pastoral systems: Rethinking trajectories of change. *Ecology and Society* 17(1): 14. http://dx.doi.org/10.5751/ES-04614-170114

Resilience Alliance. 2010. *Assessing Resilience in Social-Ecological Systems: Workbook for Practitioners.* Version 2.0. Resilience Alliance, Stockholm. http://www.resalliance.org/3871.php

Robinson, L.W. 2009. A complex-systems approach to pastoral commons. *Human Ecology* 37: 441–451.

Rosen, A.M., and I. Rivera-Collazo. 2012. Climate change, adaptive cycles, and the persistence of foraging economies during the late Pleistocene/Holocene transition in the Levant. *Proceedings of the National Academy of Sciences* 109: 3640–3645.

Sadler, R.C., and A.R. Highsmith. 2016. Rethinking Tiebout: The contribution of political fragmentation and racial/economic segregation to the Flint Water Crisis. *Environmental Justice* 9(5): 143–151.

Salvia, R., and G. Quaranta. 2015. Adaptive cycle as a tool to select resilient patterns of rural development. *Sustainability* 7: 11114–11138.

Sim, B. 2016. Poor and African American in Flint: The water crisis and its trapped population. In *The State of Environmental Migration 2016: A Review of 2015*, F. Gemenne, C. Zickgraf, and D. Ionesco, 77–101. Presses Universitaires de Liège, Liège, Belgium.

Soane, I.D., R. Scolozzi, A. Gretter, and K. Hubacek. 2012. Exploring panarchy in alpine grasslands: An application of adaptive cycle concepts to the conservation of a cultural landscape. *Ecology and Society* 17(3).

Stroink, M.L., and C.H. Nelson. 2013. Complexity and food hubs: Five case studies from northern Ontario. *Local Environment* 18: 620–635.

Sundkvist, Å., R. Milestad, and A. Jansson. 2005. On the importance of tightening feedback loops for sustainable development of food systems. *Food Policy* 30: 224–239.

Thompson, J., and I. Scoones. 2009. Addressing the dynamics of agri-food systems: An emerging agenda for social science research. *Environmental Science & Policy* 12: 386–397.

Thorén, H., and L. Olsson. 2017. Is resilience a normative concept? *Resilience* 3293: 1–17.

University of Minnesota. 2019. *IPUMS USA*. https://usa.ipums.org/usa/.

U.S. Bureau of Labor Statistics. 2019. CPI inflation calculator. https://data.bls.gov/cgi-bin/cpicalc.pl

U.S. Bureau of Labor Statistics. 2020. CPI inflation calculator. https://www.bls.gov/data/inflation_calculator.htm

U.S. Census Bureau. 2019. U.S. Census Bureau QuickFacts, Peoria city, Illinois; Green Bay city, Wisconsin; South Bend city, Indiana; Lansing city, Michigan; Flint charter township, Genesee County, Michigan; Flint city, Michigan. https://www.census.gov/quickfacts/fact/table/peoriacityillinois,greenbaycitywisconsin,southbendcityindiana,lansingcitymichigan,flintchartertownshipgeneseecountymichigan,flintcitymichigan/PST045218

Walker, B.H., N. Abel, J.M. Anderies, and P. Ryan. 2009. Resilience, adaptability, and transformability in the Goulburn-Broken Catchment, Australia. *Ecology and Society* 14(1): 12. http://www.ecologyandsociety.org/vol14/iss1/art12/

Winkel, T., P. Bommel, M. Chevarría-Lazo, G. Cortes, C. Castillo, P. Gasselin, F. Leger, J.-P. Nina-Laura, S. Rambal, M. Tichit, J.-F. Tourrand, J.-J. Vacher, A. Vassas-Toral, M. Vieira-Pak, and R. Joffre. 2016. Panarchy of an indigenous agroecosystem in the globalized market: The quinoa production in the Bolivian Altiplano. *Global Environmental Change* 39: 195–204.

CHAPTER 12

Panarchy and the Governance of Social-Ecological Systems

Brian C. Chaffin

In *Panarchy*, Gunderson and Holling (2002) clearly articulated a theory of change at the core of resilience: that rhythmic dynamics of change occur within systems, from growth and decay, to collapse and renewal; that systems are characterized by different processes and structures at different scales; and that cross-scale and cross-level interactions exert substantial influence on system dynamics, which can result in bottom-up change. The four phases of change are represented in an adaptive cycle, which are linked across scales in the concept of panarchy (Gunderson and Holling 2002). Panarchy concepts arose from historical reviews of the interaction between ecological dynamics and institutional dynamics in managed resource systems, such as the Everglades (Gunderson et al. 1995). Social scientists and ecologists have since embraced the concept of adaptive cycles and panarchy to describe nonlinear dynamics in social-ecological systems (SESs) (Allen et al. 2014).

Governance of SESs, including all aspects of governing, not just government, such as institutions (rules, laws, policies, social norms), organizations, individuals, networks, and markets, is a critical interface where human agency and biophysical processes collide (Lemos and Agrawal 2006; Bevir 2009). Governance is a key intervention point whereby human action can be adjusted to address system changes in an attempt to shape sustainable trajectories for SESs.

Applying panarchy to environmental governance is important in two ways. First, SESs are complex systems, and panarchy illuminates critical cross-scale interactions that influence systemic change and thus can be the focus of policy and management actions and interventions. Second, viewing the environmental governance regime itself as a dynamic panarchy "clearly highlights the importance of historical and political contexts as key cross-scale interactions that influence critical periods of collapse and rebirth of governance towards forms with an increased capacity to function amidst complexity and uncertainty" (Chaffin and Gunderson 2016, 82). In short, panarchy as a metaphor and model is useful to describe both patterns of nonlinear changes in SESs and the rhythms of governance that attempt to shape SESs.

Although much empirical work has been done to advance panarchy in ecological science (Allen and Holling 2008; Angeler et al. 2013; Nash et al. 2014; Herrmann et al. 2016), scholars may have only scratched the surface in leveraging panarchy theory to understand emergent scales and cross-scale interactions relevant for environmental governance in the Anthropocene. For example, governance researchers can apply panarchy as a framework for testing hypotheses about processes of transformation and resilience: When a system is disturbed, what specific aspects of higher- and lower-scale systems engage and why? How do these feedbacks differ across space, time, jurisdictions, or institutions? Researchers can apply panarchy to frame investigations that attempt to characterize the potential for (or the mechanisms of) transformation toward local and global sustainability through a more developed understanding of cross-scale and cross-level interactions (Chaffin et al. 2016).

Despite and perhaps in large part because of the need to identify sustainable futures for SESs (and avoid unsustainable trajectories), governance research framed by resilience scholarship generally, and panarchy in particular, has long been associated with outcomes related to avoiding, mitigating, adapting to, or navigating global change (e.g., Eakin et al. 2009; Garmestani and Benson 2013; Folke 2016; Winkel et al. 2016). Thus, theories of social-ecological resilience (including adaptive cycles and panarchy) have appealed to researchers for their power to identify dynamics that could lead to discoveries of effective interventions for avoiding challenges to human well-being, including social and environmental injustice, collapse of services (ecosystem and built) and economies, and a loss of critical resources (e.g., food, water) (Petrosillo et al. 2010; Holdschlag and Ratter 2013; Jacques 2015; Chaffin et al. 2016). One of the major challenges to furthering the application of panarchy in environmental governance research stems from the difficulty in capturing social dynamics such as politics, power, and human agency in resilience-framed analyses (Leach 2008; Brown 2014; Chaffin and Gunderson 2016). Despite some criticism (e.g., Brown 2014), this challenge appears surmountable, and there is much work to be done to further explore the relevance and strength of panarchy in environmental governance research.

In this chapter, I briefly review panarchy as applied in governance research, and I present both challenges and opportunities for further leveraging panarchy as a conceptual framework to increase analytical exploration of SES governance as we head deeper into the uncertainties of human-induced environmental change that define the Anthropocene. Impacts of panarchy dynamics include both disruption and entrenchment of critical system drivers, as well as resulting changes in system processes and structures. These dynamics are visible across a range of scales important to governance scholars, including at the individual (psychological), community or group (social), and societal (cultural) scales, as well as across a myriad of emergent scales that are a manifestation of biological and social patterns superimposed on the physical processes of the Earth. As influential actors within these systems, humans—individually and collectively—both drive and react to cross-scale interactions in an attempt to shape SES trajectories. In other words, humans, through processes of environmental governance, use foresight and planning to mitigate impacts of disturbances and regime shifts, adapt to major systemic changes, but also pursue transformative responses that fundamentally alter systems through induced regime shifts (Moore et al. 2014; Boyd et al. 2015; Chaffin et al. 2016). Panarchy can serve as a guide to systematic analysis of these dynamics and may help reveal patterns of scale-dependent and scale-invariant governance phenomena (including individual human behaviors and social processes) essential to better understanding how to achieve sustainable and just SES trajectories, including more effectively integrating the role of science and traditional knowledges in governance.

Panarchy and Governance Research

Panarchy describes interactions and feedbacks within and across scales of space and time, addressing how small-scale processes affect larger-scale ones and how larger-scale processes influence smaller ones. Two major contributions of this conceptual framing are the recognition that an adaptive cycle can generate novelty and change at a particular scale (Allen and Holling 2010) and the existence of cross-scale interactions that act on the dynamics of a set of adaptive cycles operating at different scales (Gunderson and Holling 2002). Through studies of ecosystem dynamics, Holling et al. (2002, 64–68) synthesized four salient patterns of scales and scaling: "(1) as scale increases, distinct objects appear and persist across a range of scales; (2) abrupt breaks will appear in these patterns across scales; (3) the degree of human impact depends on the scale and landscapes affected, i.e., land, water, atmosphere; and (4) human-derived issues, problems, and opportunities can have causes linked to processes at several scales." Empirical evidence of these patterns discovered by ecologists and social scientists (e.g., levels of a panarchy discontinuities separating scales) suggests that scale-dependent groups exhibit similarities in size, speed, and function within each discrete aggregate level (e.g., body mass distributions, cities, buildings) (Holling 1992; Allen et al. 1999, 2014; Garmestani et al. 2009b). Adaptive cycles are the engine for experimentation at a given scale; their cycles of growth, collapse, and renewal allow novelty to emerge (Holling et al. 2002). Adaptive cycles can be nested within larger, slower adaptive cycles but are not organized in strict hierarchies across scales, as they are sometimes conceptualized (e.g., in U.S. government, county, state, federal systems). There are two-way feedbacks of interactions between adaptive cycles: Influences from smaller, faster levels of adaptive cycles may feed novelty and innovation during key moments of opportunity, whereas memory from larger, slower adaptive cycles constrains the dynamics of the nested, faster scales (figure 1.3 in chapter 1). The dynamics of a panarchy are therefore both "creative and conserving" (Holling 2001). Although most of the initial examples of adaptive cycle dynamics were ecological systems (Holling 1986), subsequent investigations have used panarchy concepts in environmental and SES governance (Gunderson et al. 1995; Garmestani et al. 2019, 2020).

Environmental Governance of SESs

Governance as a term has risen in use and popularity since the 1980s as a means of capturing the diversity and messiness of governing processes beyond those of governments (Bevir 2012). Bevir (2009, 2012) described a shift away from the strong legitimacy given to central state governments, which also parallels a loss of faith in a unified sovereign state. Bevir used the term *new governance* to include markets, nongovernment organizations and cross-scale networks of individuals who play critical roles in the processes of governing. Bevir (2012, 11) wrote that "these processes increasingly involve organizational hybrids that cross hierarchy, market, and network, and embrace multiple actors from the public, private, and voluntary sectors." As scientific progress shed light on the complexities and uncertainties inherent in the governance of environmental changes, governments were unable to respond at the scales necessary to address global issues such as climate change, prompting a shift in focus toward the multiple, overlapping processes of environmental governance (Lemos and Agrawal 2006).

A holistic approach to governance of SESs is much more aligned with the theory of panarchy than the hierarchical mechanisms of governments alone. Government institutions and bureaucracies can limit the cross-scale experimentation in governance needed to understand and adapt to complex SES dynamics and trajectories of change (Gunderson and Light 2006; Garmestani et al. 2009a). Rigid regulatory frameworks of government limit the application of adaptive management and thus limit societal ability to learn and adapt, creating a pathology of bureaucratic management that ignores the rich, scale-dependent, complex dynamics of SESs (Craig 2010; Allen and Gunderson 2011; Garmestani et al. 2019). Panarchy allows researchers and practitioners to see governance regimes as embedded in broader dynamics, interacting across space and time—information and memory from levels above, but also feedbacks up from lower levels of a panarchy in the form of innovation and disturbance (Garmestani and Benson 2013).

Adaptive and Transformative Governance

Over the past two decades, concepts of resilience (*sensu* Holling 1973) and panarchy have been incorporated into scholarship on environmental governance (e.g., Garmestani et al. 2009a). Garmestani and Benson (2013) described such configurations as resilience-based governance that supports social-ecological resilience in SESs. Before integration of these concepts, governance approaches to complex environmental problems were focused on development of formal policies, regulations, or legislation that used a command-and-control approach to solving the problems (Holling and Meffe 1996). Such top-down, command-and-control approaches can work well for solving discrete problems (such as controlling pollution via end-of-pipe regulations under the U.S. Clean Water Act) but are wholly inadequate for addressing complex problems such as climate change (Ruhl 2005; Green et al. 2015). To address these problems, a different approach is needed, one that incorporates experimentation and learning as well as processes for evaluating progress toward general goals as opposed to achieving fixed outcomes. The foundation of this alternative approach to governing complex problems is to manage for a dynamic *resilience*, the amount of disturbance that can be absorbed by a system before it fundamentally changes to a new configuration (Holling 2001).

Resilience-based governance in SESs would require three critical changes to governance approaches. First, we need to recognize that SESs are made up of adaptive cycles that exist across scales, as a result of the complexity and interconnectedness inherent in SESs. Thus, a narrow focus on environmental statutes, regulations, and policies alone is inadequate for managing resilience. Second, resilience is not a static quality of systems but is ever changing based on internal dynamics of systems (described by adaptive cycles) and external influences (cross-scale interactions across panarchies). The one-size-fits-all solutions so common to command-and-control governance will not effectively manage resilience in SESs. Governance approaches must be context dependent, appropriate to a particular situation and focused on process before outcomes (Rijke et al. 2012). Third, there is much we do not know about the complex interactions between SESs in our global systems. Thus, approaches to governing resilience must embrace learning and experimentation and build into governance the capacity to capture new knowledge in a structured, iterative manner.

Adaptive governance represents the confluence of several distinct strands of scholarship related to these ideas. Brunner et al. (2005) rejected assumptions of maximum sustainable yield as an environmental management paradigm in their description of adaptive governance. Folke et al. (2005) described SES resilience as a goal for adaptive governance and management of complex systems. Dietz et al. (2003) and Ostrom (2009) argued that human institutions must match the complexity of natural systems in order to govern resources sustainably. Adaptive governance was identified as necessary to foster adaptive management (Gunderson and Light 2006) but more recently used to describe the many emergent, dynamic, learning-based forms of governance that anticipate and change while achieving sustainability in SESs (Olsson et al. 2004, 2006; Folke et al. 2005; Chaffin et al. 2014; Boyd et al. 2015; Cosens et al. 2018). Scholars have suggested many conditions "necessary" for adaptive governance to function, including polycentricity, or the multiple centers of power, authority, and legitimacy in governance (Morrison et al. 2019); experimentation and learning vis-à-vis adaptive management (Allen et al. 2011); a focus on fit between governance institutions and biophysical aspects of SESs (Huitema et al. 2009; Rijke et al. 2012); participation in governance beyond the structures of government, both the capacity of groups and individuals to participate and the legitimacy of the venue (Cosens 2013); and the integration of multiple types of knowledge, not just science but traditional knowledge, and the meaningful coproduction of governance approaches therefrom (Wyborn et al. 2019).

Adaptive governance represents a wide range of governance approaches that emerge to enable management of resilience in SESs. Adaptive governance itself is a complex system, and the degree to which it emerges and governs effectively is a product of its own dynamics and interactions with both nested and legacy approaches to governance across time and space (Chaffin and

Gunderson 2016). Often adaptive governance is sought to maintain or build resilience in an SES. However, there are times when a system exists untenably (has undergone a regime shift) or society deems a SES trajectory as undesirable (e.g., irreversible environmental degradation, economic devastation, social and environmental injustice); in these cases, the processes of adaptive governance are not particularly useful for the governance of SESs (Chaffin et al. 2016). In turn, Chaffin et al. (2016) proposed the phrase *transformative governance* to describe proactive and reactive governance that actively works to transition an SES into a more socially and ecologically desirable regime. Transformative governance requires a unique set of capacities, opportunities, and political will to emerge successfully (Chaffin et al. 2016). Such processes are transient and controlled by the internal dynamics and external influences in which SES governance is embedded.

The theories of adaptive and transformative governance follow directly from the development of panarchy. In addition, a better understanding of these governance approaches can be gained through an analytical approach that identifies the cross-scale interactions illustrated by flows to and from adaptive cycles at a particular scale. This type of investigation can aid governance scholars in more clearly identifying potential barriers or lack of capacity for SES change and performing adaptive management of governance itself, that is, identifying possible opportunities or intervention points to experiment and test hypotheses on the salience and effectiveness of elements of adaptive and transformative governance (Chaffin and Gosnell 2015).

Challenges and Opportunities for Panarchy and Governance

Panarchy is useful in the process of conceptualizing, identifying, and analyzing SESs as important aggregations that arise at particular scales, as well as the cross-scale interactions that span time and space and cross levels of jurisdictions, institutions, and human cognition or perceptions. Although quantifying the social and ecological indicators of panarchy can be challenging, there is substantial progress and potential (Garmestani et al. 2009b; Allen et al. 2014). Analyzing social indicators through a panarchy lens, particularly in the social sciences, is partial and incomplete because panarchy was not developed to be an all-encompassing explanation or panacea for SESs. Berkes and Ross (2016, 190), in their review of panarchy's application to community resilience specifically, warned that some of the ecological science assumptions implicit in resilience theory "cannot capture social capital and power relations, agency, historical and cultural contexts within which social relationships are experienced and negotiated among the levels in a panarchy." Similarly, Fraser (2003) argued that panarchy as a framework should not be used alone to analyze social systems but instead combined with other social theories to better explain the scale-constrained nature of social elements such as vulnerability. Through an analysis of historical SES dynamics surrounding the nineteenth-century Irish potato famine, Fraser asserted that panarchy theory can help clarify dynamics of vulnerability and resilience in ecological systems, but other frameworks of social theory yield valuable, complementary insights. Gotts (2007) and Stone-Jovicich (2015) suggested potentially compatible social theory frameworks such as world systems analysis as a means to better understand the historic and geographic linkages between coupled SES processes including environmental degradation, social inequity, and poverty.

Other critical scholars have also specifically called out the ecological and complex system foundations of resilience theory as a culprit for depoliticizing the "social" in SESs (Cote and Nightingale 2012; Brown 2014). Resilience, as a nonnormative property of complex systems, "does not take account of the institutions within which practices and management are embedded" (Brown 2014, 109). Pelling and Manuel-Navarrete (2011) and Holdschlag and Ratter (2016) are among the few resilience scholars who recognize how politics, power, and issues such as marginalization and legacies of colonization play out as cross-scale dynamics and how the discourse of resilience itself is powerful and can be wielded by those in positions of power and advantage for outcomes that are unreflexive at best, malicious at worst. Thus, the critiques that resilience scholarship omits key elements of social systems are red herrings, because resilience theory and panarchy were not developed to explain every aspect of SESs. Herein lies one of the major emergent challenges to a panarchy-based approach to governance research.

Challenge of the Human Element

A major critique of resilience theory (and, implicitly, panarchy) over the past decade has come from social scientists concerned that the strong normative salience of terms such as *resilience* and *transformation* for individuals, networks, organizations, and governments engaged in environmental governance and international development have been coopted on the global stage to promote a neoliberal agenda (Brown 2014; Blythe et al. 2018). Not just "resilience of what to what" (Carpenter et al. 2001, 765) but "resilience *of whom and for whom*" have become important questions to ask to better understand motivations and power dynamics concealed in applications of the terms (Brown 2014; Cutter 2016). Brown noted a general omission of analysis (or even mention) of social, political, and cultural dynamics in early resilience scholarship and little analysis of social difference and resilience. This omission is potentially attributable to the swift adoption of resilience concepts outside the scientific research community, specifically by the international environmental development community, what Brand and Jax (2007) referred to as applying resilience more as a perspective or way of thinking than as a descriptive (or analytical) concept (Brown 2014). MacKinnon and Derickson (2013) took a step further to argue that resilience had been coopted by neoliberal modes of global environmental governance, rendering it unfit as a social science concept. Such uses of *resilience* helped justify large government and nongovernment organization development programs that directed unprecedented sums of capital to untested or dubious environmental governance schemes such as REDD+ (Fletcher et al. 2016) but failed to address the question of resilience for whom (Brown 2014). Beymer-Farris et al. (2012) and Blythe et al. (2018) continued to argue that when applied thus, resilience overlooks resource conflicts and power asymmetries embedded in these conflicts. Often, such applications of the engineering definition of resilience (Allen et al. 2019) narrowly focus on external disturbances to bounded, immutable, local-scale systems. As a result, such incomplete applications never fully investigate or elucidate the nested internal dynamics at a focal scale or capture the cascading cross-scale changes and cross-level linkages within a panarchy.

Another important challenge for pushing the synergy of panarchy and SES governance research further is that most initial explorations of resilience-based governance have taken place in the developed, mostly democratic regions of the Global North (Olsson et al. 2004; Chaffin et al. 2016). Case studies on the emergence and institutionalization of adaptive governance have been primarily northern, creating a void in the understanding of adaptive governance processes in relation to the unique contexts of the Global South (Karpouzoglou et al. 2016; Yasmin et al. 2019). This challenge is probably reinforced by the potential failure of resilience-based governance features to emerge or be implemented in the Global South due to lack of capacity and coordination, persistent corruption, or other challenging governance contexts, and myriad other pressing development and human rights challenges that render holistic approaches to SES governance a secondary concern.

As an example of how further framing SES governance research in terms of panarchy can illuminate additional insights and act on the opportunities mentioned above, I describe a case of contemporary water governance in the western United States in the next section.

Pursuing Panarchy Governance: A U.S. Watershed Example

Two basic systems of water law govern surface water use and allocation in the United States, based on different hydrologic settings. A riparian rights system exists in many of the wetter eastern states of the country (Craig et al. 2017). In the drier landscapes of the western United States, a scheme called prior appropriation is applied (Craig et al. 2017). Prior appropriation is based on the idea of "first in time, first in right," thereby privileging the first person to put a water source to use through development (e.g., mining, irrigation infrastructure) above others (Tarlock 2000). Prior appropriation was developed to encourage Euro-American colonization and settlement of arid lands west of the longitudinal divide on the North American continent known as the hundredth meridian. This meridian divides lands that can sustain crops without irrigation from lands that cannot, and it is slowly advancing east because of climate change (Seager et al. 2018). The administration of prior appropriation has shaped both the formal and informal governance of water in the western United States, serving as a powerful backdrop for decisions about water use and conservation across scales, and specifically influenced by the overlay of a variety of jurisdictions and institutions (Tarlock 2000, 2001).

The primary scale at which the processes of western U.S. water governance are administered is defined by U.S. state boundaries. Jurisdictionally, U.S. states operate with a strong amount of autonomy within the bounds of prior appropriation and even provide feedbacks to the higher-level institution of prior appropriation through differences in administration from state to state that often become accepted as part of the doctrine of prior appropriation itself. In the mid- to late twentieth century, western states began to adopt procedures for adjudicating the prior appropriation doctrine in watersheds or basins within each state. The role of the state was to investigate and determine which water users had rights to use specific water sources within a watershed and the characteristics of those rights (e.g., quantity, timing, purposes of use, place of use) (Craig et al. 2017).

Most early prior appropriation water rights were simply claimed by Euro-American settlers colonizing the western United States. Such claims established a right to use water. All a potential user had to do was divert the water

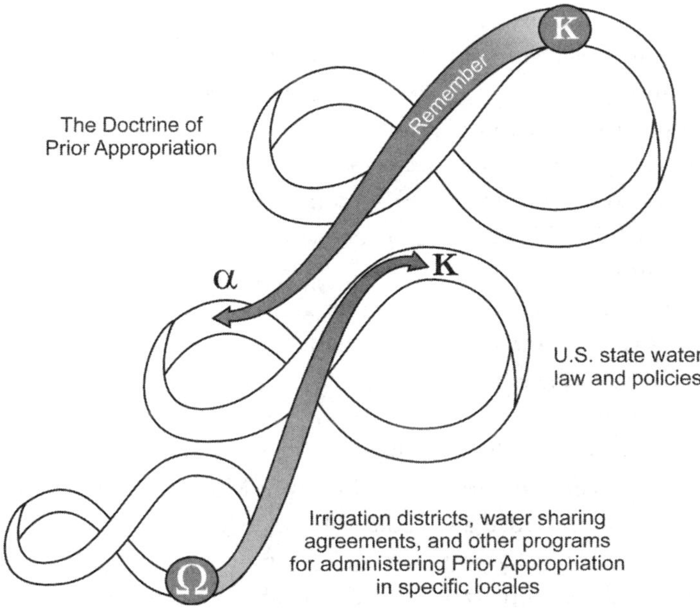

Figure 12.1. Panarchy of water governance in the western United States. At the largest level, the prior appropriation doctrine provides memory for settlement of administrative challenges to state water law. Policy failure, judicial appeal, and local agreements have led to reevaluation or modification of state water law and administration.

and put it to what was considered a beneficial use, such as irrigation, mining, or personal consumption (Montana Water Policy Interim Committee [MT WPIC] 2018). Today, state-based basin adjudications of prior appropriation rights are the norm, part of the overall concept of western water governance, despite the variation in approach from state to state. This method demonstrates a clear feedback across the panarchy of water governance from states to the institution of prior appropriation itself (figure 12.1).

Water governance levels below the state level are often emergent, based on a mediation of values placed on physical boundaries such as watersheds and institutional borders such as counties or irrigation districts. Institutions that arise at these emergent levels can guide implementation of prior appropriation to evolve and react more quickly to social and environmental change than they can at the state level, or certainly than at the level of prior appropriation itself. A clear example can be found in states such as Montana, where the state-level administration of prior appropriation often forces adherence to another major tenet of prior appropriation, the "use it or lose it" principle, which essentially requires users with an extractive water right to actually use water for the specified purpose in the specified amount (Craig et al. 2017). If a water right goes unused for a specified period of time, the user could be subject to forfeiture of those rights through state or court enforcement (MT WPIC 2018). Although this generally occurs only in overallocated basins with strong competition between users of limited water sources, it is a dominant institution at the state level.

At jurisdictional levels and geographic scales within a state (local, watershed or basin), a series of institutions and organizations have emerged that create solutions for allocating water that are both more flexible across temporal scales (within water years or irrigation seasons) and less formal (figure 12.2). Many watersheds or basins within Montana contain irrigation districts, quasigovernmental groups of users who rely on a common source of water and are governed by locally formalized rules and overseen by an elected body (Mullin 2009). For many irrigation districts in Montana, water rights at the state level are not recorded and codified for individual water users; instead, a water right is attributable to the entire irrigation district. This arrangement allows locally elected district leaders and members to craft rules and enforcement mechanisms for water use such as water sharing, transfers, and leases. All of this happens at emergent scales that are not a priori defined, within the bounds of both prior appropriation and the state, but allowing for much faster and more responsive reaction to local conditions. Another innovation at this level in Montana is the presence of volunteer water sharing agreements between water users in a particular watershed, generally organized and curated by one or more environmental or conservation organizations in areas where no irrigation district exists or the district does not adequately fit or represent the watershed under pressure, such as in the Blackfoot River watershed (Blackfoot Challenge 2019).

Currently, legacies of historically exploitive actions such as colonization are colliding with processes of prior appropriation administered at the state level, leading to emergence of new components and relationships with traditional governmental hierarchies. Montana is home to eight federally recognized Native American tribes, yet their current reservation lands represent fractions of their historically occupied territories. Historically, tribes were forcibly removed from their territories or forced into signing treaties with the U.S. government that established reservation boundaries and a federal–tribal trust relationship. Tribes are domestic-dependent sovereign nations, exercising a government-to-government relationship with the United States (Anderson

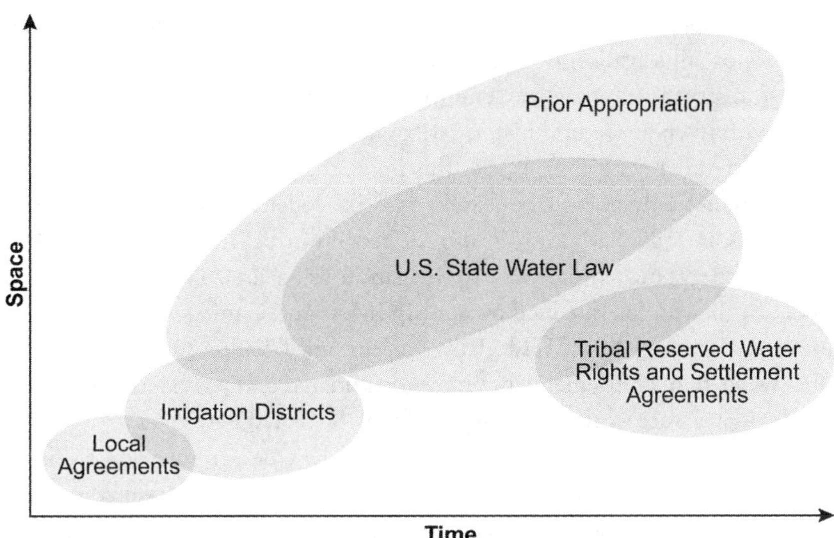

Figure 12.2. Scales of components of water governance in the U.S. West. They range from local agreements, which change more rapidly and are in effect over smallest areas (e.g., voluntary drought mitigation agreements between private landowners and a local nongovernment organization). Irrigation districts are quasigovernment groups of users who rely on a common source of water and are governed by locally formalized rules and overseen by an elected body. State water laws govern all water allocation and use within the bounds of a given U.S. state. Tribal reserved water rights are federally held water rights asserted in within state systems of water rights administration that quantify the amount of water available to tribes; reserved water rights settlement agreements are often required to formalize these water rights and concurrent agreements (payments, release of liability, water conservation programs) between states, tribes, and the federal government. The Doctrine of Prior Appropriation includes the overarching principles of water governance that guide water law and policy in most western U.S. states.

2006; Fletcher 2006). Through this relationship, the United States has a fiduciary responsibility to manage tribal resources (including land and water) in trust and advocate on behalf of the best interests of the tribe. This arrangement or relationship has led to a series of legal challenges and decisions over the past hundred years with respect to water rights. These decisions have determined that tribes and reservations have legal rights to water not previously considered within a state-based processes of prior appropriation. Moreover, priority dates for tribal water rights are based on the time at which a treaty was signed or the time of Indigenous occupation, that is, "time immemorial" (Colby et al. 2005,

Anderson 2006). As tribes and their federal trustees work through state-based processes of adjudication or litigation to quantify and gain legal recognition of these reserved water rights, seniority of water uses in many western U.S. basins with reservations and historic tribal homelands have been upended and disturbed (Anderson 2010). It is at this point that new effective scales of water governance emerge as tribes, states, and the federal government negotiate potential solutions that aim to fulfill the federal–tribal trust relationship but also protect entrenched, non-Native irrigation water users in arid basins with a mix of tribal and private land ownership (Montana-Confederated Salish and Koontenai Tribes [MT-CSKT] 2015; Cosens and Chaffin 2016). A series of tribal water rights settlements have been negotiated and passed as federal and state legislation across the West, some of which created new governance institutions that more closely align to the scales of the emergent challenges (Stern 2017). New institutions interact with established water governance institutions across levels of a water governance panarchy but emerge at a unique scale, as they both shape and are shaped by all levels of the established system of water governance. The emergent scale of these water governance institutions is a result of driving forces across space and time that include historical contexts such as the legacies of colonization, as well as contemporary power, politics, and economic dynamics that influence water governance and resource management from the global (Indigenous rights movements) to the federal and state (tribal reserved water rights settlements) to more local levels in the United States (comanagement of water and aquatic resources) (MT-CSKT 2015; Cosens and Chaffin 2016; Stern 2017).

Exploring this and similar examples of governance through the lens of panarchy can illuminate the potential for research to identify innovations that could cause cascading change across a panarchy. For example, as climate change shifts the timing and volume of spring runoff in many snowpack-driven agricultural areas of the western United States (Hamlet et al. 2005; Whitlock et al. 2017), the seemingly rigid tenets of prior appropriation will be tested (Craig 2010). Innovations such as new institutional arrangements for water sharing at the more local level or across multiple jurisdictions may serve as a template for experimentation in adjustments to water law at the state level and programs, guidance, or support from the federal level. A panarchy lens allows the research analyst to better understand the barriers to this scaling up (institutional, jurisdictional, geographic) and the potential feedbacks to and from the biophysical environment across space and time. Such action is under way in many states in the U.S. West in the form of experiments in prior appropriation policy such as split-season leasing of water rights and the ability of a water user to temporarily or permanently convert an irrigation water right into a water right for instream flow to protect fishery resources, aquatic habitat, and waterbody integrity—for both environmental and cultural values. In addition, as water governance experiments such as these proceed, arguably another coupled scale of investigation emerges as fertile ground for research: the cognitive or perceptual scale. As prior appropriation changes institutionally and jurisdictionally, the collective understanding of and beliefs about what is and is not acceptable under the tenets of prior appropriation are challenged both at the level of the individual and at the level of society. The cross-scale interactions of this shift over time and space (and jurisdiction) have arguably been a missing component of research to understand why strong legal doctrines such as prior appropriation are unlikely to explicitly shift in anticipation of the hydrologic implications of a changing climate and associated societal navigation.

Opportunities: Panarchy as a Research Agenda for Governance in the Anthropocene

A renewed focus on the dynamics of panarchy within SES governance research, with attention to the general and specific critiques of resilience, has strong potential to overcome the challenges presented above. I am not suggesting business as usual for panarchy and resilience science; instead, I suggest that governance scholars reengage and potentially reimagine panarchy by integrating novel methods and theories used in modern environmental and sustainability sciences. Many scholars are already engaged in this work, as evidenced in this volume (Olsson et al., chapter 8; Craig et al., chapter 9; Farley and Egler, chapter 10; Hodbod and Wentworth, chapter 11; and Chapin et al., chapter 13). Specifically, the socialization and politicization of panarchy warrant additional attention from social scientists who can bring specific approaches, theories, methods, and tools to couple with the conceptual foundations of panarchy and expand the breadth of cross-scale and within-scale interactions (e.g., politics, power, historical legacies) considered in governance systems (Berkes and Ross 2016).

This brief review of panarchy's influence in environmental governance research also emphasizes a series of important opportunities that parallel the challenges articulated above. Here I outline a handful of these opportunities as a research agenda to inspire environmental governance scholars engaged in SES and sustainability research to further apply and experiment with panarchy.

- *Test a range of spatial and temporal scales for panarchy dynamics in SES governance.* Develop an empirical base of more observations at a range of resolutions, especially in the Global South (Karpouzoglou et al. 2016) and among marginalized populations, such as Indigenous peoples (Cosens and Chaffin 2016). The panarchy metaphor is ripe for hypothesis testing, whether directly through quantitative methods or indirectly through more qualitative approaches to SES governance research.
- *Investigate the potential for discontinuous scaling relationships between governance variables and processes across levels of a panarchy.* Are scale breaks and clusters of attributes observable in governance dynamics, as some previous work has suggested (Garmestani et al. 2009b; Allen et al. 2014)? Do self-organized patterns emerge across scales, or are governance patterns predominantly scale invariant?
- *Apply panarchy to better understand the role of slow variables in SES governance change.* "Both the management of ecosystems and the development of nations require that attention be focused on the slow variables while encouraging experiments that engage fast ones. A critical number of levels of the panarchy need to be involved in order to satisfy minimal needs for understanding and action" (Holling et al. 2002, 88).
- *Empirically address the lag in analyses of power and politics across panarchies.* It is time for critical scholars to engage with panarchy and assist their more positivist environmental governance colleagues with analyses that couple critical theories of social dynamics with panarchy to reveal interactions (stabilizing, destabilizing) across scales. Panarchy holds the potential to assist environmental social scientists with expanding analyses of the impact of actions across scales, such as helping to more clearly illuminate chains of explanation, as proposed by Rocheleau (2008). Some of this work is already under way via efforts to develop and apply theories such as critical institutionalism (Cleaver and Whaley 2018).
- *Apply panarchy in attempts to model the future and make governance predictions.* Given rapid environmental change is further entrenching grand SES governance challenges such as poverty and severe environmental degradation, we need more than just retrospective or historical analysis of SES dynamics. It is imperative that we use what we know—clearly and explicitly recognizing uncertainty—to model and forecast

potential futures that address grand global challenges and make progress toward managing social-ecological resilience (Chaffin et al 2016).

Conclusions

Panarchy, as a heuristic and model, has been useful for guiding empirical discoveries and theoretical advances in environmental and SES governance research, but there is more to be explored on both fronts. Empirically, applying panarchy to governance research may illuminate scales at which critical, previously unknown elements of SES governance emerge and catalyze change through cross-scale interactions (Chaffin et al. 2016). This is especially likely in places such as the Global South, where institutions such as traditional or indigenous land and resource tenure systems do not match but instead coexist with more western conceptions of government hierarchy. Theoretically, panarchy can push scholars to deepen the politicization and socialization of resilience, integrating social theory to learn more about power struggles between institutions, organizations, and individuals and the social-ecological impact of these struggles across space and time. Applying panarchy to critically analyze the cross-scale nature of SES dynamics, and the systems that govern them, is an important tool in any research agenda for governing social-ecological resilience.

Acknowledgments

This material is based on work supported by the National Science Foundation under grant nos. 1738857 and 1920938. Any opinions, findings, and conclusions or recommendations expressed in this material are those of the author and do not necessarily reflect the views of the National Science Foundation.

Literature Cited

Allen, C.R., D.G. Angeler, B.C. Chaffin, D. Twidwell, and A.S. Garmestani. 2019. Resilience reconciled. *Nature Sustainability* 2: 898–900.

Allen, C.R., D.G. Angeler, A.S. Garmestani, L.H. Gunderson, and C.S. Holling. 2014. Panarchy: Theory and application. *Ecosystems* 17: 578–589.

Allen, C.R., J.J. Fontaine, K.L. Pope, and A.S. Garmestani. 2011. Adaptive management for a turbulent future. *Journal of Environmental Management* 92: 1339–1345.

Allen, C.R., E.A. Forys, and C.S. Holling. 1999. Body mass patterns predict invasions and extinctions in transforming landscapes. *Ecosystems* 2: 114–121.

Allen, C.R., and L.H. Gunderson. 2011. Pathology and failure in the design and implementation of adaptive management. *Journal of Environmental Management* 92: 1379–1384.

Allen, C.R., and C.S. Holling (Eds.). 2008. *Discontinuities in Ecosystems and Other Complex Systems*. Columbia University Press, New York.

Allen, C.R., and C.S. Holling. 2010. Novelty, adaptive capacity, and resilience. *Ecology and Society* 15: 24.

Anderson, R.T. 2006. Indian water rights: Litigation and settlements. *Tulsa Law Review* 42: 23–35.

Anderson, R.T. 2010. Indian water rights, practical reasoning, and negotiated settlements. *California Law Review* 98(4): 1133–1164.

Angeler, D.G., C.R. Allen, and R.K. Johnson. 2013. Measuring the relative resilience of subarctic lakes to global change: Redundancies of functions within and across temporal scales. *Journal of Applied Ecology* 50: 572–584.

Berkes, F., and H. Ross. 2016. Panarchy and community resilience: Sustainability science and policy implications. *Environmental Science & Policy* 61: 185–193.

Bevir, M. 2009. *Key Concepts in Governance*. Sage Publications, Thousand Oaks, CA.

Bevir, M. 2012. *Governance: A Very Short Introduction*. Oxford University Press, Oxford, UK.

Beymer-Farris, B.A., T.J. Bassett, and I. Bryceson. 2012. Promises and pitfalls of adaptive management in resilience thinking: The lens of political ecology. In *Resilience in the Cultural Landscape*, ed. T. Plieninger and C. Bieling, 283–300. Cambridge University Press, Cambridge, UK.

Blackfoot Challenge. 2019. The Blackfoot drought response plan. https://blackfootchallenge.org/portfolio/water/

Blythe, J., J. Silver, L. Evans, D. Armitage, N.J. Bennett, M.L. Moore, T.H. Morrison, and K. Brown. 2018. The dark side of transformation: Latent risks in contemporary sustainability discourse. *Antipode* 50: 1206–1223.

Boyd, E., B. Nykvist, S. Borgström, and I.A. Stacewicz. 2015. Anticipatory governance for social-ecological resilience. *Ambio* 44: 149–161.

Brand, F.S., and K. Jax. 2007. Focusing the meaning (s) of resilience: Resilience as a descriptive concept and a boundary object. *Ecology and Society* 12(1): 23.

Brown, K. 2014. Global environmental change I: A social turn for resilience? *Progress in Human Geography* 38: 107–117.

Brunner, R.D., T.A. Steelman, L. Coe-Juell, C.M. Cromley, D.W. Tucker, and C.M. Edwards. 2005. *Adaptive Governance: Integrating Science, Policy, and Decision Making*. Columbia University Press, New York.

Carpenter, S., B. Walker, J.M. Anderies, and N. Abel. 2001. From metaphor to measurement: Resilience of what to what? *Ecosystems* 4: 765–781.

Chaffin, B.C., R.K. Craig, and H. Gosnell. 2014. Resilience, adaptation, and transformation in the Klamath River Basin social-ecological system. *Idaho Law Review: Natural Resources & Environmental Law Edition* 51(1): 157–193.

Chaffin, B.C., A.S. Garmestani, L. Gunderson, M. Harm Benson, D.G. Angeler, C.A. Arnold, B. Cosens, R.K. Craig, J.B. Ruhl, and C.R. Allen. 2016. Transformative environmental governance. *Annual Review of Resources and Environment* 41: 399–423.

Chaffin, B.C., and H. Gosnell. 2015. Measuring success in adaptive management projects. In *Adaptive Management of Social-Ecological Systems*, ed. C.R. Allen and A.S. Garmestani, 85–105. Springer, Dordrecht, The Netherlands.

Chaffin, B.C., and L.H. Gunderson. 2016. Emergence, institutionalization and renewal: Rhythms of adaptive governance in complex social-ecological systems. *Journal of Environmental Management* 165: 81–87.

Cleaver, F., and L. Whaley. 2018. Understanding process, power, and meaning in adaptive governance. *Ecology and Society* 23(2): 49.

Colby, B.G., J.E. Thorson, S. Britton, and D.H. Getches. 2005. *Negotiating Tribal Water Rights: Fulfilling Promises in the Arid West.* University of Arizona Press, Tucson.

Cosens, B.A. 2013. Legitimacy, adaptation, and resilience in ecosystem management. *Ecology and Society* 18(1): 3.

Cosens, B., and B.C. Chaffin. 2016. Adaptive governance of water resources shared with Indigenous peoples: The role of law. *Water* 8(3): 97.

Cosens, B., L. Gunderson, and B.C. Chaffin. 2018. Introduction to the special feature practicing panarchy: Assessing legal flexibility, ecological resilience, and adaptive governance in regional water systems experiencing rapid environmental change. *Ecology and Society* 23(1): 4.

Cote, M., and A.J. Nightingale. 2012. Resilience thinking meets social theory: Situating social change in socio-ecological systems (SES) research. *Progress in Human Geography* 36: 475–489.

Craig, R.K. 2010. Stationarity is dead—long live transformation: Five principles for climate change adaptation law. *Harvard Environmental Law Review* 34: 9–73.

Craig, R.K., A.S. Garmestani, C.R. Allen, C.A.T. Arnold, H. Birgé, D.A. DeCaro, A.K. Fremier, H. Gosnell, and E. Schlager. 2017. Balancing stability and flexibility in adaptive governance: An analysis of tools available in US environmental law. *Ecology and Society* 22(2): 1–3.

Cutter, S.L. 2016. Resilience to what? Resilience for whom? *Geographical Journal* 182: 110–113.

Dietz, T., E. Ostrom, and P.C. Stern. 2003. The struggle to govern the commons. *Science* 302: 1907–1912.

Eakin, H., A. Winkels, and J. Sendzimir. 2009. Nested vulnerability: Exploring cross-scale linkages and vulnerability teleconnections in Mexican and Vietnamese coffee systems. *Environmental Science & Policy* 12: 398–412.

Fletcher, M.L.M. 2006. Politics, history, and semantics: The federal recognition of Indian tribes. *North Dakota Law Review* 82: 487–518.

Fletcher, R., W. Dressler, B. Büscher, and Z.R. Anderson. 2016. Questioning REDD+ and the future of market-based conservation. *Conservation Biology* 30: 673–675.

Folke, C. 2016. Resilience. In *Subject: Framing Concepts in Environmental Science. Oxford Research Encyclopedias, Environmental Science*, ed. H. Shugart. Oxford University Press, New York.

Folke, C., T. Hahn, P. Olsson, and J. Norberg. 2005. Adaptive governance of social-ecological systems. *Annual Review of Environment and Resources* 30: 441–473.

Fraser, E.D. 2003. Social vulnerability and ecological fragility: Building bridges between social and natural sciences using the Irish Potato Famine as a case study. *Conservation Ecology* 7(2): 9.

Garmestani, A.S., C.R. Allen, and H. Cabezas. 2009a. Panarchy, adaptive management and governance: Policy options for building resilience. *Nebraska Law Review* 87: 1036–1054.

Garmestani, A.S., C.R. Allen, and L. Gunderson. 2009b. Panarchy: Discontinuities reveal similarities in the dynamic system structure of ecological and social systems. *Ecology and Society* 14(1): 15.

Garmestani, A.S., and M.H. Benson. 2013. A framework for resilience-based governance of social-ecological systems. *Ecology and Society* 18(1): 9.

Garmestani, A.S., J.B. Ruhl, B.C. Chaffin, R.K. Craig, H.F. van Rijswick, D.G. Angeler, C. Folke, L. Gunderson, D. Twidwell, and C.R. Allen. 2019. Untapped capacity for resilience in environmental law. *Proceedings of the National Academy of Sciences* 116: 19899–19904.

Garmestani, A.S., D. Twidwell, D.G. Angeler, S. Sundstrom, C. Barichievy, B.C. Chaffin, T. Eason, N. Graham, D. Granholm, L. Gunderson, and M. Knutson. 2020. Panarchy: Opportunities and challenges for ecosystem management. *Frontiers in Ecology and the Environment.*

Gotts, N.M. 2007. Resilience, panarchy, and world-systems analysis. *Ecology and Society* 12(1): 24.

Green, O.O., A.S. Garmestani, C.R. Allen, L.H. Gunderson, J.B. Ruhl, C.A. Arnold, N.A.J. Graham, B. Cosens, D.G. Angeler, B.C. Chaffin, and C.S. Holling. 2015. Barriers and bridges to the integration of social-ecological resilience and law. *Frontiers in Ecology and the Environment* 13: 332–337.

Gunderson, L.H., and C.S. Holling (Eds.). 2002. *Panarchy: Understanding Transformations in Human and Natural Systems.* Island Press, Washington, DC.

Gunderson, L.H., C.S. Holling, and S.S. Light. 1995. *Barriers and Bridges to the Renewal of Ecosystems and Institutions.* Columbia University Press, New York.

Gunderson, L., and S.S. Light. 2006. Adaptive management and adaptive governance in the Everglades ecosystem. *Policy Sciences* 39: 323–334.

Hamlet, A.F., P.W. Mote, M.P. Clark, and D.P. Lettenmaier. 2005. Effects of temperature and precipitation variability on snowpack trends in the western United States. *Journal of Climate* 18(21): 4545–4561.

Herrmann, D.L., K. Schwarz, W.D. Shuster, A. Berland, B.C. Chaffin, A.S. Garmestani, and M.E. Hopton. 2016. Ecology for the shrinking city. *BioScience* 66: 965–973.

Holdschlag, A., and B.M. Ratter. 2013. Multiscale system dynamics of humans and nature in The Bahamas: Perturbation, knowledge, panarchy and resilience. *Sustainability Science* 8: 407–421.

Holling, C.S. 1973. Resilience and stability of ecological systems. *Annual Review of Ecology and Systematics* 4: 1–23.

Holling, C.S. 1986. The resilience of terrestrial ecosystems: Local surprise and global change. In *Sustainable Development of the Biosphere*, ed. W. C. Clark and R.E. Munn, 292–317. Cambridge University Press, Cambridge, UK.

Holling, C.S. 1992. Cross-scale morphology, geometry, and dynamics of ecosystems. *Ecological Monographs* 62: 447–502.

Holling, C.S. 2001. Understanding the complexity of economic, ecological, and social systems. *Ecosystems* 4: 390–405.

Holling, C.S., L.H. Gunderson, and G.D. Peterson. 2002. Sustainabilities and panarchies. In *Panarchy: Understanding Transformations in Human and Natural Systems*, ed. L.H. Gunderson and C.S. Holling, 63–102. Island Press, Washington, DC.

Holling, C.S., and G.K. Meffe. 1996. Command and control and the pathology of natural resource management. *Conservation Biology* 10: 328–337.

Huitema, D., E. Mostert, W. Egas, S. Moellenkamp, C. Pahl-Wostl, and R. Yalcin. 2009. Adaptive water governance: Assessing the institutional prescriptions of adaptive (co-) management from a governance perspective and defining a research agenda. *Ecology and Society* 14(1): 26.

Jacques, P.J. 2015. Are world fisheries a global panarchy? *Marine Policy* 53: 165–170.

Karpouzoglou, T., A. Dewulf, and J. Clark. 2016. Advancing adaptive governance of social-ecological systems through theoretical multiplicity. *Environmental Science & Policy* 57: 1–9.

Leach, M. (Ed.). 2008. *Reframing Resilience: A Symposium Report.* STEPS Working Paper 13. STEPS Centre, Brighton, UK.

Lemos, M.C., and A. Agrawal. 2006. Environmental governance. *Annual Review of Environment and Resources* 31: 297–325.

MacKinnon, D., and K.D. Derickson. 2013. From resilience to resourcefulness: A critique of resilience policy and activism. *Progress in Human Geography* 37: 253–270.

Montana-Confederated Salish and Koontenai Tribes (MT-CSKT). 2015. *Water Compact.* Montana Department of Natural Resources & Conservation, Helena, MT. http://dnrc.mt.gov/divisions/reserved-water-rights-compact-commission/confederated-salish-and-kootenai-tribes-compact

Montana Water Policy Interim Committee (MT WPIC). 2018. *Water Rights in Montana.* Water Policy Interim Committee, Helena, MT.

Moore, M.-L., O. Tjornbo, E. Enfors, C. Knapp, J. Hodbod, J.A. Baggio, A. Norström, P. Olsson, and D. Biggs. 2014. Studying the complexity of change: Toward an analytical framework for understanding deliberate social-ecological transformations. *Ecology and Society* 19(4): 54.

Morrison, T.H., W.N. Adger, K. Brown, M.C. Lemos, D. Huitema, J. Phelps, L. Evans, P . Cohen, A.M. Song, R. Turner, and T. Quinn. 2019. The black box of power in polycentric environmental governance. *Global Environmental Change* 57: 101934.

Mullin, M. 2009. *Governing the Tap: Special District Governance and the New Local Politics of Water.* MIT University Press, Cambridge, MA.

Nash, K.L., C.R. Allen, D.G. Angeler, C. Barichievy, T. Eason, A.S. Garmestani, N.A. Graham, D. Granholm, M. Knutson, R.J. Nelson, and M. Nyström. 2014. Discontinuities, cross-scale patterns, and the organization of ecosystems. *Ecology* 95: 654–667.

Olsson, P., C. Folke, and T. Hahn. 2004. Social-ecological transformation for ecosystem management: The development of adaptive co-management of a wetland landscape in southern Sweden. *Ecology and Society* 9(4): 2.

Olsson, P., L.H. Gunderson, S.R. Carpenter, P. Ryan, L. Lebel, C. Folke, and C.S. Holling. 2006. Shooting the rapids: Navigating transitions to adaptive governance of social-ecological systems. *Ecology and Society* 11(1): 18.

Ostrom, E. 2009. A general framework for analyzing sustainability of social-ecological systems. *Science* 325: 419–422.

Pelling, M., and D. Manuel-Navarrete. 2011. From resilience to transformation: The adaptive cycle in two Mexican urban centers. *Ecology and Society* 16(2): 11.

Petrosillo, I., N. Zaccarelli, and G. Zurlini. 2010. Multi-scale vulnerability of natural capital in a panarchy of social–ecological landscapes. *Ecological Complexity* 7: 359–367.

Rijke, J., R. Brown, C. Zevenbergen, R. Ashley, M. Farrelly, P. Morison, and S. van Herk. 2012. Fit-for-purpose governance: A framework to make adaptive governance operational. *Environmental Science & Policy* 22: 73–84.

Rocheleau, D.E. 2008. Political ecology in the key of policy: From chains of explanation to webs of relation. *Geoforum* 39: 716–727.

Ruhl, J.B. 2005. Regulation by adaptive management: Is it possible? *Minnesota Journal of Law, Science & Technology* 7: 21–57.

Seager, R., J. Feldman, N. Lis, M. Ting, A.P. Williams, J. Nakamura, H. Liu, and N. Henderson. 2018. Whither the 100th Meridian? The once and future physical and human geography of America's arid–humid divide. Part II: The meridian moves east. *Earth Interactions* 22: 1–24.

Stern, C.V. 2017. *Indian Water Rights Settlements*. Congressional Research Service (CRS), Washington, DC.

Stone-Jovicich, S. 2015. Probing the interfaces between the social sciences and social-ecological resilience: Insights from integrative and hybrid perspectives in the social sciences. *Ecology and Society* 20(2): 25.

Tarlock, A.D. 2000. Prior appropriation: Rule, principle, or rhetoric. *North Dakota Law Review* 76: 881–910.

Tarlock, A.D. 2001. The future of prior appropriation in the new west. *Natural Resources Journal* 41: 769–793.

Whitlock, C., W. Cross, B. Maxwell, N. Silverman, and A.A. Wade. 2017. *2017 Montana Climate Assessment*. Montana State University and University of Montana, Montana Institute on Ecosystems, Bozeman and Missoula. doi:10.15788/m2ww8w

Winkel, T., P. Bommel, M. Chevarría-Lazo, G. Cortes, C. Del Castillo, P. Gasselin, F. Léger, J.P. Nina-Laura, S. Rambal, M. Tichit, and J.F. Tourrand. 2016. Panarchy of an indigenous agroecosystem in the globalized market: The quinoa production in the Bolivian Altiplano. *Global Environmental Change* 39: 195–204.

Wyborn, C., A. Datta, J. Montana, M. Ryan, P. Leith, B.C. Chaffin, C. Miller, and L. van Kerkhoff. 2019. Co-producing sustainability: Reordering the governance of science, policy, and practice. *Annual Review of Environment and Resources* 44: 319–346.

Yasmin, T., M. Farrelly, and B.C. Rogers. 2019. Adaptive governance: A catalyst for advancing sustainable urban transformation in the global South. *International Journal of Water Resources Development* 36(5): 818–838.

CHAPTER 13

Cross-Scale Social-Ecological Stewardship for Navigating toward More Sustainable and Just Futures

F. Stuart Chapin III, Reinette Biggs, Nadia Sitas, Carl Folke, and Gary P. Kofinas

Perhaps the greatest challenge humanity faces today is the accelerating degradation of the capacity of our planet to meet the needs of nature and society for current and future generations (Folke et al. 2016; Steffen et al. 2018). There is growing evidence that the cumulative effects of human activities around the world are pushing the Earth system beyond the planetary boundaries within which the Earth has functioned for the past 10,000 years (Rockström et al. 2009; Steffen et al. 2018). The relative climate stability of the Holocene Epoch contributed to the beginning of agriculture and the development of complex settled societies. Exceeding these planetary boundaries could have profound consequences for human societies by jeopardizing crop production, affecting the security of housing and infrastructure, and undermining livelihoods and economies. However, these effects will be experienced differently in different parts of the world, because of differences in biophysical conditions and in the capacity of people to recover from shocks, adapt to ongoing stressors, or migrate to new places (Adger et al. 2020).

These impacts are not the result of deliberate attempts to degrade Earth but a side effect of short-term goals that are pursued across a wide range of scales and places. The failure of society to recognize the panarchical nature of the Earth system (Allen et al. 2014) and its increasing interconnectivity with human society has resulted in fragmented efforts to meet society's needs. Well-intended efforts to solve problems at a particular place and scale often create externalities that emerge at other places and scales and are now accumulating in ways that threaten nature and society globally (Díaz et al. 2019). For example, society's efforts to produce material goods and services at scales of an industry or nation release greenhouse gases that are radically altering Earth's climate system. Such changes strongly affect people in other places, often the poorest and most vulnerable people who have benefited least from these activities (O'Brien and Leichenko 2000; Denton 2002; IPCC 2018). For example, nitrogen added by individual farmers to maximize productivity of their local crops moves across the landscape and accumulates in the estuaries of the world's major rivers (Boyer et al. 2006). The resulting nearshore dead zones persist in oceans hundreds to thousands of miles from where farming takes place.

Efforts to improve environmental quality in wealthier regions can result in resource use, degradation, and pollution in poorer, less well-regulated regions. For example, decreasing wood supply due to forest protection in wealthy countries can inadvertently fuel deforestation in poorer regions in response to continued market demand (Lang and Chan 2006). Changes in land use through hydrocarbon development may benefit a nation while degrading wildlife resources needed by those who rely on local subsistence harvesting. Moreover, these cross-scale interactions often exacerbate existing inequities, so efforts to meet the needs of nature and society must address linkages across multiple scales if equitable outcomes are to emerge (ISSC et al. 2016; Leach et al. 2018). This chapter introduces and discusses social-ecological stewardship across multiple interlinked scales as an approach to addressing these profound, interlinked challenges.

Stewardship across Scales

The word *stewardship* traces its origins back to medieval England, where it referred to a guard (*weard*) who had a responsibility, on behalf of others, to care for a house or hall (*stig*) (Chapin et al. 2015). The concept is deeply rooted in Christian religious thought (Pope Francis 2015), with parallel concepts in other religions. It also captures the spirit of Aldo Leopold's writings about a land ethic of respect and care for nature (Leopold 1949). However, it has come under scrutiny for its historical associations with colonialism and patriarchy (Welchman 2012). Despite these diverse connotations, stewardship has become an important theme in modern conservation and resource management as a strategy to shape the health of ecosystems and society's long-term quality of life (Chapin et al. 2009). Stewardship can provide common ground for conversations about environmental and social justice and ways that environmental decisions benefit or burden different actor groups within and between nations (Cockburn et al. 2018; Enqvist et al. 2018).

Historically, the stewardship concept was applied largely to land management decisions at local scales, as managers cared for their lands, based on local knowledge, to sustain ecosystem properties and processes (Olsson and Folke 2001; Chapin et al. 2009). For example, farmers may manage their lands and crops to sustain soil fertility by building soil organic matter and minimizing nutrient loss. Land managers may seek to sustain the productivity or conservation value of the lands that they manage. Stewardship also guides national and regional institutions, such as Natura 2000 in the European Union, the National Park Service in the United States, or international planning through international conventions (e.g., Biodiversity Convention and the Convention on the Conservation of Antarctic Marine Living Resources), international research planning (e.g., Diversitas, Future Earth, Circumarctic Flora and Fauna), coordinated nongovernment organization activities (e.g., the Water Funds program of the Nature Conservancy), and management of supply chains by some transnational corporations (Mathevet et al. 2016; Williams et al. 2019).

Stewardship efforts that focus at a single scale tend to treat cross-scale interactions as external influences and consequences rather than as integral components of system dynamics. Incorporating these cross-scale dynamics is important to understanding cross-scale interactions (i.e., a panarchy) (Gunderson and Holling 2002). At particular scales, some system components may be stable, others change rapidly, while others are on the verge of a regime shift, potentially precipitating complex and unanticipated changes in dynamics at other scales. For example, national and international efforts that address threats to the biodiversity and ecosystem services to benefit society generally can cause sudden catastrophic changes for a local community that is denied access to its traditional lands (Pinkerton 1989; Berkes et al. 2009).

Many cross-scale interactions are accelerating. These include information mobilization (especially through the internet), the globalization of goods via international trade, long-range transport of pollutants (e.g., carbon dioxide), and movement of people (through migration). Stewardship must therefore be reframed to address these accelerating cross-scale interdependencies, all of which have the potential to behave as panarchies, that is, to cause unexpected changes in dynamics at other scales (Balvanera et al. 2017).

Social-Ecological Stewardship

Historically, stewardship initiatives focused primarily on improving ecosystem function (Chapin et al. 2009). In recent years there has been a more explicit focus on interconnected social-ecological processes. Social-ecological stewardship (SE stewardship) addresses the shaping of physical, biological, and social processes to benefit both people and nature (Chapin et al. 2010). This requires careful attention to reciprocal interactions between people and nature (Barthel et al. 2010), as grounded in many indigenous ideologies (Berkes 2012). Key features of SE stewardship are proactive interventions to shape changes rather than to respond reactively, a focus on coupled systems in which people are part of nature rather than considered separate from it, an emphasis on the benefits to both nature and people, and an intent to foster equitable benefits among people, with a particular focus on poor or marginalized people, who are most disadvantaged by environmental degradation and least empowered or included in decision making.

We suggest that SE stewardship develops most readily at a single (usually local) scale and emerges across scales when there is sufficient motivation, respectful engagement, dialogue, and power sharing (Young 1994, 2002). Development of cross-scale SE stewardship often requires new configurations of social networks and institutions and collaborations between groups that have not traditionally interacted. When these new interactions foster trust, share power, and coproduce new knowledge and commitment to action, this can trigger innovations in policy and practice at multiple scales, leading to complex panarchical dynamics (Folke et al. 2005; Hahn et al. 2006; Olsson et al. 2008). However, politics and unequal access to decision-making power often complicate these new interactions and their consequences for distribution of ecosystem services among segments of society (Pascual et al. 2017), within and across scales (Leach et al. 2018; Kashwan et al. 2019). In general, cross-scale SE stewardship requires improved understanding of ways that power mediates cross-scale integration, interdependencies, and decision-making processes across cultures, sectors, jurisdictions, and disciplines. Social and economic planning that ignores these power dynamics often prevents marginalized people from mobilizing, planning, and acting to shape their future by creating unique modes of discrimination or privilege of certain groups of people, institutions, or individuals (Hankivsky et al. 2014).

Cross-scale SE stewardship can bring a diversity of perspectives to inform understanding of a system's dynamics. Diversity in planning approaches can, in turn, contribute to the creative processes inherent in innovation (Pereira et al. 2019). Power sharing in these multiscale institutional arrangements helps avoid unintended consequences that are typical when macro-level governance dictates broad solutions that ignore their impacts at smaller scales (Kofinas et al. 2007, 2020). In this way, multiscale SE stewardship can implement an intent to sustain the long-term well-being of all members of society by shaping the dynamics of ecosystem services and their distribution to sustain well-being for all.

Navigating toward More Sustainable and Just Futures

Policymakers and the public often assume that many aspects of the environment, such as climate and sea level, will not change and that, in the future, people will behave and power will be distributed in ways that are similar to those of the present or recent past. This assumption is clearly inconsistent with accelerating changes such as climate warming, rising inequalities, technological change, social connectivity, biodiversity loss, greater frequency of extreme events such as fires, floods, and storms, increasing pesticide resistance, and changes in regional hydrology (Steffen et al. 2011, 2018; IPCC 2018). Future ecological, social, political, and economic dynamics and their interactions are likely to give rise to new cross-scale social-ecological dynamics and vulnerabilities, as observed with the COVID-19 pandemic that emerged in China in late 2019 and rapidly affected most of the planet. Inclusion of known mechanisms linking human behavior to environmental and social changes and their cascading, interlinked biophysical and societal consequences will provide a stronger basis for projecting and exploring scenarios of future changes. More importantly, it will better enable us to prepare for known and unknown futures and to proactively shape changes in power and political structures that influence behavior, technology, economies, and institutions to achieve more sustainable and just futures.

Path dependence adds an additional challenge to managing complex systems over time (North 1990) and the identification and pursuit of pathways toward sustainability. Ongoing system trajectories are often determined by past events when a tipping point alters the time course and trajectory of the system. Such trajectories are maintained by stabilizing (negative) feedback mechanisms that maintain the current trajectory or path. For example, events linked to political struggles over different interests and distributions of power often influence public policies or institutions (often through coercive measures), which, once set, are difficult to change in the absence of a new crisis (Gunderson and Holling 2002). Moving beyond past patterns of behavior and establishing new pathways toward sustainability requires adopting forward-looking processes that draw on multiple knowledge sources and methods and which capture the nonlinearities of system change. For example, models that incorporate thresholds and tipping points can explore and identify risks and opportunities where interventions can either reduce the likelihood of undesirable trajectories or foster opportunities for desired changes. Such models, if designed and implemented to engage decision makers, have been useful in managing pest and disease outbreaks such as the COVID-19 pandemic, fisheries, and social behaviors such as smoking (Holling et al. 2002; Nyborg et al. 2016). In such path-dependent systems, SE stewardship plays at least three distinct roles:

Stewardship can nudge systems toward pathways that society views as desirable and just and then sustain these desired states. This requires sustaining or enhancing ecological slow variables (e.g., soils, climate, disturbance regime, and functional groups of organisms) that support the desired state of ecosystems and sustaining or enhancing social slow variables (e.g., altruistic values, education, well-being, equity, and empowerment) that promote a just society. Changes in ecological and social slow variables often reflect changes in society's underlying cultural beliefs and values, individual behaviors, and the institutions that shape these behaviors (Chapin 2020). When there are differing values and views on the future, such efforts require respectful communication efforts with high transaction costs (North 1990).

Stewardship seeks to understand and shape the dynamics of systems that are undergoing exponential changes resulting from amplifying feedbacks. Exponential change occurs when stabilizing (negative) feedbacks (constraints) are weakened or amplifying (positive) feedbacks (subsidies) are established or strengthened. For example, climate warming can be minimized by increasing carbon capture by forests and wetlands (strengthening a feedback that stabilizes atmospheric CO_2) and by reducing fossil fuel emissions (weakening a feedback that accelerates the increase in atmospheric CO_2) (IPCC 2018). Feedbacks can also be manipulated to accelerate change toward a more desirable state. For example, shifting subsidies from fossil fuels to renewable energy weakens the feedbacks that sustain fossil fuel consumption and strengthens feedbacks that would accelerate a shift to renewable energy. Key to this process is assessing the resilience that determines the sensitivity of key components to environmental and social changes and charting pathways that increase resilience of desired states and reduce resilience of undesired states.

Identifying and changing the likelihood of events that could trigger transformation toward either a desirable or an undesirable state. Triggers for transformative change are events that lead to a bifurcation of pathways that shape dramatically different outcomes. In many cases, identification of these transformative events requires learning from other situations with different contexts where transformative change occurred or failed to occur (Baker and Chapin 2018). Institutional changes that alter the distribution of power, for example, often lead to transformative change. Triggering desirable transformations requires a strategy to prepare for a transformation, navigate during transformation (often at a time of crisis when the previous state of the system is fragile), and stabilizing a new system configuration (Olsson et al. 2008; Chaffin et al. 2016). Preventing undesirable transformations (e.g., outbreak of war) requires reducing the likelihood of events that might trigger these transformations.

Navigating Transformative Change in a Panarchical World

There is growing recognition that we need to address increasing global environmental and societal risks by transforming the ways we live and interact with our environment and one another at local and regional levels (Olsson et al. 2004; Westley et al. 2011). There is a growing body of theory on how transformative change happens (Chaffin et al. 2016; Herrfahrdt-Pähle et al. 2020). It typically involves an interaction between bottom-up and top-down processes, and the panarchy model provides a useful framework for understanding how and when change may cascade upward or downward across scales. The Seeds of Good Anthropocenes project (https://goodanthropocenes.wordpress.com/) illustrates ways that mostly local-scale innovations can contribute to building a more sustainable and just future, often through innovative stewardship initiatives (Bennett et al. 2016). The project has developed a theory of change that combines insights from the social-ecological transformations and social–technical transition literatures to understand how small-scale initiatives may be supported to generate larger systemic change (Pereira et al. 2018) (figure 13.1).

In the preparation phase (figure 13.1), the environmental and social problems at the global level generate emerging awareness and metanarratives about the need for larger systemic change. These narratives are widely shared both locally and across scales through news, social media, and the published literature. This communication, in turn, inspires micro-scale experimentation of new ways to do things on the ground (Olsson et al. 2004): various rewilding projects, such as the reintroduction of wolves in Europe (Reinhardt et al. 2019); the introduction of grassroots community currencies in a range of

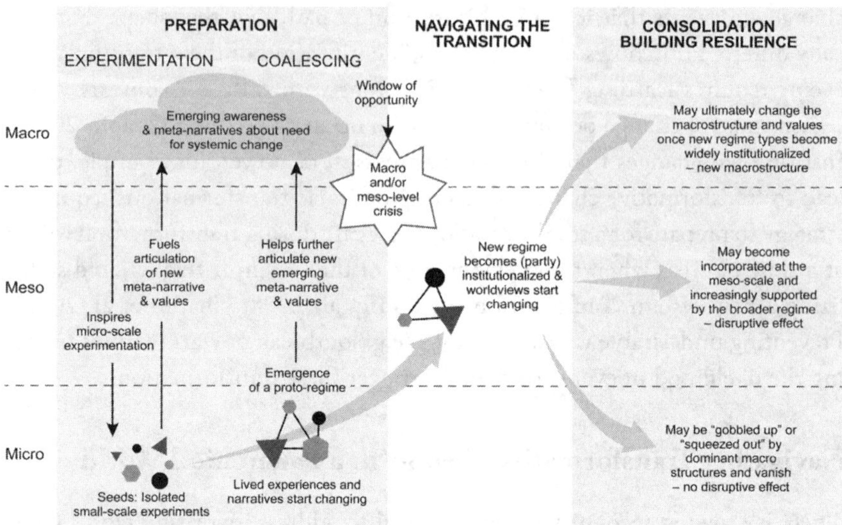

Figure 13.1. Model showing the role of cross-scale interactions (between micro, meso, and macro scales) in fostering three stages of transformation (preparation, navigation, and consolidation) in ways that shift long-term system dynamics. (Adopted from Pereira et al. 2018.)

different contexts that work alongside national currencies (Collom 2011); and the Health in Harmony initiative, which works with local organizations to provide healthcare in exchange for a commitment to protect natural resources in neighboring rainforests (Buck et al. 2020), for example. Experiments such as these further fuel and articulate emerging metanarratives at macro scales. In time, different micro-scale initiatives, such as combining innovative financing solutions with healthcare and environmental stewardship, may connect to form a protoregime—an alternative potential way in which the system might work. Transformative change can also propagate upward from local initiatives when networks of communities or regions share experiences, capture public attention, and become social movements, such as collaborations between Indigenous Peoples (Lauderdale 2008).

In the transitional phase (figure 13.1), the fate of a protoregime often depends on windows of opportunity, institutional contexts, and innovative leadership. The novel protoregime may remain stuck at the margins of a routinized dominant system (i.e., no larger-scale systemic transformation) unless an opportunity arises where the alternative system can become institutionalized (Gelcich et al. 2010). Panarchy provides a useful framework for understanding how this opportunity context might arise and be fostered. Two key strategies include creating an awareness of larger systemic problems that can inform bottom-up innovation and providing support for such initiatives, as well as creating an openness to experiment with alternative institutional arrangements, such as the incorporation of smart grids for local-scale renewable electricity generation into national grids. In other cases, the institutionalization of an alternative system may be triggered by a crisis.

The outcome of transformation (figure 13.1) depends on the alignment of local dynamics with those at other scales. In some cases, national or regionally institutionalized innovations can be adopted more widely (Westley et al. 2017). In other cases, they may become integrated into and modify the existing regime or may be squeezed out. Importantly, in many cases there may be long lag times (i.e., a long and uncertain transition) where an alternative regime exists at micro or meso levels but is not widely adopted because of a lack of opportunity contexts or misalignment of phases of adaptive cycles at different scales. In these cases, institutional entrepreneurs sometimes work across scales to influence system dynamics to create new opportunity contexts, such as engaging in dialogues or coproduction processes with national governments or transnational corporations to make them aware of the need for change and the existence of initiatives that could address apparent barriers (Westley et al. 2011; Rosen and Olsson 2013). Some transformations, such as the institutionalization of knowledge coproduction as a legitimate process, also require culture change among nongovernment actors, such as academics. However, culture is typically a slow variable, especially when underpinned by economic self-interest. Macro-level institutions can provide resources or institutional capacity that facilitates the collaboration of groups that are focused at and across scales, such as funding by nongovernment organizations of networks of climate-resilient cities.

Within the Good Anthropocenes theory of change (figure 13.1; Pereira et al. 2018), institutional entrepreneurs are assumed to work toward more just and sustainable trajectories. However, these opportunity contexts can easily be captured by individuals and institutions (whether these be multinational corporations, nation states, or government departments), resulting in further environmental degradation and inequitable benefit sharing (Abel et al. 2006; Phillips 2017; Folke et al. 2019). In these cases, the panarchy model is useful for understanding interactions between system variables that create maladaptive systems and rigidity traps and for understanding where power is held and used to determine where detrimental transformation can be blocked and positive transformations facilitated.

Conclusions

The Anthropocene is increasingly characterized by global connectivity and by rapid, high-impact human actions and reactions. In a globally intertwined social-ecological world, local experiments and opportunity contexts that enable these experiments to be adapted and taken up at larger scales arise in new ways and in new constellations. The theory of change that we present (Pereira et al. 2018) provides a framework to prepare strategies and create and navigate conditions that increase the possibility of just and sustainable futures at local to global scales. The social-ecologically intertwined Anthropocene context highlights the necessity of new forms of SE stewardship across levels and scales to foster new visions, meanings, and narratives that actively reconnect and reinvent human actions and developments to the biosphere in support of sustainability.

Literature Cited

Abel, N., D.H.M. Cumming, and J.M. Anderies. 2006. Collapse and reorganization in social-ecological systems: Questions, some ideas, and policy implications. *Ecology and Society* 11(1): 17. http://www.ecologyandsociety.org/vol11/iss1/art17/

Adger, W.N., A.-S. Crépin, C. Folke, D. Ospina, F.S. Chapin III, K. Segerson, K.C. Seto, J.M. Anderies, S. Barrett, E.M. Bennett, G. Daily, T. Elmqvist, J. Fischer, N. Kautsky, S.A. Levin, J.F. Shogren, J. van den Bergh, B. Walker, and J. Wilen. 2020. Urbanisation, migration and adaptation to climate change. One-Earth D3(4): 396–399.

Allen, C.R., D.G. Angeler, A.S. Garmestani, L.H. Gunderson, and C.S. Holling. 2014. Panarchy: Theory and application. *Ecosystems* 17(4), 578–589. doi:10.1007/s10021-013-9744-2

Baker, S., and F.S. Chapin III. 2018. Going beyond "It depends": The role of context in shaping participation in natural resource management. *Ecology and Society* 23(1): 20. https://doi.org/10.5751/ES-09868-230120

Balvanera, P., R. Calderón-Contreras, A.J. Castro, M.R. Felipe-Lucia, I.R. Geijzendorffer, S. Jacobs, B. Martín-López, U. Arbieu, C.I. Speranza, B. Locatelli, N.P. Harguindeguy, I.R. Mercado, M.J. Spierenburg, A. Vallet, L. Lynes, and L. Gillson. 2017. Interconnected place-based social-ecological research can inform global sustainability. *Current Opinion in Environmental Sustainability* 29: 1–7. http://doi.org/10.1016/j.cosust.2017.09.005

Barthel, S., C. Folke, and J. Colding. 2010. Social-ecological memory in urban gardens: Retaining the capacity for management of ecosystem services. *Global Environmental Change* 20: 255–265.

Bennett, E.M., M. Solan, R. Biggs, T. McPhearson, A.V. Norström, P. Olsson, L. Pereira, G.D. Peterson, C. Raudsepp-Hearne, F. Biermann, S.R. Carpenter, E.C. Ellis, T. Hichert, V. Galaz, M. Lahsen, M. Milkoreit, B. Martin López, K.A. Nicholas, R. Preiser, G. Vince, J.M. Vervoort, and J. Xu. 2016. Bright spots: Seeds of a Good Anthropocene. *Frontiers in Ecology and the Environment* 14(8): 441–448.

Berkes, F. 2012. *Sacred Ecology*. Routledge, New York.

Berkes, F., G.P. Kofinas, and F.S. Chapin III. 2009. Conservation, community, and livelihoods: Sustaining, renewing, and adaptive cultural connections to the land. In *Principles of Ecosystem Stewardship: Resilience-Based Natural Resource Management in a Changing World*, ed. F.S. Chapin III, G.P. Kofinas, and C. Folke, 129–147. Springer, New York.

Boyer, E.W., R.W. Howarth, J.N. Galloway, F.J. Dentener, P.A. Green, and C.J. Vörösmarty. 2006. Riverine nitrogen export from the continents to the coasts. *Global Biogeochemical Cycles* 20(1). https://doi.org/10.1029/2005GB002537

Buck, J.C., S.B. Weinstein, G. Titcomb, and H.S. Young. 2020. Conservation implications of disease control. *Frontiers in Ecology and the Environment* 18(6): 329–334.

Chaffin, B.C., A.S. Garmestani, L.H. Gunderson, M.H. Benson, D.G. Angeler, C.A. Arnold, B. Cosens, R.K. Craig, J.B. Ruhl, and C.R. Allen. 2016. Transformative environmental governance. *Annual Review of Environment and Resources* 41: 399–423.

Chapin, F.S. III. 2020. *Grassroots Stewardship: Sustainability Within Our Reach*. Oxford University Press, New York.

Chapin, F.S. III, S.R. Carpenter, G.P. Kofinas, C. Folke, N. Abel, W.C. Clark, P. Olsson, D.M. Stafford-Smith, B.H. Walker, O.R. Young, F. Berkes, R. Biggs, J.M. Grove, R.L. Naylor, E. Pinkerton, W. Steffen, and F.J. Swanson. 2010. Ecosystem stewardship: Sustainability strategies for a rapidly changing planet. *Trends in Ecology and Evolution* 25: 241–249.

Chapin, F.S. III, G.P. Kofinas, and C. Folke (Eds.). 2009. *Principles of Ecosystem Stewardship: Resilience-Based Natural Resource Management in a Changing World*. Springer, New York.

Chapin, F.S. III, S.T.A. Pickett, M.E. Power, S.L. Collins, J.S. Baron, D.W. Inouye, and M.G. Turner. 2015. Earth Stewardship: An initiative by the Ecological Society of America to foster engagement to sustain planet Earth. In *Earth Stewardship*, ed. R. Rozzi et al., 173-194. Springer, New York.

Cockburn, J., G. Cundill, S. Shackleton, and M. Rouget. 2018. Towards place-based research to support social-ecological stewardship. *Sustainability* 10(5): 1434. https://doi.org/10.3390/su10051434

Collom, E. 2011. Motivations and differential participation in a community currency system: The dynamics within a local social movement organization. *Sociological Forum* 26(1): 144–168.

Denton, F. 2002. Climate change vulnerability, impacts, and adaptation: Why does gender matter? *Gender & Development* 10(2): 10–20.

Díaz, S., J. Settele, E.S. Brondízio, H.T. Ngo, J. Agard, A. Arneth, P. Balvanera, K.A. Brauman, S.H.M. Butchart, K.M.A. Chan, L.A. Garibaldi, K. Ichii, J. Liu, S.M. Subramanian, G.F. Midgley, P. Miloslavich, Z. Molnár, D. Obura, A. Pfaff, S. Polasky, A. Purvis, J. Razzaque, B. Reyers, R.R. Chowdhury, Y.-J. Shin, I. Visseren-Hamakers, K.J. Willis, and C.N. Zayas. 2019. Pervasive human-driven decline of life on Earth points to the need for transformative change. *Science* 366(6741): eaax3100. doi:10.1126/science.aax3100

Enqvist, J.P., S. West, V.A. Masterson, L.J. Haider, U. Svedin, and M. Tengö. 2018. Stewardship as a boundary object for sustainability research: Linking care, knowledge and agency. *Landscape and Urban Planning* 179: 17–37.

Folke, C., R. Biggs, A.V. Norström, B. Reyers, and J. Rockström. 2016. Social-ecological resilience and Biosphere-based sustainability science. *Ecology and Society* 21(3): 41. http://dx.doi.org/10.5751/ES-08748-210341

Folke, C., T. Hahn, P. Olsson, and J. Norberg. 2005. Adaptive governance of social-ecolog ical systems. *Annual Review of Environment and Resources* 30: 441–473.

Folke, C., H. Österblom, J.-B. Jouffray, E. Lambin, M. Scheffer, B.I. Crona, M. Nyström, S.A. Levin, S.R. Carpenter, W.N. Adger, J.M. Anderies, F.S. Chapin III, A.-S. Crépin, A. Dauriach, V. Galaz, L.J. Gordon, N. Kautsky, B.H. Walker, J.R. Watson, J. Wilen, and A. de Zeeuw. 2019. Transnational corporations and the challenge of Biosphere stewardship. *Nature Ecology & Evolution* 3: 1396–1403.

Gelcich, S., T.P. Hughes, P. Olsson, C. Folke, O. Defeo, M. Fernández, S. Foale, L.H. Gunderson, C. Rodríguez-Sieker, M. Scheffer, R. Steneck, and J.C. Castilla. 2010. Navigating transformations in governance of Chilean marine coastal resources. *Proceedings of the National Academy of Sciences* 107: 16794–16799.

Gunderson, L.H., and C.S. Holling (Eds.). 2002. *Panarchy: Understanding Transformations in Human and Natural Systems*. Island Press, Washington, DC.

Hahn, T., P. Olsson, C. Folke, and K. Johansson. 2006. Trust building, knowledge generation and organizational innovations: The role of a bridging organization for adaptive co-management of a wetland landscape around Kristianstad, Sweden. *Human Ecology* 34: 573–592.

Hankivsky, O., D. Grace, G. Hunting, M. Giesbrecht, A. Fridkin, S. Rudrum, O. Ferlatte, and N. Clark. 2014. An intersectionality-based policy analysis framework: Critical reflections on a methodology for advancing equity. *International Journal for Equity in Health* 13: 119.

Herrfahrdt-Pähle, E., M. Schlüter, P. Olsson, C. Folke, S. Gelcich, and C. Pahl-Wostl. 2020. Sustainability transformations: Socio-political shocks as opportunities for governance transitions. *Global Environmental Change* 63: 102097. https://doi.org/10.1016/j.gloenvcha.2020.102097

Holling, C.S., L.H. Gunderson, and G.D. Peterson. 2002. Sustainability and panarchies. In *Panarchy: Understanding Transformations in Human and Natural Systems*, ed. L.H. Gunderson and C.S. Holling, 63–102. Island Press, Washington, DC.

IPCC. 2018. Summary for policy makers. In *Global Warming of 1.5°C above Pre-Industrial Levels and Related Global Greenhouse Gas Emission Pathways, in the Context of Strengthening the Global Response to the Threat of Climate Change, Sustainable Development, and Efforts to Eradicate Poverty*, ed. V. Masson-Delmotte et al. World Meteorological Organization, Geneva, Switzerland.

ISSC, IDS, & UNESCO. 2016. *World Social Science Report 2016: Challenging Inequalities: Pathways to a Just World*. UNESCO Publishing, Paris.

Kashwan, P., L.M. MacLean, and G.A. García-López. 2019. Rethinking power and institutions in the shadows of neoliberalism. *World Development* 120: 133–146.

Kofinas, G., S. BurnSilver, and A.N. Petrov. 2020. Methodological challenges and innovations in arctic community sustainability research. In *Arctic Sustainability, Key Methodologies and Knowledge*, ed. A.N. Petrov and J.K. Graybill. Routledge, New York.

Kofinas, G., S.J. Herman, and C. Meek. 2007. Novel problems require novel solutions: Innovation as an outcome of adaptive co-management. In *Adaptive Co-Management: Collaboration, Learning, and Multi-Level Governance*, ed. F. Berkes, N. Doubleday, and D. Armitage, 249–467. UBC Press, Vancouver.

Lang, G., and C.H.W. Chan. 2006. China's impact on forests in Southeast Asia. *Journal of Contemporary Asia* 36(2): 167–194.

Lauderdale, P. 2008. Indigenous peoples in the face of globalization. *American Behavioral Scientist* 51: 1836–1843.

Leach, M., B. Reyers, X. Bai, E.S. Brondizio, C. Cook, S. Díaz, G. Espindola, M. Scobie, M. Stafford-Smith, and S.M. Subramanian. 2018. Equity and sustainability in the Anthropocene: A social–ecological systems perspective on their intertwined futures. *Global Sustainability* 1: e13. https://doi.org/10.1017/sus.2018.12

Leopold, A. 1949. *A Sand County Almanac*. Oxford University Press, Oxford, UK.

Mathevet, R., J.D. Thompson, and C. Folke. 2016. Protected areas and their surrounding territory: Socioecological systems in the context of ecological solidarity. *Ecological Applications* 26: 5–16.

North, D.C. 1990. A transaction cost theory of politics. *Journal of Theoretical Politics* 2: 355–367.

Nyborg, K., J.M. Anderies, A. Dannenberg, T. Lindahl, C. Schill, M. Schlüter, W.N. Adger, K.J. Arrow, S. Barrett, S. Carpenter, F.S. Chapin III, A.-S. Crépin, G. Daily, P. Ehrlich, C. Folke, W. Jager, N. Kautsky, S.A. Levin, O.J. Madsen, S. Polasky, M. Scheffer, B. Walker, E.U. Weber, J. Wilen, A. Xepapadeas, A. de Zeeuw. 2016. Social norms as solutions. *Science* 354(6308): 42–43.

O'Brien, K.L., and R.M. Leichenko. 2000. Double exposure: Assessing the impacts of climate change within the context of economic globalization. *Global Environmental Change* 10: 221–232.

Olsson, P., and C. Folke. 2001. Local ecological knowledge and institutional dynamics for ecosystem management: A study of Lake Racken watershed, Sweden. *Ecosystems* 4: 85–104.

Olsson, P., C. Folke, and T. Hahn. 2004. Social-ecological transformation for ecosystem management: The development of adaptive co-management of a wetland landscape in southern Sweden. *Ecology and Society* 9(4): 2. http://www.ecologyandsociety.org/vol9/iss4/art2

Olsson, P., C. Folke, and T.P. Hughes. 2008. Navigating the transition to ecosystem-based management of the Great Barrier Reef, Australia. *Proceedings of the National Academy of Sciences* 105: 9489–9494.

Pascual, U., I. Palomo, W.M. Adams, K.M. Chan, T.M. Daw, E. Garmendia, E. Gómez-Baggethun, R.S. De Groot, G.M. Mace, B. Martín-López, and J. Phelps. 2017. Off-stage ecosystem service burdens: A blind spot for global sustainability. *Environmental Research Letters* 12: 075001.

Pereira, L., E. Bennett, R. Biggs, G.D. Peterson, T. McPhearson, A. Norström, P. Olsson, R. Preiser, C. Raudsepp-Hearne, and J Vervoort. 2018. Seeds of the future in the present: Exploring pathways for navigating towards "Good" Anthropocenes. In *Urban Planet: Knowledge towards Sustainable Cities*, ed. T. Elmqvist X. Bai, N. Frantzeskaki, C. Griffith, D. Maddox, T. McPhearson, S. Parnell, P. Romero-Lankao, D. Simon, and M. Watkins, 327–350. Cambridge University Press, Cambridge, UK.

Pereira, L., N. Sitas, F. Ravera, A. Jimenez-Aceituno, and A. Merrie. 2019. Building capacities for transformative change towards sustainability: Imagination in intergovernmental science-policy scenario processes. *Elementa: Science of the Anthropocene* 7(1): 35.

Phillips, N. 2017. Power and inequality in the global political economy. *International Affairs* 93: 429–444.

Pinkerton, E. 1989. Introduction: Attaining better fisheries management through co-management: Prospects, problems, and propositions. *Co-operative Management of Local Fisheries: New Directions for Improved Management and Community Development*, ed. E. Pinkerton, 3–33. UBC Press, Vancouver Canada.

Pope Francis. 2015. *Laudato Si', The Vatican.* http://w2.vatican.va/content/francesco/en/encyclicals/documents/papa-francesco_20150524_enciclica-laudato-si.html

Reinhardt, I., G. Kluth, C. Nowak, C.A. Szentiks, O. Krone, H. Ansorge, and T. Mueller. 2019. Military training areas facilitate the recolonization of wolves in Germany. *Conservation Letters* 12(3): e12635.

Rockström, J., W. Steffen, K. Noone, Å. Persson, F.S. Chapin III, E. Lambin, T.M. Lenton, M. Scheffer, C. Folke, H. Schellnhuber, B. Nykvist, C.A. De Wit, T. Hughes, S. van der Leeuw, H. Rodhe, S. Sörlin, P.K. Snyder, R. Costanza, U. Svedin, M. Falkenmark, L. Karlberg, R.W. Corell, V.J. Fabry, J. Hansen, B. Walker, D. Liverman, K. Richardson, P. Crutzen, and J. Foley. 2009. Planetary boundaries: Exploring the safe operating space for humanity. *Ecology and Society* 14(2): 32. http://www.ecologyandsociety.org/vol14/iss2/art32/

Rosen, F., and P. Olsson. 2013. Institutional entrepreneurs, global networks, and the emergence of international institutions for ecosystem-based management: The Coral Triangle Initiative. *Marine Policy* 38: 195–204. doi:10.1016/j.marpol.2012.05.036

Steffen, W., J. Rockström, K. Richardson, T.M. Lenton, C. Folke, D. Liverman, C.P. Summerhayes, A.D. Barnosky, S.E. Cornell, M. Crucifix, J.F. Donges, I. Fetzer, S.J. Lade, M. Scheffer, R. Winkelmann, and H.J. Schellnhuber. 2018. Trajectories of the Earth System in the Anthropocene. *Proceedings of the National Academy of Sciences* 115(33): 8252–8259.

Steffen, W., Å. Persson, L. Deutsch, J. Zalasiewicz, M. Williams, K. Richardson, C. Crumley, P. Crutzen, C. Folke, L. Gordon, M. Molina, V. Ramanathan, J. Rockström, M. Scheffer, H.J. Schellnhuber, and U. Svedin. 2011. The Anthropocene: From global change to planetary stewardship. *Ambio* 40: 739–761.

Welchman, J. 2012. A defence of environmental stewardship. *Environmental Values* 21: 297–316.

Westley, F., K. McGowan, and O. Tjörnbo. 2017. *The Evolution of Social Innovation: Building Resilience through Transitions*. Edward Elgar, Northampton, MA.

Westley, F., P. Olsson, C. Folke, T. Homer-Dixon, H. Vredenburg, D. Loorbach, J. Thompson, M. Nilsson, E. Lambin, J. Sendzimir, B. Banarjee, V. Galaz, and S. van der Leeuw. 2011. Tipping towards sustainability: Emerging pathways of transformation. *Ambio* 40: 762–780.

Williams, A., G. Whiteman, and J.N. Parker. 2019. Backstage interorganizational collaboration: Corporate endorsement for the sustainable development goals. *Academy of Management Discoveries* 5: 367–395.

Young, O.R. 1994. The problem of scale in human/environment relationships. *Journal of Theoretical Politics* 6(4): 429–447.

Young, O.R. 2002. Institutional interplay: The environmental consequences of cross-scale interactions. In The Drama of the Commons, ed. E. Ostrom, T. Deitz, N. Dolsak, P.C. Stern, S. Stonich, and E.U. Weber. National Academy Press, Washington DC.

PART IV

SUMMARY, SYNTHESIS, AND FUTURE ADVANCES

Chapter 14

Applications and Diffusion of Panarchy Theory

Lance H. Gunderson, Craig R. Allen, and Ahjond Garmestani

> I made my most recent fundamental discovery almost twenty long years ago, and yet still I muse about its meaning and significance. I called it the lumpy world, because I was testing a Panarchy hypothesis that implied that all ecosystems, and indeed all the living and non-living components of the planet, had to be organized in distinct lumps across scales from small and fast to big and slow. The panarchy hypothesis still exists as a fruitful way to explore structure and patterns and their significance. That language using the word lumps is all right as a crude statistical descriptor. But what I see, hear and feel and intuit needs poetic imagery. The lumpy world really lays out the patterns or frequencies of rhythms whose structures coalesce into whirling patterns of evolutionary creation where elements emerge and prosper, then collapse and renew. Lumps as rhythms creates a kind of music above all . . . and each system has its own distinctive tune. —*C. S. (Buzz) Holling (2017)*

The preceding chapters in this volume have provided different perspectives on the ways in which humans attempt to explain, understand, and interact with our environment. At a coarse and abstract level, humans collectively interact, make decisions, and intervene in environmental systems based on assumptions and conceptualizations or heuristic representations. Heuristics are general formulations that help guide inquiry and complex problem solving. Holling (1978a) was among the first to suggest that these heuristics consist of constructs or ideas that are oversimplifications of a much more complex reality. Holling (1978b) described them as myths or cartoons that help characterize different and distinct archetypes that structure human foresight, intervention, and expectations of dynamic systems of people and their environment. Holling and Gunderson (2002) outlined five such cartoons, random, balanced, anarchic, resilient, and evolutionary, to describe the interaction between persistence and change over time in systems. Panarchy (Gunderson and Holling 2002) was developed to explain nature's rules across both spatial and temporal dimensions in social-ecological systems (SESs).

In the two decades since the publication of the original *Panarchy* volume, scholarship has proceeded on at least three distinct research trajectories. One is the linkage between resilience-related concepts (Holling 1973; Gunderson et al. 1995; Folke et al. 2004; Folke 2016) and panarchy theory. The first *Panarchy* volume (Gunderson and Holling 2002) was produced as a result of a project challenging economic, political, and social theory with ecological theories of nonlinear change and resulted in dozens of books and tens of thousands of articles published on resilience in scientific journals (Folke 2016). The second thrust of panarchy research is based on scale and cross-scale dynamics and has achieved significant attention as well as numerous journal articles testing the theory and applying panarchy theory to resource management (Garmestani et al. 2009a, 2009b; Angeler et al. 2016). Panarchy theory also informs a growing body of scholarship in which the concepts have diffused into other scholarly areas. This third research trajectory includes governance and management of natural resources, law and policy, Indigenous and traditional ecological knowledge, interdisciplinary approaches to agriculture and land management, and innovation, creativity, and novelty (Allen et al. 2014).

The primary goals of this volume are to document and characterize how panarchy has been modified as a result of interdisciplinary scholarship over time, how panarchy has diffused and modified other fields of scientific inquiry, and how panarchy has influenced the management of SESs.

Theory and Concepts

Although proposed as a set of integrative concepts around the cross-scale dynamics of SESs, panarchy has been applied to a wide range of disciplines. For example, Delcourt and Delcourt (2004) described the cross-scale changes that occurred as Indigenous North Americans influenced vegetation composition and distributions within a bioregion over several decades, and Thoms et al. (2018) applied cross-scale resilience concepts to describe how geomorphologic processes operate across scales consistent with cross-scale adaptive cycles. Abel et al. (2006) used a panarchy framework but applied political and economic theories to explain how cross-scale transfers of resources can promote or set back reorganization in grazing systems after collapses. Garmestani et al.

(2009a) used panarchy to examine cross-scale patterns and discontinuities in urban and regional economic systems. Holdschlag and Ratter (2013) examined multiscale dynamics in three SESs in the Bahamas, including hurricanes and disaster recovery, coastal degradation due to land development, and impact of a nonnative species (lionfish) on local fisheries. Moen and Keskitalo (2010) characterized the history of multiuse forests in Sweden as multiple, interlocking panarchies. Finally, Petrosillo et al. (2010) examined the effects of multiscale disturbances on landcover patterns and vulnerability on biodiversity, and panarchy has also been used to assess agroecosystems in the Netherlands (van Apeldoorn et al. 2011).

The use of panarchy in these different types of SESs has greatly expanded the core set of concepts presented in the original volume. These include ideas of self-organization and design, as well as furthering the understanding of relationships between resilience, transformation, and panarchies. Each of these is discussed in turn.

Self-organization and design

SESs have emergent, discontinuous dynamics across scales that comprise a panarchy (Gunderson and Holling 2002). Such cross-scale dynamics interact and oscillate between forces and actions that generate stability and persistence and those that generate instability and crisis (Holling 1992; Gunderson and Holling 2002). The tendency toward stability can occur in two different ways, depending on the degree of human intervention. One of those mechanisms is self-organization within the system, the other by human foresight, ingenuity, design, and coercion. SESs have both at play, which contributes to the inherent unpredictability of these systems.

Self-organization is the process by which structures and processes emerge from interactions between components in complex adaptive systems in the absence of centralized control. Self-organization of panarchies has been shown to occur in ecological systems, as described in forests (Holling 1992), grasslands (chapters 5 and 7), and atmospheric systems (Gunderson and Holling 2002). Sundstrom and Allen (chapter 3) describe how regimes develop toward self-organized criticality, driven by external inputs of energy and materials, and Allen et al. (chapter 5) describe how quasistable spatial regimes are the result of self-organization due to multiple, nonlinear interactions across scales.

SESs have elements of human foresight and intentionality that can be positive but also negative (Holling et al. 2002). For example, the design, operations, and management that engineered the great river systems in the United States in the twentieth century were aimed at achieving human objectives of water supply and flood control (Cosens and Gunderson 2018). However, attempts at controlling variation in water flow have also led to the loss of social-ecological resilience through time (Holling and Meffe 1996; Gunderson and Holling 2002; Cosens and Gunderson 2018). Indeed, degrees of coercion and resilience appear to be inversely related.

SESs show characteristics of both self-organized and designed panarchical dynamics, which manifest at regional scales to national scales to the planetary scale. On the regional scale of the U.S. Great Plains, the transitions between regimes of grassy and woody vegetation (Twidwell et al., chapter 7) are a combination of self-organized landscape vegetation and fire dynamics, as well as intentional policies and management (Garmestani et al. 2020). At the scale of the planet, human interventions of technology, fossil fuels, and global economies have altered self-organized biogeochemical cycles to the extent that we now live on a human-dominated planet in a phase of Earth's existence called the Anthropocene (Angeler et al., chapter 3; Steffen et al. 2018; Folke et al. 2021). All of these systems demonstrate the dynamic nature of SESs, as elaborated in the following sections that discuss resilience, transformations, and panarchies.

Resilience and panarchies

Resilience (*sensu* Holling 1973) depends on the distribution of functions within and across scales of a panarchy (Peterson et al. 1998; Allen et al. 2005; Allen et al., chapter 5). In ecosystems, species functional traits maintain system processes and feedbacks. Similarly, functional differences define social and organizational systems at different scales (Chaffin, chapter 12; Olsson et al., chapter 8; Hodbod and Wentworth, chapter 11). Craig and colleagues (chapter 9) describe how a resilient system will have diverse functions present within a scale domain and redundant functions distributed across scale domains, which buffers against scale-specific disturbances (Peterson et al. 1998). Redundancy of function is also critical, because redundant functions occur at different spatial and temporal scales, thus filling different niches because of their separation by scale. This represents an important aspect of resilience and panarchy: the ability to limit the potential cascade of collapse given perturbations that could otherwise scale up, such as insect outbreaks, and the ability to provide compensatory function when some species' populations are reduced

(Holling 1987; Winfree and Kremen 2009; Mouillot et al. 2013; Sundstrom et al. 2018).

The cross-scale aspects of resilience can be studied through a variety of methods. Discontinuity analysis, time series modeling, spatial analysis, and early warning indicators have all been proposed as methods of quantifying aspects of resilience based on relevant data (Angeler and Allen 2016). Discontinuity analysis is an attractive option for assessing SESs because the method uses proxies (e.g., body size, city size) to identify scaling structure in complex systems and the breaks that identify scale domains (Angeler et al. 2016). Several methods have been used to conduct discontinuity analysis, and analyses can use both temporal and spatial data (Spanbauer et al. 2016). Time series modeling and spatial analysis allow the use of data reflecting abundance, presence or absence, and remotely sensed data. When monitoring data are present, early warning indicators (e.g., increased variance and autocorrelation, skewness, critical slowing down, Fisher information) can be used to assess whether the resilience of a system is eroding and a tipping point is near (Scheffer et al. 2012; Eason et al. 2016), and these approaches might be applicable to detecting the regime shifts that occur when one changes scale (Allen and Holling 2010); when one changes scale in a panarchy, structure changes because the structuring processes—the process and disturbance regimes responsible for structure—also change.

The assessment of resilience, regime shifts, and leading indicators of regime shifts has occurred mostly at the mesoscales identified as ecosystems, such as shallow lakes (Scheffer and Carpenter 2003). However, resilience theory, the cross-scale resilience model, and panarchy all suggest that within scale domains there is cross-scale structure and an adaptive cycle and that the transition to a new scale domain is in essence a regime shift (Gunderson and Pritchard 2002). A lake can have multiple possible regimes, each characterized by a different set of processes and functions at the scale of the lake but also at *each* scale domain within the lake. Fish, like all the other taxa tested, have discontinuous body mass distributions that reflect a series of scale domains within the boundary of the lake (Havlicek and Carpenter 2001). Each scale domain has its own adaptive cycle and is defined by a regime. Although most research has focused on the phenomena of a regime shift and adaptive cycles at one discrete scale, such as a lake, the cross-scale hierarchical nature of these systems has been largely ignored.

Spatial resilience

Spatial resilience is at the forefront of efforts to operationalize and quantify resilience concepts in landscapes. Landscapes are characterized by dynamics that are spatially and temporally complex but not random. Spatial resilience is a concept that captures the resilience and transformability of heterogeneous and dynamic systems that are spatially (and otherwise) nonstationary, that is, they may move in response to exogenous or endogenous factors. Spatial approaches related to resilience have also identified leading indicators of spatial thresholds (Kéfi et al. 2014), assessed the role of structural and functional spatial components of systems in relation to their resilience (Allen et al. 2014; Angeler et al. 2016), determined the relationship between connectivity, dispersal, and resilience (Underwood et al. 2009; Uden et al. 2014), assessed the influence of network membership on resilience (Keitt et al. 1997; Maciejewski et al. 2015), evaluated the relationship of landscape metrics to resilience (Uden et al. 2014) and developed approaches for understanding cross-scale interactions within complex systems (Cumming et al. 2015).

Spatial boundaries are rarely stationary, and the concept of spatial resilience captures this nonstationarity. Much early work on resilience was within bounded, physically stationary systems (e.g., lakes). This simplifies analyses, but most SES boundaries move in space, generally in a nonrandom manner driven by path dependency, climate change, or a number of other factors. Spatial boundaries may be defined in social, economic, or ecological terms. For example, corn seed is part of a spatially segregated economic supply chain; corn is harvested in one location, processed in another, sold in another, and consumed in yet another (figure 14.1). Shared elements, such as the movements of individuals between patches within a metapopulation, can lead to nonscaled regime shifts. For example, eastern redcedar spreads unevenly into rangeland grass patches based on patch soil nutrients, soil moisture, other species, disturbance history, and distance from other patches. A spatial resilience lens can therefore be used to consider how the underlying spatial heterogeneity from historical land use and disturbance legacies influence resilience.

Transformation and panarchies

Transformation is a human construct and a human-mediated process and occurs when humans purposefully degrade the resilience of a system in an

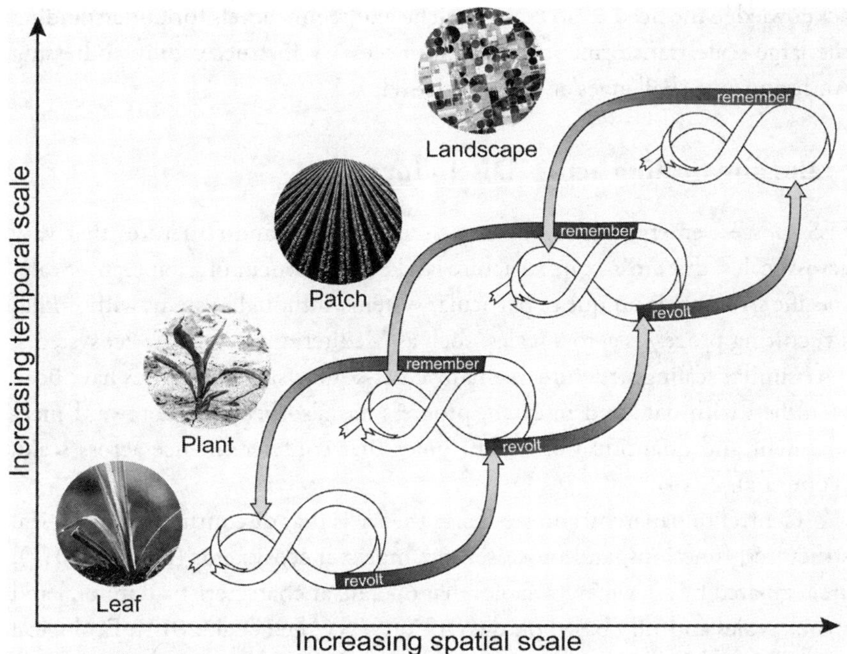

Figure 14.1. Cross-scale (panarchy) structures in a coerced agricultural system. Such panarchies are prevalent in many types of complex systems.

undesirable state in order to collapse it and guide reorganization to a more desired state. Transformations do not necessarily involve the collapse of one panarchical system replaced by a new system. This challenges the notion that transformations occur in a linear fashion (Moore et al. 2016; Chaffin et al. 2016; Preiser et al. 2018).

Chaffin et al. (2016) proposed *transformative governance* to describe proactive governance that actively works to transition an SES into a more socially and ecologically desirable regime. Transformative governance requires a unique set of capacities, opportunities, and political will to emerge successfully (see Chaffin et al. 2016). Olsson et al. (chapter 8) use panarchy to develop a model that might help increase understanding of complex interactions across scales and generate insight on transformations toward sustainable and equitable futures. Panarchy is central to the model they developed and used to explore cross-scale interactions and transformations in natural resource governance regimes in Chile, Uzbekistan, and South Africa. Olsson et al. (chapter 8) acknowledge the need to go beyond niche and regime levels for understanding the large-scale transformations that are necessary for successfully addressing Anthropocene challenges at the global level.

Scales and Scaling across Disciplines

SESs possess emergent properties such as resilience and structures that vary across scales; this cross-scale structure is a key component of a panarchy. Scale-specific structure is unique to particular systems, although systems with similar structuring processes across scales, such as Mediterranean climate ecosystems, have similar scaling structure. As many cross-scale system structures have been identified with data and research, progress has also been made toward measurement and quantification of attributes that confer resilience across scales (Pope et al. 2014).

Central to panarchy and resilience theory is that discontinuous ecological structures, functions, and processes, the music of ecosystems (Holling 2017), are regulated by a few key variables that operate at characteristic temporal and spatial scales and vary based on the type of SES (Allen et al. 2014). Ecological resilience emerges from the interaction of ecological structure and processes within and across these scales. For example, community change in subarctic lakes is scale-specific, with a subgroup of littoral invertebrates responding to slow changes of regional environmental conditions, whereas other subgroups respond to faster-changing processes (Angeler et al. 2013). In this example, resilience emerges from the reinforcement of functional traits of invertebrates within and across the scale-specific temporal dynamics observed.

In SESs, humans are often the ultimate drivers of regime shifts. Anthropogenic stressors, including biological invasions, soil degradation, habitat loss, the emergence of novel diseases, and climate change, affect the SESs on which humanity relies. In many cases, small linear changes in these stressors cause abrupt and large nonlinear changes in the SES (Scheffer et al. 2001). Transitions such as the conversion of productive rangelands to invaded woodlands are typically characterized by sudden changes in the production of valued ecosystem services (Folke et al. 2002; Birge et al. 2016).

A core hypothesis of panarchy is that the key processes that structure complex systems occur at different spatial and temporal scales, often separated by orders of magnitude—scale domains, where key processes, structure, and resources either do not change or change monotonically (Wiens 1989; Holling 1992; Nash et al. 2014). Thus, the spatiotemporal scales of pine needle turnover on a pine tree are different than the physiographic scales of the processes that drive where boreal forest occurs on Earth. Adaptive cycles occur at each scale domain in a system, resulting in hierarchical complex systems with multiple and nested domains of scales at different phases of the adaptive cycle. The presence of discrete scales of pattern and process in complex systems creates discontinuities, or scale breaks, between scale domains, and this ensures a degree of modularity that can confine disturbances to a particular scale domain. However, theory and empirical analysis suggest that these scale breaks also generate novelty and innovation as a result of inherent dynamics at these transitions.

Novelty across scales

Novelty and innovation are necessary for systems to be adaptive and thus persistent over time. Without innovation, novelty, and key sources of renewal, complex systems tend to become rigid and overly connected with capital and energy bound into existing structures (Gunderson and Holling 2002). Novelty is critical during the renewal phase of the adaptive cycle, because novelty provides new pathways for the flow of energy necessary for the development of new structures after collapses.

In a panarchy, novelty is generated at the edge of scale breaks (at the transitions between domains of scale) as a result of the highly variable distribution and occurrence of resources in space and time, which in turn is reflected in the high variability in biotic components of the system (e.g., Allen et al. 1999; Skillen and Maurer 2008; Wardwell and Allen 2009). This generation of novelty continuously creates options for systems, is critical in maintaining adaptive capacity, and serves as a reservoir of potential functions that may be needed after transformations or for adaptation (Angeler et al. 2019). Such novelty is at the heart of resilience.

When the resilience of a complex system is exceeded, a system may collapse, triggering the renewal phase (Gunderson and Holling 2002). Novelty may be added to or introduced to a system during reorganization. Here the novelty added (e.g., new animals to a community or ecosystem) may be either locally or globally novel. The addition of novelty may also occur within adaptive cycles, when individual cycles enter the omega–alpha phases, and this may also build resilience over time and in response to changing conditions.

Having a constant source of novelty is important for complex adaptive systems, both after change and during their normal dynamics. Novelty is a critical component of resilience and adaptive and transformative capacities. Scale breaks are a key source of such novelty. However, novelty may be a destructive force as well. For example, invasive species can alter basic process and structure in ecosystems and be a source of collapse (and occur, more than expected, at scale breaks; Allen et al. 1999). Thus, innovation and novelty can be beneficial or detrimental. In ecosystems, for example, in addition to being a cause of major extinctions, innovation is the prime source of recovery (Jain and Krishna 2002).

Scales of traps

System traps (Holling and Gunderson 2002; Allison and Hobbs 2004; Carpenter and Brock 2008; Chapin et al. 2009; Uden et al. 2018) are systems that are characterized by maladaptive trajectories. A trap is created when a regime is stable and persistent, but the trajectory is not desired or is maladaptive. Social examples of trapped systems such as poverty and racism are difficult to change. Recurring harmful algal blooms are another example of a trapped system. Panarchy, especially cross-scale linkages of revolt and remember, can help us understand different types of traps.

Systems with high degrees of potential, resources, and resilience can become trapped. Such traps are sustained by control exerted by large-scale processes that suppress innovation, diversity, and experimentation (Gunderson et al. 2018). For example, food insecurity increased in Flint, Michigan after the water crisis, as grassroots initiatives failed to scale up and were squelched by well-meaning attempts by state and federal governments to stem the crisis (Hodbod and Wentworth, chapter 11). A rigidity trap continues to halt progress on Everglades restoration as a result of sustained hierarchical controls in the form of fiscal resources (Gunderson et al. 2018). Coerced regimes (Angeler and Allen, chapter 4) are maintained by broad-scale inputs that focus on stability. The stability provided by law (Craig et al., chapter 9) sometimes reinforces power relations across scales such that novelty and innovation are suppressed.

In contrast, systems in poverty traps can be characterized by the lack of critical types of larger-scale inputs (memory, resources) and the inability to constrain or adapt to small-scale perturbations. Hodbod and Wentworth (chapter 11) indicate how a poverty trap for food insecure households makes it difficult for social memory to be transferred from larger and slower scales through remember mechanisms. When in a poverty trap, small-scale disturbances lead to crises and reorganizations that sustain trajectories of continued poverty conditions.

Governance and Management

Panarchy (Gunderson and Holling 2002) is the conceptual foundation of cross-scale governance and management (Garmestani and Benson 2013; Cosens and Gunderson 2018; Garmestani et al. 2020). Such approaches acknowledge the inherent uncertainty and unpredictability in self-organized ecological systems and hence have developed flexible forms of governance and management in contrast to those that are based on command and control.

In the United States, environmental governance and management are based primarily on laws that assume that dynamics of SESs are easily predicted and mitigated (Craig 2010). Environmental governance has met with mixed success because managers must make decisions in the context of laws (e.g., Endangered Species Act) that were not written considering a system perspective. Consequently, environmental governance is often based on simple metrics to assess the status of one or more endangered species and their habitats. Simple measures for SESs (e.g., number of an endangered species) neglect ecosystem-scale metrics of change (e.g., ecosystem processes and structures), contributing to management failures (Clement and Standish 2018). In addition, most environmental governance has a difficult time accounting for scale (an inherent aspect of SESs; Garmestani and Benson 2013). Managing for single variables is problematic because such approaches often do not account for potential interactions across scales in SESs (Gunderson and Holling 2002). Furthermore, the spatial and temporal scale at which management is applied is often chosen arbitrarily or limited by law or administrative jurisdictions (Green et al. 2015). Most current approaches to environmental governance and management assume that data can be directly upscaled, as if one could add up the pieces to describe the whole. However, when scale is arbitrarily delineated, it can lead to mismatches between management and the ecosystem of interest, which in turn can lead to adverse management outcomes (e.g., coral reef management that does not account for impacts from climate change). Thus, the current U.S. legal framework is not well suited for a no-analogue future or for confronting cross-scale challenges for managing SESs (Garmestani and Allen 2014).

Resilience-based approaches (*sensu* Holling 1973) that led to adaptive management (Holling 1978a; Walters 1986; Allen and Garmestani 2015) explicitly account for spatial and temporal scales at which ecological patterns and processes operate, as well as within- and cross-scale interactions (Allen et al. 2014). The identification of scale is crucial for management of SESs and allows management actions to be driven by a clear understanding of the social-ecological scales at which processes occur rather than scales that are convenient. In addition to being based on systems thinking, resilience-based approaches include several quantitative methods originally derived from the work of Holling (1973, 1992), Gunderson and Holling (2002), Allen and Holling (2008), and Allen et al. (2005). These methods were created to identify scales in SESs, with critical implications for management (e.g., cross-scale interactions) (Angeler et al. 2016). Objective methods to delineate scale are an essential component of using panarchy to assess and manage cross-scale resilience in SESs, because, for example, the effects of disturbance have scale-specific impacts (Nash et al. 2014).

Chaffin (chapter 12) argues that government institutions and bureaucracies can limit the cross-scale experimentation in governance needed to understand SES dynamics (Gunderson and Light 2006; Garmestani et al. 2009b). Rigid regulatory frameworks of government limit the application of adaptive management and thus limit societal ability to learn and adapt, creating a pathology of bureaucratic management that ignores the rich, scale-dependent dynamics of SESs (Craig 2010; Allen and Gunderson 2011; Garmestani et al. 2019). Although adaptive management has received attention from both agency scientists (Williams et al. 2009) and legal scholars (Fischman and Ruhl 2015; Craig and Ruhl 2014), the application of adaptive management to extensive landscapes has not worked well (Volkman and McConnaha 1993; Lee 1999).

These problems with management of SESs highlight the need for information sharing across levels of management to increase the likelihood of generating desirable environmental outcomes (Jacobson and Haubold 2014). The interaction between top-down and bottom-up aspects of social and ecological systems must be taken into account and in many cases harnessed to manage resilience. Social-ecological baselines are rapidly changing, and resilience-based approaches can complement traditional management approaches (e.g., assessments of extinction and invasion risks; Angeler et al. 2016). Social-ecological stewardship develops most readily at a single (usually local) scale and emerges across scales when there is sufficient motivation, respectful engagement, dialogue, and power sharing (Young 2002; Chapin et al., chapter 13).

Development of cross-scale social-ecological stewardship often requires new configurations of social networks and institutions and collaborations between groups that have not traditionally interacted. When these new interactions foster trust, share power, and coproduce new knowledge and commitment to action, this can trigger innovations in policy and practice at multiple scales with cross-scale interactions (Folke et al. 2005; Hahn et al. 2006; Olsson et al. 2008). However, politics and unequal access to decision-making power often complicate these new interactions and their consequences for distribution of ecosystem services among segments of society (Pascual et al. 2017), within and across scales (Leach et al. 2018; Kashwan et al. 2019).

Fostering innovation and experimentation

Allen et al. (chapter 5), discuss the relationships of novelty and innovation to regime shifts and how the cross-scale resilience model can help improve management of SESs. Reform to a resilience-based perspective that addresses within-scale and cross-scale interactions via an iterative, multiscale path for law and policy would require massive shifts in lawmaking and policymaking at national and international levels (Garmestani et al. 2019). These reforms are ideals and could be incredible improvements over current law and policy, because they would revolve around change originating from different levels, with creative solutions (flexibility) coupled with enforceability at each level. Experimentation is encouraged at smaller scales, with the intent that larger (national, global) levels offer resilience to the entire system (Garmestani et al. 2019). Subsidiarity principles can allow delegation of environmental decision making to the lowest appropriate level for an issue, as local, state, and non-state entities can network, build trust, and react more quickly at small scales. Legislatures and federal and state agencies operate at broader levels and provide oversight and enforcement, if necessary, for entities conducting environmental governance at lower levels (Garmestani et al. 2019). This innovation is important because ecosystems do not exhibit the same biotic components, physical and chemical properties, and ecosystem processes at every scale (Karkkainen 2006). Rather, a larger-scale ecosystem typically has spatially distinct subsystems, with considerable heterogeneity in composition, which is better suited for delegated environmental governance.

Dynamic federalism, in which federal and state governments work together to achieve shared goals, creates space for dialogue, plurality, and redundancy, and a check on interest group capture at one level and an opportunity for regulatory innovation at multiple levels (Schapiro 2006, Engel 2007). In this sense, dynamic federalism fosters communication between state and federal officials, multiple approaches for addressing environmental issues, and recourse to state or federal government if one level of government fails in its responsibilities to citizens. In essence, federal law should allow states to innovate, providing an iterative, flexible framework for interaction between levels of government (Engel 2007; Garmestani and Benson 2013). The classic example of this phenomenon is when the Clean Air Act was implemented after California established automobile tailpipe standards (Engel 2007). In the United States, states (and local governments) have taken the lead for addressing accelerating environmental change, even though the issue has broader-scale implications.

Prescriptive approaches for governance, in which overly generalized recommendations are offered up as the blueprint for social-ecological resilience, are rife with problems (Hill and Engle 2013; Lubell et al. 2017). Rather, context appears to matter more than any other factor in the type of governance appropriate for a specific environmental issue. Although there are factors that appear to foster environmental governance that generates desirable environmental outcomes (Hill Clarvis and Engle 2015), these factors (e.g., leadership, intermediaries, communication) can be offered only as suggestions or a starting point for determining the best approach for the problem presented. Furthermore, these factors may be impossible to quantify, driving home the importance of context when engaging in environmental governance.

Panarchy frontiers

Panarchy is a critical consideration for global-scale transformations in the Anthropocene (Folke et al. 2021). Angeler and Allen (chapter 3) write that management of complex systems of people and nature under rapid environmental change has been translated into concepts that directly link management challenges, such as fostering ecosystem service provisioning, to resilience theory. For example, coerced resilience has been used in production ecosystems (e.g., agriculture, silviculture, fisheries) and refers to the improvement of selected functions related to the production of food and fiber through substantial external physical and chemical subsidies. A related concept, coerced regimes, has a systemic focus and refers to the creation and maintenance of artificial, non–self-sustaining system regimes. Angeler and Allen (chapter 4) explore within- and cross-scale aspects of coercion with special attention to how coercion may be implemented within and across the distinct phases of the adaptive cycles of a panarchy. They introduce the term *coerced panarchy* to show this control of hierarchical system dynamics through management and discuss how management interventions at defined scales can propagate through other levels in a panarchy and the global climate regime.

The ongoing COVID-19 pandemic, for example, has created a window of opportunity (Walker et al. 2020) for transitions across scales. The current glacial–interglacial climate regime is undergoing profound changes that probably will result in a planetary regime shift known as hothouse Earth (Steffen et al. 2018). To avoid such a shift, the current climate regime must be maintained artificially through a series of mitigation interventions. Angeler and Allen (chapter 4) discuss how the implementation of panarchy with redundant Earth stewardship measures within and across scales can potentially coerce the current climate regime, which is probably no longer tenable without such interventions, for continued human survival.

The importance of panarchy has become increasingly apparent as we move forward in the Anthropocene. For thousands of years, Earth was able to absorb humankind's impact through Earth's inherent capacity for self-organization and compartmentalization of biophysical systems (i.e., multiscale ecosystems), in short, resilience (Holling 1973). Earth has an inherent capacity to persist in the face of perturbations, but this resilience can and has been eroded to a point where SESs shift from one configuration to a different configuration. For decades scientists have known that ecosystems have multiple regimes, but this dynamic in SESs was largely observed at small spatial scales (e.g., small lakes, individual reefs, part of a grassland biome). Now, we see that the effects of human domination of Earth have begun to scale up. What this means is that the panarchy at the global level (the globe being the largest effective scale in the global panarchy) is being challenged, which is unprecedented in human history. Now we see much larger SESs experiencing shocks that could cause shifts at the largest global scales, which are likely to be undesirable for humankind because we've coevolved with the current panarchy. For example, larger lakes are experiencing larger and larger harmful algal blooms, giving warning of impending shifts (e.g., frequency and size of algae blooms in Lake Erie in the past five to ten years); larger coral reef SESs are on the verge of collapse, including the largest living organism on Earth (e.g., mass bleaching of Great Barrier Reef in Australia); and grassland biomes are trending toward permanent conversion to forests (e.g., conversion of the Great Plains grassland in North America to red cedar). Contributors to this book addressed some of these critical issues for humankind and offered suggestions for a path forward in the Anthropocene.

Farley and Egler (chapter 10) assert that unfettered economic growth has the potential to flip the Earth into an alternative state with negative consequences for humankind and advocate for a new economic model based on current understanding of the dynamics of SESs and panarchy. Farley and Egler (chapter 10) argue that panarchy can play a critical role in the development of this economic model and introduce economic panarchy, a system that would account for adaptive economic cycles at different scales (from the family level up to the global level) as a way of managing resilience in economic systems.

Pumo and coauthors (chapter 6) discuss different definitions of resilience and how these different definitions can result in completely different management recommendations (see Allen et al. 2019). For example, resilience is defined differently for engineered systems and SESs, respectively. Accounting for panarchy is essential moving forward for dealing with linked systems of humans and nature where infrastructure (e.g., sewer and water pipes in a city) is subject to nature's rules. Pumo and co-authors (chapter 6) assert that clearly defining the type of resilience one is managing for has critical implications for environmental planning and discuss the implications of panarchy for a large federal agency, the U.S. Army Corps of Engineers, which is responsible for managing navigable waters and wetlands in the United States.

The accelerating rate of planetary changes requires a framework that accounts for cross-scale interactions, emergent properties such as resilience, consideration of nonstationarity, novelty, and tipping points (Allen et al. 2019). Panarchy is such a framework and has proven useful in multiple disciplines, as demonstrated by the contributions to this volume. Humanity has decreased system diversity across scales, which along with increased connectivity has led to the erosion of resilience at the scale of the planet. A lack of controls has allowed small-scale processes (such as carbon emissions or human reproduction) to contribute to global instabilities. Modern humans are drivers in a global regime shift, with high degrees of uncertainty regarding what alternative regime might emerge. Ideas generated by the theories associated with social-ecological resilience, discontinuities, and panarchy have influenced numerous disciplines. We hope that these ideas also improve our stewardship of planet Earth and allow humanity to help transform the planet into a more sustainable future.

Literature Cited

Abel, N., D.H.M. Cumming, and J.M. Anderies. 2006. Collapse and reorganization in social-ecological systems: Questions, some ideas, and policy implications. *Ecology and Society* 11(1): 17.

Allen, C.R., D.G. Angeler, B.C. Chaffin, D. Twidwell, and A. Garmestani. 2019. Resilience reconciled. *Nature Sustainability* 2: 898–900.

Allen, C.R., D.G. Angeler, A.S. Garmestani, L.H. Gunderson, and C.S. Holling. 2014. Panarchy: Theory and application. *Ecosystems* 17: 578–589.

Allen, C.R., E.A. Forys, and C.S. Holling. 1999. Body mass patterns predict invasions and extinctions in transforming landscapes. *Ecosystems* 2: 114–121.

Allen, C.R., and A. Garmestani. 2015. *Adaptive Management of Social-Ecological Systems*. Springer, Dordrecht, The Netherlands.

Allen, C.R., and L.H. Gunderson. 2011. Pathology and failure in the design and implementation of adaptive management. *Journal of Environmental Management* 92: 1379–1384.

Allen, C.R., L. Gunderson, and A.R. Johnson. 2005. The use of discontinuities and functional groups to assess relative resilience in complex systems. *Ecosystems* 8: 958–966.

Allen, C.R., and C.S. Holling. 2008. *Discontinuities in Ecosystems and Other Complex Systems*. Columbia University Press, New York.

Allen, C.R., and C.S. Holling. 2010. Novelty, adaptive capacity, and resilience. *Ecology and Society* 15: 24.

Allison, H.E., and R.J. Hobbs. 2004. Resilience, adaptive capacity, and the "lock-in trap" of the Western Australian agricultural region. *Ecology and Society* 9(1): 3.

Angeler, D.G., and C.R. Allen. 2016. Quantifying resilience. *Journal of Applied Ecology* 53: 617–624.

Angeler, D.G., C.R. Allen, C. Barichievy, T. Eason, A.S. Garmestani, N.A.J. Graham, D. Granholm, L. Gunderson, M. Knutson, K.L. Nash, R.J. Nelson, M. Nystrom, T. Spanbauer, C.A. Stow, and S.M. Sundstrom. 2016. Management applications of discontinuity theory. *Journal of Applied Ecology* 53: 688–698.

Angeler, D.G., C.R. Allen, and R.K. Johnson. 2013. Measuring the relative resilience of subarctic lakes to global change: Redundancies of functions within and across temporal scales. *Journal of Applied Ecology* 50: 572–584.

Angeler, D.G., H. Peterson, C.R. Allen, A. Garmestani, D. Twidwell, W. Chuang, V.M. Donovan, T. Eason, C.P. Roberts, S.M. Sundstrom, and C.L. Wonkka. 2019. Adaptive capacity in ecosystems. *Advances in Ecological Research* 60: 1–24.

Birge, H., K. Pope, A. Garmestani, and C.R. Allen. 2016. Adaptive management for ecosystem services. *Journal of Environmental Management* 183: 343–352.

Carpenter, S.R., and W.A. Brock. 2008. Adaptive capacity and traps. *Ecology and Society* 13(2): 40.

Chaffin, B.C., A.S. Garmestani, L.H. Gunderson, M.H. Benson, D.A. Angeler, C Arnold, B. Cosens, R.K. Craig, J.B. Ruhl, and C.R. Allen. 2016. Transformative environmental governance. *Annual Review of Environment and Resources* 41: 399–423.

Chapin, F.S., G. Kofinas, and C. Folke. 2009. *Principles of Ecosystem Stewardship*. Springer, New York.

Clement, S., and R.J. Standish. 2018. Novel ecosystems: Governance and conservation in the age of the Anthropocene. *Journal of Environmental Management* 208: 36–45.

Cosens, B., and L.H. Gunderson (Eds.). 2018. *Practical Panarchy for Adaptive Water Governance: Linking Law to Social-Ecological Resilience*. Springer, New York.

Craig, R.K. 2010. "Stationarity is dead"—Long live transformation: Five principles for climate change adaptation law. *Harvard Environmental Law Review* 31: 9–75.

Craig, R., and J.B Ruhl. 2014. Designing administrative law for adaptive management. *67 Vanderbilt Law Review* 1.

Cumming, G.S., C.R. Allen, N.C. Ban, D. Biggs, H.C. Biggs, D.H.M. Cumming, A. De Vos, G. Epstein, M. Etienne, K. Maciejewski, R. Mathevet, C. Moore, M. Nenadovic, and M. Schoon. 2015. Understanding protected area resilience: A multi-scale, social-ecological approach. *Ecological Applications* 25: 299–319.

Delcourt, P.A., and H.R. Delcourt. 2004. *Prehistoric Native Americans and Ecological Change: Human Ecosystems in Eastern North America since the Pleistocene*. Cambridge University Press, Cambridge, UK.

Eason, T., A.S. Garmestani, C.A. Stow, M. Alvarez-Cobelas, C. Rojo, and H. Cabezas. 2016. Managing for resilience: An information theory-based approach to assessing ecosystems. *Journal of Applied Ecology* 53: 656–665.

Engel, K.H. 2007. Harnessing the benefits of dynamic federalism in environmental law. *Emory Law Journal* 56: 159–188.

Fischman, R.L, and J.B. Ruhl. 2015. Judging adaptive management practices of U.S. agencies. *Conservation Biology* 30: 268–275.

Folke, C. 2016. Resilience. In *Subject: Framing Concepts in Environmental Science. Oxford Research Encyclopedias, Environmental Science*, ed. H. Shugart. Oxford University Press, New York.

Folke, C., S. Carpenter, T. Elmqvist, L. Gunderson, C.S. Holling, and B. Walker. 2002. Resilience and sustainable development: Building adaptive capacity in a world of transformations. *Ambio* 31: 437–440.

Folke, C., S. Carpenter, B. Walker, M. Scheffer, T. Elmqvist, L. Gunderson, and C.S. Holling. 2004. Regime shifts, resilience, and biodiversity in ecosystem management. *Annual Review of Ecology Evolution and Systematics* 35: 557–581.

Folke, C., T. Hahn, P. Olsson, and J. Norberg. 2005. Adaptive governance of social-ecological systems. *Annual Review of Environment and Resources* 30: 441–473.

Folke, C., S. Polasky, J. Rockstrom, V. Galaz, F. Westley, M. Lamont, M. Scheffer, H. Osterblom, S.R. Carpenter, F.S. Chapin III, K.C. Seto, E.U. Weber, B.I. Crona, G.C. Daily, P. Dasgupta, O. Gaffney, L.J. Gordon, H. Hoff, S.A. Levin, J. Lubchenco, W. Steffen, and B.H. Walker. 2021. Our future in the Anthropocene biosphere. *Ambio* 50: 834–869.

Garmestani, A.S., and C.R. Allen. 2014. *Social-Ecological Resilience and Law*. Columbia University Press, New York.

Garmestani, A.S., C.R. Allen, and H. Cabezas. 2009a. Panarchy, adaptive management and governance: Policy options for building resilience. *Nebraska Law Review* 87: 1036–1054.

Garmestani, A.S., C.R. Allen, and L. Gunderson. 2009b. Panarchy: Discontinuities reveal similarities in the dynamic system structure of ecological and social systems. *Ecology and Society* 14 (1): 15.

Garmestani, A.S., and M.H. Benson. 2013. A framework for resilience-based governance of social-ecological systems. *Ecology and Society* 18 (1): 9.

Garmestani, A., J.B. Ruhl, B.C. Chaffin, R.K. Craig, H.F.M.W. van Rijswick, D.G. Angeler, C. Folke, L. Gunderson, D. Twidwell, and C.R. Allen. 2019. Untapped capacity for resilience in environmental law. *Proceedings of the National Academy of Sciences* 116: 9899–19904.

Garmestani, A., D. Twidwell, D.G. Angeler, S.M. Sundstrom, C. Barichievy, B.C. Chaffin, T. Eason, N.A.J. Graham, D. Granholm, L. Gunderson, M. Knutson, K.L. Nash, M.J. Nelson, M. Nystrom, T.L. Spanbauer, C.A. Stow, and C.R. Allen. 2020. Panarchy: Opportunities and challenges for ecosystem management. *Frontiers in Ecology and the Environment* 18: 576–583.

Green, O.O., A.S. Garmestani, C.R. Allen, L.H. Gunderson, J.B. Ruhl, C.A. Arnold, N.A.J. Graham, B. Cosens, D.G. Angeler, B.C. Chaffin, and C.S. Holling. 2015. Barriers and bridges to the integration of social-ecological resilience and law. *Frontiers in Ecology and the Environment* 13: 332–337.

Gunderson, L.H., B. Cosens, and B. Chaffin. 2018. Trajectories of change in regional scale social-ecological water systems. In Practical Panarchy, Linking Law, Resilience and Adaptive Water Governance of Regional Scale Social-Ecological Systems, ed. B. Cosens and L. Gunderson. Springer, New York.

Gunderson, L.H., and C.S. Holling. 2002. *Panarchy: Understanding Transformations in Systems of Humans and Nature*. Island Press, Washington, DC.

Gunderson, L.H., Holling, C.S., and S.S. Light. 1995. *Barriers and Bridges to the Renewal of Ecosystems and Institutions*. Columbia University Press, New York.

Gunderson, L.H., and S. S. Light. 2006. Adaptive management and adaptive governance in the Everglades ecosystem. *Policy Sciences* 39: 323–334.

Gunderson, L.H., and L. Pritchard. 2002. *Resilience and the Behavior of Large-Scale Ecosystems*. SCOPE volume. Island Press, Washington, DC.

Hahn, T., P. Olsson, C. Folke, and K. Johansson. 2006. Trust-building, knowledge generation and organizational innovations: The role of a bridging organization for adaptive co-management of a wetland landscape around Kristianstad, Sweden. *Human Ecology* 34: 573–592.

Havlicek, T.D., and S.R. Carpenter. 2001. Pelagic species size distributions in lakes: Are they discontinuous? *Limnology and Oceanography* 46: 1021–1033.

Hill, M., and N.L. Engle. 2013. Adaptive capacity: Tensions across scales. *Environmental Policy and Governance* 23: 177–192.

Hill Clarvis, M., and N.L. Engle. 2015. Adaptive capacity of water governance arrangements: A comparative study of barriers and opportunities in Swiss and US states. *Regional Environmental Change* 15: 517–527.

Holdschlag, A., and B. M. W. Ratter. 2013. Multiscale system dynamics of humans and nature in the Bahamas: Perturbation, knowledge, panarchy and resilience. *Sustain Science* 8: 407–442.

Holling, C.S. 1973. Resilience and stability of ecological systems. *Annual Review of Ecology and Systematics* 4: 1–23.

Holling, C.S. 1978a. *Adaptive Environmental Assessment and Management*. Wiley and Sons: London.

Holling, C.S. 1978b. Myths of ecological stability: Resilience and the problem of failure. In *Studies in Crisis Management*, ed. in C. F. Smart and W. T. Stanbury. Butterworth & Co., Montreal.

Holling, C. S. 1987. Simplifying the complex: The paradigms of ecological function and structure. *European Journal of Operational Research* 30: 139–146.

Holling, C.S. 1992. Cross-scale morphology, geometry, and dynamics of ecosystems. *Ecological Monographs* 62: 447–502.

Holling, C.S. 2017. *Bubbles and Spirals: The Memoirs of Buzz Holling.* http://www.stockholmresilience.org/holling-memoirs

Holling, C.S., and L. H. Gunderson. 2002. Resilience and adaptive cycles. In *Panarchy: Understanding Transformations in Systems of Humans and Nature*, ed. L.H. Gunderson and C.S. Holling, 25–62. Island Press, Washington, DC.

Holling, C.S., L.H. Gunderson, and G.D. Peterson. 2002. Sustainability and panarchies. In *Panarchy: Understanding Transformations in Systems of Humans and Nature*, ed. L.H. Gunderson and C.S. Holling, 63–101. Island Press, Washington, DC.

Holling, C.S., and G.K. Meffe. 1996. Command and control and the pathology of natural resource management. *Conservation Biology* 10: 328–337.

Jacobson, C.A., and E.M. Haubold. 2014. Landscape conservation cooperatives: Building a network to help fulfill public trust obligations. *Human Dimensions of Wildlife* 19: 427–436.

Jain, S., and S. Krishna. 2002. Large extinctions in an evolutionary model: The role of innovation and keystone species. *Proceedings of the National Academy of Sciences* 99: 2055–2060.

Karkkainen, B.C. 2006. Managing transboundary aquatic ecosystems: Lessons from the Great Lakes. *Global Business and Development Law Journal* 19: 209–240.

Kashwan, P., L.M. MacLean, and G.A. García-López. 2019. Rethinking power and institutions in the shadows of neoliberalism. *World Development* 120: 133–146.

Kéfi, S., V. Guttal, W.A. Brock, S.R. Carpenter, A.M. Ellison, V.N. Livina, D.A. Seekell, M. Scheffer, E.H. van Nes, and V. Dakos. 2014. Early warning signals of ecological transitions: Methods for spatial patterns. *PLoS One* 9: e92097.

Keitt, T.H., D.L. Urban, and B.T. Milne. 1997. Detecting critical scales in fragmented landscapes. *Conservation Ecology* 1(1): 4.

Leach, M., B. Reyers, X. Bai, E. Brondizio, C. Cook, S. Díaz, and S. Subramanian. 2018. Equity and sustainability in the Anthropocene: A social–ecological systems perspective on their intertwined futures. *Global Sustainability* 1: E13.

Lee, K.N. 1999. Appraising adaptive management. *Conservation Ecology* 3(2): 3.

Lubell, M., L. Jasny, and A. Hastings. 2017. Network governance for invasive species management. *Conservation Letters* 10: 699–707.

Maciejewski, K., A. de Vos, G.S. Cumming, C. Moore, and D. Biggs. 2015. Cross-scale feedbacks and scale mismatches as influences on cultural services and the resilience of protected areas. *Ecological Applications* 25: 11–23.

Moen, J., and E.C.H. Keskitalo. 2010. Interlocking panarchies in multi-use boreal forests in Sweden. *Ecology and Society* 15(3): 17.

Moore, C., J. Grewar, and G.S. Cumming. 2016. Quantifying network resilience: Comparison before and after a major perturbation shows strengths and limitations of network metrics. *Journal of Applied Ecology* 53: 636–645.

Mouillot, D., N.A.J. Graham, S. Villéger, N.W.H. Mason, and D.R. Bellwood. 2013. A functional approach reveals community responses to disturbances. *Trends in Ecology and Evolution* 28: 167–177.

Nash, K.L., C.R. Allen, D.G. Angeler, C. Barichievy, T. Eason, A.S. Garmestani, N.A.J. Graham, D. Granholm, M. Knutson, R.J. Nelson, M. Nystrom, C.A. Stow, and S.M. Sundstrom. 2014. Discontinuities, cross-scale dynamics, and ecosystem organization. *Ecology* 95: 654–667.

Olsson, P., C. Folke, and T.P. Hughes. 2008. Navigating the transition to ecosystem-based management of the Great Barrier Reef, Australia. *Proceedings of the National Academy of Sciences* 105: 9489–9494.

Pascual, U., P. Balvanera, S. Díaz, G. Pataki, E. Roth, M. Stenseke, R. Watson, E.B. Dessane, M. Islar, E. Kelemen, V. Maris, M. Quaas, S. Subramanian, H. Wittmer, A. Adlan, S. Ahn, Y.S. Al-Hafedh, E. Amankwah, S. Asah, P. Berry, A. Bilgin, S. Breslow, C. Bullock, D. Cáceres, H. Daly-Hassen, E. Figueroa, C. Golden, E. Gómez-Baggethun, D. González-Jiménez, J. Houdet, H. Keune, R. Kumar, K. Ma, P. May, A. Mead, P. O'Farrell, R. Pandit, W. Pengue, R. Pichis-Madruga, F. Popa, S. Preston, D. Pacheco-Balanza, H. Saarikoski, B. Strassburg, M. van den Belt, M. Verma, F. Wickson, and N. Yagi. 2017. Valuing nature's contributions to people: The IPBES approach. *Current Opinion in Environmental Sustainability* 26–27: 7–16.

Peterson, G., C.R. Allen, and C.S. Holling. 1998. Ecological resilience, biodiversity and scale. *Ecosystems* 1: 6–18.

Petrosillo, I., N. Zaccarelli, and G. Zurlini. 2010. Multi-scale vulnerability of natural capital in a panarchy of social–ecological landscapes. *Ecological Complexity* 7: 359–367.

Pope, K.L., D. Angeler, and C.R. Allen. 2014. Fishing for resilience. *Transactions of the North American Fisheries Society* 143: 467–478.

Preiser, R., R. Biggs, A. De Vos, and C. Folke. 2018. Social-ecological systems as complex adaptive systems: Organizing principles for advancing research methods and approaches. *Ecology and Society* 23(4): 46.

Schapiro, R.A. 2006. From dualist federalism to interactive federalism. *Emory Law Journal* 56: 1–18.

Scheffer, M., and S.R. Carpenter. 2003. Catastrophic regime shifts in ecosystems: Linking theory to observation. *Trends in Ecology and Evolution* 18: 648–656.

Scheffer, M., S.R. Carpenter, J.A. Foley, C. Folke, and B. Walker. 2001. Catastrophic shifts in ecosystems. *Nature* 413: 591–596.

Scheffer, M., S.R. Carpenter, T.M. Lenton, J. Bascompte, W. Brock, V. Dakos, J. Van De Koppel, I.A. Van De Leemput, S.A. Levin, E.H. Van Nes, and M. Pascual. 2012. Anticipating critical transitions. *Science* 338: 344–348.

Skillen, J.J., and B.A. Maurer. 2008. The ecological significance of discontinuities in body mass distributions. In *Discontinuities in Ecosystems and Other Complex Systems*, ed. C.R. Allen and C. S. Holling, 193–218. Columbia University Press, New York.

Spanbauer, T.L., C.R. Allen, D.G. Angeler, T. Eason, S.C. Fritz, A.S. Garmestani, K.L. Nash, J.R. Stone, C.A. Stow, and S.M. Sundstrom. 2016. Body size distributions signal a regime shift in a lake ecosystem. *Proceedings of the Royal Society B* 283: 20160249.

Steffen, W., J. Rockstrom, K. Richardson, T.M. Lenton, C. Folke, D. Liverman, and H.J. Schellnhuber. 2018. Trajectories of the Earth system in the Anthropocene. *Proceedings of the National Academy of Sciences* 115: 8252–8259.

Sundstrom, S.M., D.G. Angeler, C. Barichievy, T. Eason, A.S. Garmestani, L.H. Gunderson, M. Knutson, K.L. Nash, T. Spanbauer, C. Stow, and C.R. Allen. 2018. The distribution and role of functional abundance in cross-scale resilience. *Ecology* 99: 2421–2432.

Thoms, M.C., H. Piégay, and M. Parsons. 2018. What do you mean, "resilient geomorphic systems"? *Geomorphology* 305: 8–19.

Uden, D., C.R. Allen, F. Munoz-Arriola, G. Ou, and N. Shank. 2018. A framework for tracing social–ecological trajectories and traps in intensive agricultural landscapes. *Sustainability* 10(5): 1646.

Uden, D.R., M.L. Hellman, D.G. Angeler, and C.R. Allen. 2014. The role of reserves and anthropogenic habitat for functional connectivity and resilience of ephemeral wetlands. *Ecological Applications* 24: 1569–1582.

Underwood, J.N., L.D. Smith, M.J.H. van Oppen, and J.P. Gilmour. 2009. Ecologically relevant dispersal of corals on isolated reefs: Implications for managing resilience. *Ecological Applications* 19: 18–29.

Van Apeldoorn, D.F., K. Kok, M.P.W. Sonneveld, and T.(A.) Veldkamp. 2011. Panarchy rules: Rethinking resilience of agroecosystems, evidence from Dutch dairy-farming. *Ecology and Society* 16(1): 39.

Volkman, J.M., and W.E. McConnaha. 1993. Through a glass, darkly: Columbia River salmon, the Endangered Species Act and adaptive management. *Environmental Law* 23: 1249–1272.

Walker, B., S.R. Carpenter, C. Folke, L. Gunderson, G.D. Peterson, M. Scheffer, M. Schoon, and F.R. Westley. 2020. Navigating the chaos of an unfolding global cycle. *Ecology and Society* 25(4): 23.

Walters, C.J. 1986. *Adaptive Management of Renewable Resources*. Blackburn Press, Caldwell, NJ (2002).

Wardwell, D., and C.R. Allen. 2009. Variability in population abundance is associated with thresholds between scaling regimes. *Ecology and Society* 14(2): 42.

Wiens, J. 1989. Spatial scaling in ecology. *Functional Ecology* 3: 385–397.

Williams, B.K., R.C. Szaro, and C.D. Shapiro. 2009. *Adaptive Management: The U.S. Department of the Interior Technical Guide*. Adaptive Management Working Group, U.S. Department of the Interior, Washington, DC.

Winfree, R., and C. Kremen. 2009. Are ecosystem services stabilized by differences among species? A test using crop pollination. *Proceedings of the Royal Society of London* 276: 229–237.

Young, O.R. 2002. *The Institutional Dimensions of Environmental Change: Fit, Interplay, and Scale*. MIT Press, Cambridge, MA.

ABOUT THE EDITORS

Craig R. Allen is a professor in the School of Natural Resources and director of the Center for Resilience in Agricultural Working Landscapes at the University of Nebraska in Lincoln. He serves on the board of directors for the Resilience Alliance and is coeditor-in-chief of the journal *Ecology and Society*. He is a fellow of the American Association for the Advancement of Science and the National Strategic Research Institute. His research focuses on the cross-scale dimensions and linkages between land use and land cover change, biological invasions, and extinctions and has resulted in more than 200 peer-reviewed publications.

Ahjond Garmestani is a research scientist at the U.S. Environmental Protection Agency, Office of Research and Development, Gulf Breeze, Florida, and is a fellow of the Utrecht Centre for Water, Oceans and Sustainability Law, Utrecht University, Utrecht, The Netherlands. He received his PhD in policy studies from Clemson University in 2006, a JD from Florida State University College of Law in 2001, a MS in wildlife ecology from the University of Florida in 1997, and a BA in anthropology from Emory University in 1993. His transdisciplinary research is focused on governance and management of social-ecological systems and has resulted in more than 100 peer-reviewed publications, two books, and three journal special features.

Lance H. Gunderson is professor and chair in the Department of Environmental Sciences at Emory University in Atlanta, Georgia. He is a founding board member of the Resilience Alliance and a fellow of the Beijer Institute of Ecological Economics of the Royal Swedish Academy of Sciences. His scholarly work has addressed the application of ecological understanding to the policy and practice of managing natural resources.

CONTRIBUTORS

Brady W. Allred is an associate professor of rangeland ecology in the W.A. Franke College of Forestry and Conservation, University of Montana. He works on dynamics, structures, and processes of rangeland ecology and conservation.

David G. Angeler is an associate professor in the Department of Aquatic Sciences and Assessment, Swedish University of Agricultural Sciences Uppsala, Sweden, and an adjunct professor in the School of Natural Resources, University of Nebraska–Lincoln. His research is in the areas of resilience in complex systems, transdisciplinary applications of ecological theory, and ways of combining art and science to inform sustainability challenges.

Reinette (Oonsie) Biggs is a South African sustainability scientist whose research focuses on food, water, and the benefits people receive from nature. Biggs is the codirector of the Centre for Complex Systems in Transition at Stellenbosch University, South Africa, and a researcher at Stockholm Resilience Centre, Stockholm University, Sweden.

Brian C. Chaffin is an associate professor of water policy in the W.A. Frank College of Forestry and Conservation, University of Montana. His research focuses on complex questions of water policy and government. He works closely with ecologists and hydrologists, spanning boundaries between disciplinary science, interdisciplinary synthesis, and environmental decision making.

F. Stuart Chapin III is professor emeritus of ecology at the Institute of Arctic Biology, University of Alaska Fairbanks. Terry's ongoing scholarship on Earth stewardship led to receipt of the 2019 Volvo Environmental Prize. He has a long career linking ecology, ethics, and the management of natural resources.

Barbara Cosens is a university distinguished professor emerita at the University of Idaho College of Law. She continues to work at the interface of law and social-ecological resilience to improve governance of water resources.

Robin Kundis Craig is the Robert C. Packard Trustee Chair in Law at the Gould School of Law, University of Southern California. Robin is a leading environmental law scholar on social-ecological resilience and law, as well as sustainability of freshwater, coastal, and marine resources.

Megan Egler is a Leadership for the Ecozoic PhD fellow at the Gund Institute for Environment and Rubenstein School of Environment and Natural Resources at the University of Vermont. She works on issues of economics, political ecology, and resource development.

Joshua Farley is a professor of community development and applied economics and a fellow at the Gund Institute for Ecological Economics, University of Vermont. He is an ecological economist who works on integrating social, human, and natural capital.

Carl Folke is a professor and chair of the board at the Stockholm Resilience Centre, Stockholm University, and director of the Beijer Institute of Ecological Economics of the Royal Swedish Academy of Sciences. He is a transdisciplinary scientist who publishes on the economic, human, and environmental dynamics of social-ecological systems.

Stephanie Galaitsi is a scientist with the Risk and Decision Science Team of the New England District of the U.S. Army Corps of Engineers. She is a scientist focusing on the interaction between water issues and human prosperity.

Jennifer Hodbod is an assistant professor in the Department of Community Sustainability, College of Agriculture and Natural Resources at Michigan State University. Her scholarly work is on resilient food systems, food security, livelihoods, and sustainable agriculture.

Matthew O. Jones is a research scientist in the W.A. Franke College of Forestry and Conservation, University of Montana. He is a systems ecologist and uses remote sensing to monitor and model land cover and vegetation responses to changing climate.

Gary P. Kofinas is a professor emeritus of resource policy and management at the University of Alaska Fairbanks. His research addresses resilience of indigenous rural communities, community-based resource stewardship, and the use of local ecological knowledge in resource management.

Margaret Kurth is a research scientist with the Risk and Decision Science Team of the U.S. Army Corps of Engineers, New England District. Her work involves engaging stakeholders to improve environmental decision making.

Igor Linkov is a research scientist with the Environmental Laboratory of the U.S. Army Engineer Research and Development Center in Concord, Massachusetts. He works on assessing risks and resilience in sectors of critical infrastructure, food security, transportation, homeland security and defense, and translating those assessments to resilience management approaches.

Michele-Lee Moore is an associate professor at the University of Victoria and Strategic Advisor of Transdisciplinary Education at the Stockholm Resilience Centre, Stockholm. Her research is on social innovation and water governance, transnational networks, and resilience.

David E. Naugle is a professor of large-scale wildlife ecology in the W.A. Franke College of Forestry and Conservation at the University of Montana. His research is on understanding relationships between organisms and habitats at different scales.

Per Olsson is an associate professor and principle researcher at the Stockholm Resilience Centre of Stockholm University, Stockholm. He currently researches entrepreneurship, social-ecological innovations, and transformations to sustainability.

Ian Pumo is chief of the Technical Services Branch of the U.S. Army Corps of Engineers in Seattle, Washington. His work attempts to incorporate resilience concepts into the planning of civil works projects.

Caleb P. Roberts is the assistant unit leader of the USGS Arkansas Cooperative Fish and Wildlife Research Unit at the University of Arkansas–Fayetteville. His research interests are in landscape ecology, invasive species, resilience, and complexity theory.

J. B. Ruhl is the David Daniels Allen Distinguished Chair of Law, director of the Program on Law and Innovation, and codirector of the Energy, Environment, and Land Use Program at Vanderbilt University Law School. He is an internationally recognized expert in environmental, natural resource, and property law and also studies the legal industry and legal technology.

Nadia Sitas is a senior researcher at the Centre for Complex Systems in Transition at Stellenbosch University. She works at the science–policy interface on social-ecological resilience with a specific focus on equity and gender.

Shana M. Sundstrom is a research scientist at the School of Natural Resources, University of Nebraska–Lincoln. Her scholarly work focuses on understanding cross-scale structures and processes in ecosystems.

Dirac Twidwell is an associate professor in the Large-Scale Rangeland Conservation Lab at the University of Nebraska–Lincoln. He works on understanding resilience of rangelands to fire and weather and applies that understanding to the management of natural resources.

Dan R. Uden is an assistant professor in the School of Natural Resources, University of Nebraska–Lincoln. He is a resilience spatial scientist, studying ecosystems and disturbances and how the spatial properties of each influence social-ecological resilience.

Chelsea Wentworth is a research assistant professor in the Department of Community Sustainability, affiliated core and GJEC faculty for the Center for Gender in a Global Context, and affiliated faculty with the Institute for Public Policy and Social Research at Michigan State University. Her research examines hunger and food security, food as a human right, and sustainable food systems through a gendered lens.

INDEX

Page numbers for figures *(fig)* and tables *(t)* are noted in *italics*.

Island Press | Board of Directors